INDUSTRIAL ENZYMES AND THEIR APPLICATIONS

CONTRIBUTORS

Dr. D. Frank, Former Vice President, Research, AKZO Chemical Co. (Sub-chapter 4.4.5)

Dr. K. Gottman, Former Director, Analytical Laboratory, Röhm GmbH. (Chapter 2)

Dr. K. Köhler, Lebadd, Inc. (Chapter 5.2)

Dr. E. M. Linsmaier-Bednar, In Vitro Associates (translator, editor, contributor, including updating; all chapters)

Dr. E. Pfleiderer, Former Director, Leather Technology Customer Service, Röhm GmbH. (Chapter 5.9)

Dr. H. Plainer, Patent Department, Röhm GmbH. (Chapter 4)

Dr. G. Richter, Genencor International Germany, GmbH. (Chapter 5.1)

K. Schmitt, Enzyme Technologist, Fruits and Vegetable Processing, Röhm GmbH. (Chapter 5.5)

Dr. W. G. Spring, Business Development, F. Hoffmann-La Roche, Ltd. (Subchapter 5.13.8)

Dr. H. Uhlig, Former Director, Enzyme R & D, Röhm GmbH. (Chapters 1; 3; 5.3; 5.4; 5.6; 5.7; 5.8; 5.10–5.15; 6; 7; 8; 9 and 10)

INDUSTRIAL ENZYMES AND THEIR APPLICATIONS

Helmut Uhlig, Ph.D.

Translated and Updated by
Elfriede M. Linsmaier-Bednar, Ph.D.

A WILEY-INTERSCIENCE PUBLICATION

JOHN WILEY & SONS, INC.

New York • Chichester • Weinheim • Brisbane • Singapore • Toronto

Copyright © 1998 by John Wiley & Sons, Inc. All rights reserved.

Published simultaneously in Canada.

Library of Congress Cataloging-in-Publication Data:

Uhlig, Helmut.
    Industrial enzymes and their applications / Helmut Uhlig :
translated and updated by Elfriede M. Linsmaier–Bednar.
        p.   cm.
    Includes index.
    This is a revised translation of: Enzyme arbeiten für uns.
    ISBN 0-471-19660-6 (cloth : alk. paper)
    1. Enzymes—Industrial applications. 2. Enzymes—Biotechnology.
3. Enzymes. I. Enzyme arbeiten für uns. II. Title.
TP248.E5U37   1998
660′.634—dc21   97-38099
                                                                               CIP

Printed in the United States of America.

10 9 8 7 6 5 4 3 2

CONTENTS

PREFACE

Until quite recently, industrial enzymes were of interest only to scientists, engineers, and others directly involved in enzyme research and development; now, with the widespread commercial advertisement (in television and other media) of the relative efficiency and safety of enzymes in everyday products such as household detergents and foods, enzymes are of increasing concern to the general public. The consumer's decision whether to buy a specific product is often based on close perusal of the contents or of the ingredients' list on the package label; for instance, one may question whether eating a food treated with a certain ingredient (viz., enzyme) might pose a health risk or whether that ingredient will improve the quality of the food.

This book describes the current state of technology in industrial enzyme preparations and applications, especially those in which enzymes have replaced chemical reagents. It also discusses how enzymes can be used to reduce costs in low-energy processes, improve raw materials utilization, and enhance the quality of foods and other products. The book begins with an overview of the most important industrial enzymes, describing their microbiological origin and such parameters as pH, temperature dependence, and reaction stability of each specific enzyme. Throughout the book, the frequent descriptions of enzyme properties and activities are intended to help food technologists, engineers, students, and others find new applications for industrial enzymes, and should be of value to microbiologists in the development of enzymes with modified characteristics.

During my 30 years of work on enzymes, I have often been asked by students, clients, technicians, and marketing representatives to recommend an easily understandable book on industrial enzymes. Since a comprehensive treatise on this subject was not available in German, my Röhm colleagues and I collected our practical experience and observations on enzymes in a book entitled *Enzyme arbeiten für uns,* which was published in 1991 by Carl Hanser in Munich and was intended as a reference source for industrial enzyme chemists at all levels, from beginning students to R&D technologists, as well as for nonchemists involved in industrial enzyme marketing.

The previous manuscript has been updated by including discussion of some recently developed industrial enzymes and progress in enzyme applications. Especially updated are the chapters on baking and fruit juice; in particular, new amylases and xylanases have been added. Chapter 4, on enzyme immobilization, was completed by Dr. D. Frank. New references have been added, and errors in the literature references of the German edition were eliminated by Dr. Linsmaier-Bednar.

I thank my colleague Mr. Reinhold Schmitt, a world-class computer expert, for producing the figures and for his patience in teaching me how to handle a computer. I also thank Prof. Dr. W. Pilnik, Wageningen; Prof. Dr. K. Girschner, Hohenheim; Prof. Dr. K. Wünscher, Lemgo; and Prof. Dr. H. Ruttloff, Potsdam-Rehbrücke for critical discussions and information. Dr. Don Scott, of Scott Biotechnology Inc., Chicago, helped greatly with his long experience in the enzyme field. He provided me with the motivation for this English edition. I am grateful to the enzyme-producing companies Alko, Amano, Cultor, Daiwa, Gist Brocades, Grindsted, Meiji, Novo, Röhm, Solvey, Shin Nihon, Stern-Enzyme, and Ueda for permission to use their technical information.

HELMUT UHLIG, PH.D.

1 Introduction

1.1 WHAT ENZYMES DO

Life on earth is based on constant change. Inorganic matter forms the complicated structures that make up the living world, which will again decay into lifeless matter. The lives of plants, animals, and human beings are included in this great cycle. The inert substances that form the foundation of life have a very simple structure. Water, carbon dioxide, and nitrogen are the basic ingredients that, with the energy supplied by sunlight, permit plants to synthesize the molecular moieties that support life. These compounds are then used to provide more elaborate compounds required by animal life.

Chemical experiments have shown how specific compounds, namely, the essential amino acids, can be generated from simple mixtures of carbon, nitrogen, oxygen, and hydrogen when subjected to extreme pressures and temperatures. Once formed, less extreme conditions can cause the degradation of these complex structures as well. Under the current global conditions, such processes proceed only very slowly.

What, then, causes the rapid synthesis of these compounds that are essential for life, and the degradation of the high-molecular-weight and highly organized structures that make up life? The catalysts for this assembly and disassembly were for a long time thought to be life itself, or rather a fundamental principle of living cells.

The first observation of an enzymatic degradation reaction was in 1783 by Spallanzani (1729–1799), a priest and naturalist from Padua, Italy. After placing meat in small porous capsules, he examined the regurgitated pellets of the hawks who had eaten this material and found the capsules to be empty, proving that the meat had been rapidly liquefied by the stomach juices of these birds of prey.

In 1814, Kirchhoff (1815) found that barley contained a substance that was capable of liquefying starch paste into sugar. He assumed that the reaction was caused by the gluten protein of the barley. Subsequently, Payen and Persoz (1833) termed the working principle of this saccharification *diastase* (Greek for *separation*), a term still used for the amylases in the brewing industry. The degradation of starch, which is composed of sugar moieties, and the subsequent process of alcoholic fermentation of the sugar into carbon dioxide and alcohol occupied much of this nineteenth-century chemist's time.

In 1857 Pasteur showed that fermentation is closely associated with live yeast. He distinguished between the actions of "organized ferments" (cellular) and the "unorganized ferments" (soluble). These soluble "ferments," which are not bound to the living cell, were labeled *enzymes* by Kühne (1878). This term is derived from the

1

Greek *en zyme,* meaning "in sour dough." Concrete evidence for this assumption was provided by E. Buchner in 1897, as he showed that the cell-free extract from yeast cells could also produce alcohol from sugars (Buchner, 1897, 1898). The enzyme active in fermentation was termed *zymase*. The first book summarizing the "wonderful mechanism" of the enzymes known at that time and outlining the history of their discovery was written by Green (1901) and is still read with interest today.

1.2   ENZYMES DEFINED AS CATALYSTS

Friederich Wilhelm Ostwald first coined the now-familiar definition of a catalyst, specifically, that a catalyst is a substance that alters the rate of a chemical reaction without being present in the reaction products. Thus, Ostwald recognized enzymes as catalysts.

Today, *catalysts* and *catalysis* are familiar terms. The catalysts in internal-combustion engines accelerate the conversion of the pollutants carbon monoxide and nitrogen oxides into the less polluting carbon dioxide and nitrogen, without any change in the catalysts during the process. According to the basic research of Tammann (1892) and Van't Hoff (1898), a catalyst must be able to regenerate the initial substrate from its products. Enzymatic reactions also have an equilibrium. An example is the formation and scisson of an ester that forms on reaction between an alcohol with an acid (Ammon and Dirscherl, 1959):

$$\text{Acetic acid} + \text{alcohol} \underset{\text{hydrolysis}}{\overset{\text{synthesis}}{\rightleftharpoons}} \text{acetate ester} + \text{water}$$

If acetic acid and alcohol are mixed in equimolar amounts, some of the molecules will react to form an ester and water (synthesis). Yet, there still will be significant quantities of acetic acid and alcohol in the reaction mixture as well as the products generated, ester and water. A mixture of the same composition can also be obtained when starting with water and an ester (hydrolysis). Even in the absence of a catalyst, the reactions reach equilibrium between ester and synthesis and ester hydrolysis.

When forming and splitting butyl ester, it was found that the equilibrium was rapidly established from either direction on the addition of porcine pancreatic powder (pancreatic lipase). Thus, for life to exist, the biochemical equilibria of essential and exceedingly complex reaction systems must be established rapidly. Therefore, enzymes are the vital catalysts in all life processes.

1.3   SOME HISTORICAL USES OF ENZYMES

The use of enzymes and microorganisms in processing raw materials from plants and animals has been practiced for a long time. At one time, living microorganisms were used predominantly. Traditional processes, such as the production of alcoholic beverages and yeast-fermented doughs in baking bread, are displayed in Egyptian wall paintings. Further examples are the processes for preserving food, such as vegetable conservation by fermentation with lactobacilli (Sauerkraut), or preserving milk by making cheese.

"Bread and wine in ancient times" is a contribution by Reisner (1981). He provides an interesting description of the historical development of food processing biotechnology:

> The oldest baking utensils and ovens are 5000 years old and were used in Jericho (Canaan), one of the most ancient cities in the world. The simplest way of baking was to place a flat loaf of dough in glowing ashes. However, around 3500 BC, the variety of breads produced at that time which were mainly flat breads, were baked in the first cylindrical ovens. A dough was made from coarse milled kernels without addition of leavening agents. Such breads were tasty only when fresh. Sourdough was rarely used except when extended storage was desired. Ovens for baking bread were installed in the kitchens of the temple area of Ur, the capital of the Sumerians. Near Abu Kemal, where the Damaskus–Mossul line crosses the Euphrates, lie the high walls of the palace of Mari. Here the cuneiform writings relate the marshalling of men for military conscription. On such occasions, government officials distributed free beer and bread.
>
> A large bakery dating from around 2000 BC has been excavated in the palace of Sargon in Babylon. Clay pots were found in which pieces of dough were still stuck to the inner walls and apparently were baked in a quickly kindled fire. In Nebuchadnezzar's Babylon, baking with sourdough was already known. In the palace bakery of Ramses III, cylindrical baking ovens were used, as shown in a wall drawing in the side-chamber of El Amarna [Fig. 1.1].
>
> In this picture the dough is not hand kneaded but trodden on while employing two plots to help balance the worker. Above this scene a basket with sourdough is suspended. Two other bakers carry liquids in jugs to a table upon which a firm dough is formed. In the middle row, a heated oil bath with lid can be seen. By utilizing tongs, a spiral shaped pastry is fried in the oil. A baker removes the finished, round baked goods from a cooled baking oven. In the bottom row, flour is run through a sieve and on the right, a drum containing fresh sourdough can be seen. It is not known whether the Egyptians sometimes used malt in baking. They were familiar with the production of malt by wetting and sprouting barley.
>
> Wine had already been mentioned in the book of Genesis as a special fermentation product. "And Noah, a farmer, started to till the soil and planted a vineyard. And then he drank the wine and became drunk." In Babylon drinking was a favorite pastime; 16 different beer varieties were known, of which black beer was the most popular. The first word deciphered from the hieroglyphics by Champollion was "Erep" meaning wine. At least six types of wine could be distinguished, including white, red, black and lower Egyptian. After the grape harvest, the baskets were carried to a wine press, in this case a long stone trough, above which a man-sized wooden scaffold was erected [Figs. 1.2 and 1.3].
>
> The men supported themselves with the upper beams of the scaffolding and crushed the grapes with their feet. Later, round wine presses were built; the workers held onto ropes and the juice was discharged from the trough into a container [as shown in Fig.

Figure 1.1   Court bakery of pharaoh Ramses III. Tomb painting from El Amarna. [*Source:* I. Rosellini, *I Monumenti dell'Egitto,* Pisa, Italy, 1832 (Reisner, 1981).]

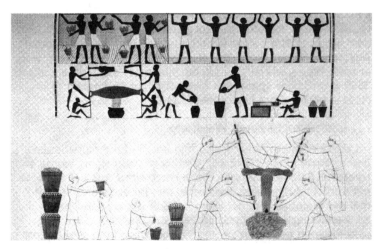

Figure 1.2   Harvesting and making wine. From a Theban tomb painting. [*Source:* I. Rosellini, *I Monumenti dell'Egitto,* Pisa, Italy, 1832 (Reisner, 1981).]

Figure 1.3   Advanced Egyptian technology in winemaking. From a Theban tomb painting. [*Source:* I. Rosellini, *I Monumenti dell'Egitto,* Pisa, Italy, 1832 (Reisner, 1981).]

1.3]. This would correspond to the present run-off from a preliminary dejuicer. Because the grape pomace still contained juice, it was emptied into a large sack and the juice extracted by twisting the sack.

A tip for wine enthusiasts: There is a wine museum in Torgiano, a small village south of Perugia. In the palace of Baglioni, a number of displays describe the process and treatment of wine from the fourth millennium through the Etruscan and Roman periods until the Middle Ages. The use of yeast and sourdough belongs to the oldest biological processes of our culture. Today, they are still used in a somewhat altered form often involving enzyme technology.

1.4   A HISTORY OF THE DEVELOPMENT OF TECHNICAL ENZYMES

Enzyme preparations, from extracts of plants or animal tissue, were used well before much was known about the nature and properties of enzymes. The development of today's industrially important microbial enzymes began with the Japanese Jokichi Takamine. Born in Japan and later a U.S. immigrant, Takamine studied the production of enzymes from molds. In 1894, he obtained a process patent for making a diastatic enzyme preparation from fungi (Takamine, 1894). He named the product *Takadiastase,* a mixture of carbohydrases and proteolytic enzymes. The production occurred either via a surface culture, a semisolid culture, or the *koji* process on moist wheat bran with nutrient salts and buffers added. The sterile culture media, mixed with spores from *Aspergillus oryzae,* was poured onto metal trays to a thickness of a few centimeters and incubated in climate-controlled culture

Figure 1.4 Fungal semisolid culture at Röhm GmbH.

chambers. The culture media, permeated with fungal mycelia, are ground and extracted with water. In a modified form, this process is still used today by various manufacturers, although it has largely been replaced by the development of the submerged fermentation process carried out in large fermenters (Fig. 1.4).

In 1895, Auguste Boidin discovered a new process for the manufacture of alcohol from cereals. Known as the *amyloprocess,* this method involves the extraction of cereal polysaccharides by boiling, then inoculation with a fungus that, in turn, produces saccharifying enzymes and subsequently yeast fermentation of the sugars. Boidin and Professor Jean Effront (Boidin and Effront, 1917), who had studied enzymes involved in alcohol production since 1900, founded the Société Rapidase in 1920.

Enzyme preparations from animal organs also continue to play an important industrial role. In 1907, Otto Röhm discovered the effectiveness of pancreatic proteases in liming and bating of hides in leather manufacture (Röhm, 1907a, 1907b). Until then, slurries of dog feces had been used in bating. Röhm concluded that the

Figure 1.5   Patent, Röhm GmbH, 1979.

bating effect of dog feces was due to excreted pancreatic proteases and subsequently developed the first standardized enzymatic bating agent that, marketed under the name "Oropon," offered the tanner greater control and reliability in the sensitive leather manufacturing process (Figs. 1.5–1.7).

Otto Röhm developed additional uses for the pancreatic enzymes he manufactured in Darmstadt, including the degumming of raw silk. The importance of these enzymes for the silk industry at that time can be shown by the following cost computation: "One kilogram Degomma S (pancreatic protease) with a value of 4 [Reichsmark] can degum the same amount of silk as ten kilograms of expensive Marseilles soap" (at that time 2–3 reichsmarks per kilogram).

In 1913, Röhm introduced the first commercially enzymatic detergent. The product, named "Burnus," rapidly became known as a good prewash and soaking agent. In addition to pancreatic enzymes, it contained sodium carbonate and sodium bicarbonate, which held the pH in the wash water below 9. The pancreatic proteases are sufficiently active and stable at low temperatures in this pH range. However, pancreatic enzymes are inactive in new detergent formulations that contain sequestrants, anionic detergents, and oxidizing agents.

In 1959, Jaag in Switzerland developed a detergent that contained, as active components, proteases from *Bacillus subtilis* (Jaag, 1968). The Novo Company

Figure 1.6   The Röhm GmbH Oropon label.

[NO] in Copenhagen deserves credit for the subsequent development of this enzyme. This will be discussed in more detail later. Plant enzymes, especially the proteases from papaya fruits (i.e., papain), are industrially important. Wallerstein (1911) was the first to use papain for stabilizing beer. Beer stabilization, which prevents protein flocculation or hazing on refrigeration, is particularly important in the United States, where beer is cold-stored.

Boidin and Effront (1917) obtained a process patent for manufacturing an enzyme preparation from cultures of *B. subtilis,* grown on the surface of a liquid nutrient medium. The preparation primarily contained amylase, which was able to replace the malt that previously had been used in textile sizing.

Figure 1.7   The Röhm GmbH Burnus label.

Beginning in 1932, Röhm and Haas began producing fungal proteases with semisolid cultures. These enzymes produced gentle bating preparations suited for special applications in leather manufacture. About 1934, the first pectinase preparations for clarifying fruit juices and improving juice extraction from grapes were produced by the same method.

In 1958, Underkofler et al. published a list of commercial enzyme preparations. The most important products were the amylases that were used in bakeries and breweries and in the manufacture of starch hydrolysates. Second were the microbial proteases, which were used in leather manufacture and in removing hazes from cold beer. Other enzymes such as pectinases, lactases, invertases, lipases, and cellulases were, however, produced only in small quantities. The years following World War II saw a rapid development of submerged fermentation technology, triggered by the then-extant antibiotic manufacturing methods. Bacterial and fungal amylases could now be manufactured inexpensively. This had a great influence on the commercial development of enzymatic starch conversion to dextrins and glucose.

In 1960, Novo Nordisk [NO] began production of alkaline bacterial proteases from *Bacillus licheniformis* for use in detergents. A flurry of activity followed that led to a wave of innovations in the entire field of industrial enzymes.

The commercial development of alkaline bacterial proteases was briefly interrupted when cases of lung disease, due to allergic reactions, were reported in the United States. These allergies occurred in workers who had direct contact with dusty enzyme concentrates during detergent manufacture. Granulating the concentrates helped avoid these risks. In 1971, The National Academy of Sciences (USA) confirmed that the use of detergents containing enzymes posed no health hazard to the consumer.

Proteases from *B. licheniformis* are currently the commercially most important microbial enzymes. They represent approximately 60% of all the industrial enzymes, followed by the carbohydrases making up about 30%. The latter include the amylases, glucose isomerases, pectinases, and cellulases.

The development of other industrial enzymes will be discussed in subsequent chapters.

1.5 APPLICATIONS OF MODERN ENZYME TECHNOLOGY

Major targets of modern enzyme technology continue to be preservation of foods and food components (e.g., vitamins), more efficient use of raw materials, and improvement of food quality such as texture and taste. Enzymes are also used in the food industry for process optimization to reduce process costs, including energy requirements. Other current objectives include the utilization of new raw materials for feeding humans and animals, the manufacture of dietetic foods, and eliminating antinutritive substances from certain nutritional raw materials.

Biotechnological methods have also replaced some of the traditional chemical processes. Enzymatic methods, which include the most recent procedures with immobilized enzymes and immobilized microorganisms, constitute only one aspect of modern biotechnology. Biotechnology encompasses many processes for producing biologically active compounds, such as amino acids and antibiotics as well as processing wastewater.

Examples of new biotechnological processes include:

- The use of enzymes in nonaqueous media for the production of chiral compounds and the synthesis of special polymers (Dordick, 1992).
- The use of enzymes for the synthesis of amino acids, peptides, and antibiotics (Soda and Yonaha, 1987).
- The use of enzymes for recycling food wastes and in wastewater treatment (Shoemaker, 1986).
- The production of sweeteners, such as aspartame, using combined microbiological and enzymatic methods.

- The production of cyclodextrins from starch with specific enzymes.
- The use of cellulases and lipases as active components of detergents.
- The production of rare and expensive mammalian enzymes by microorganisms. For example, pure calf stomach rennet is currently not available in sufficient quantities.
- The production of tailored enzymes to serve as specific, process-adapted catalysts. These procedures are part of the so-called protein engineering, which means the design of proteins with specific functions. For more information, see Goodenough (1991, 1995).
- Large-scale production of enzymes by gene technology is to be expected in future years; recombinant chymosin was a pioneering first. It has since overcome the world shortage of chymosin in the cheese industry. Thus, many limited or costly traditional enzyme sources will be replaceable.

In the past, obtaining the desired metabolites and improved performance depended on the conventional genetic methods of inducing mutations and hybridizing sexually propagating microorganisms. In asexually reproducing microorganisms, physical and chemical mutagenesis produced the necessary variability. The sought-after characteristic or product was subsequently found by screening large numbers of progeny. Mutations randomly alter the genetic or hereditary information. Success depends on whether one of the random mutations occurs at the correct site within the genome; only then will it be possible to develop better microbial strains. The extensive screening required makes it very expensive to determine whether such random mutations have produced the changes sought in the production of the desired substance. Now, gene technology permits site-specific changes and yields products of great purity under economically feasible conditions. For those who wish to read further on this subject, the book on gene technology by Gassen et al. (1986) is recommended.

REFERENCES

Ammon, R., and Dirscherl, W., *Fermente, Hormone, Vitamine,* Georg Thieme, Stuttgart, 1959.

Boidin, A., and Effront, J., "Process manufacturing diastases and toxins by oxidizing ferments," U.S. Patent 1,227,525 (1917).

Buchner, E., *Ber Dtsch. Chem. Ges.* **30,** 117 (1897).

Buchner, E., *Ber. Dtsch. Chem. Ges.* **31,** 586 (1898).

Dordick, J. S., in Heineman, R., and Wolnak, B. (eds.), *Opportunities with Industrial Enzymes,* Wolnak, Chicago, 1992, p. 40.

Gassen, H.-G., Martin, A., and Sachse, G., *Der Stoff, aus dem die Gene sind,* Schweitzer, Munich, 1986.

Goodenough, P. W., in Tucker, G. A., and Woods, L. F. J. (eds.), *Enzymes in Food Processing,* Blackie, Glasgow, 1991, p. 35.

Goodenough, P. W., in Tucker, G. A., and Woods, L. F. J. (eds.), *Enzymes in Food Processing,* 2nd ed., Blackie Academic & Professional (imprint of Chapman & Hall), New York, 1995, p. 41.

Green, J. R., *Die Enzyme,* Parey, Berlin, 1901.

Jaag, H., "Über proteolytische Enzyme, deren Prüfmöglichkeit in der Waschmittelindustrie," *Eur. Hochschulschr. Reihe VIII Chemie, Sec. B, Biochem.,* Herbert Lang, Bern, 1968.

Kirchoff, K., *Schweig's J.* **14,** 389 (1815).

Kühne, W. F., *Erfahrungen und Bemerkungen über Enzyme und Fermente,* Physiol. Inst., Univ. Heidelberg, 1878.

Ostwald, F. W., in *Enzymologie,* Hoffmann-Osteuhot, O. (ed.), Springer Wien, 1954, p. 9.

Payen, F. W., and Persoz, J. F., *Ann. Chim. Phys.* **53,** 73 (1833).

Reisner, W., *Röhm Spektrum* **27,** 48 (1981).

Röhm, O., "Preparation of hides for the manufacture of leather," U.S. Patent 886,411 (1908a).

Röhm, O., "Aufbereitung der Häute für die Lederherstellung," German Patent 200519 (1907b).

Shoemaker, S., in Harlander, S. K., and Labuza, T. P. (eds.), *Biotechnology in Food Processing,* Noyes, Parkridge, NJ, 1986, p. 259.

Soda, K., and Yonaha, K., in Rehm, H.-J., and Reed, G. (eds.), *Biotechnology,* Vol. 7a, VCH, Weinheim, 1987, p. 605.

Takamine, J., "Preparing and making taka-koji," U.S. Patent 525,820 (1894).

Tammann, G., *Ztschr. Physiol. Chemie* **16,** 271 (1892).

Underkofler, L. A., Barton, R. R., and Rennert, S. S., *Appl. Microbiol.* **6,** 212 (1958).

Van't Hoff, J. H., *Ztschr. anorg. allg. Chemie* **18,** 1 (1898).

Wallerstein, L., "Treating beer or ale," U.S. Patent 995,825 (1911).

2 General Characteristics of Technical Enzymes

2.1 HOW ENZYMES WORK

A wealth of scientific literature addresses this issue. The current state of knowledge on enzymes can be found under captions such as "enzyme structure," "enzymatic activity," "active site," "mechanism," and "enzyme technology." Presently, about 50,000 publications on enzymes appear each year. As any attempt to cover the total field of enzyme technology would burst this book at the seams, only those factors that deal with the general understanding and the commercial applications of enzymes will be covered. A short summary of enzymatic activity and kinetics is addressed for the practitioner and taken from a publication by Sprössler (1978). For more in-depth reading on enzyme structure, mechanisms of action, activity, and parameters that influence enzymatic activity, kinetics and enzyme analysis, the reader is referred to standard textbooks on this subject (Cornish-Bowden and Wharon, 1988; Dixon and Webb, 1979; Segel, 1975).

We saw earlier that enzymes as catalysts are able to rapidly adjust the equilibria of chemical reactions. For example, ester formation between an acid and an alcohol shows that reaction equilibrium is also reached, but extremely slowly, in the absence of an enzyme. Experience has shown, however, that the rate of a chemical reaction depends on the temperature. This also holds true for chemical reactions in food. Butter, for example, becomes rancid more rapidly at room temperature than when refrigerated. At very low temperatures the acid–alcohol system can be regarded as approximately stable in that there will be no reactions between the components.

2.1.1 Catalysis: A Lowering of the Energy Barrier

Change in a stable condition can be achieved only by supplying energy. An energy barrier must be surmounted. This can be done in one of two ways:

- The necessary activation energy can be generated. The energy required is inversely proportional to the temperature; that is, the activation energy decreases with increasing temperature and vice versa. This means however, that very high temperatures are required for certain reactions. Where extremely high temperatures must be avoided, or where the delivery of large amounts of energy is economically infeasible, another possibility remains.

• The energy barriers may be removed with enzymes that can reduce energy barriers.

2.1.2   Reduction of Energy Barriers by Contact with Active Sites

Enzymes are large, three-dimensional protein molecules with an active site at a defined location on the folded surface. This part of the surface can be envisioned as a pocket that will permit entry only to a specific substrate for a reaction to occur. Emil Fischer (1884) stipulated that enzyme and substrate must fit like a lock and key.

The temporary bonds between the enzyme and substrate that form the enzyme-substrate complex will loosen the bonds that hold the substrate together. Thus, the energy barrier for cleaving is lowered and the reaction can proceed and reach equilibrium at room temperature. Products A and B are formed from the substrate and the enzyme is liberated again ready to catalyze the next reaction (Fig. 2.1). For the

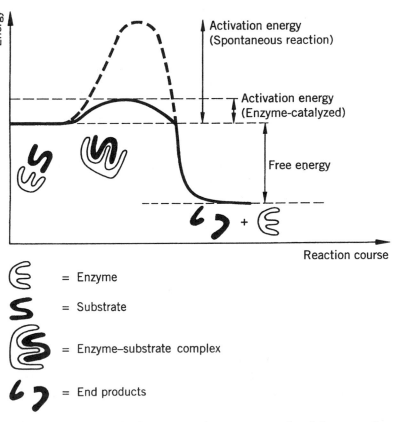

Figure 2.1   Lowering the activation energy of an enzyme-catalyzed decay reaction compared to spontaneous decay.

fastest reactions, the contact time between enzyme and substrate is only about $\frac{1}{85,000\text{th}}$ of a second.

2.1.3   Enzyme Specificity for Substrate Binding

According to Emil Fischer, the substrate binds to the enzyme only when it matches the active site. In other words, enzymes possess a high degree of specificity. A protease can only recognize and cleave proteins but will not react with starch molecules.

However, Fischer's lock and key analogy can be modified in that the substrate, and especially the enzyme, can undergo conformational changes. Zech and Domagk (1986) stated that "the garment can change to fit the figure and the figure to fit the garment." Thus, the enzyme exhibits not only substrate specificity but also, in some cases, a certain flexibility (Fig. 2.2).

2.1.4   Enzymatic Activity: Conversion Rate per Unit Time

Enzymes are evaluated according to their activities. The following can serve as a simplified analogy. Group A with 10 workers saw 10 cubic meters (m^3) of wood in one hour. Group B, consisting of 20 workers, requires the same time. Thus, work group B is only half as active as group A. Enzymatic activity is determined in a similar way. An enzyme is more active than another when a smaller quantity is required for a specific conversion. A measure of conversion per unit time is the amount of product formed per minute under well-defined, standardized conditions.

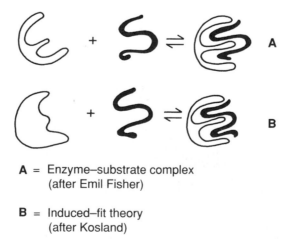

A = Enzyme–substrate complex
(after Emil Fisher)

B = Induced–fit theory
(after Kosland)

Figure 2.2   Lock-and-key concept; conformational change of enzymes.

2.1.5   Optimal Conditions for Enzymatic Activity

Enzymes need an optimal supply of substrate. The substrate should saturate the enzyme. Again, using our example, this means that the workers can reach their maximum performance only when there is sufficient wood available to be sawed.

Enzyme and substrate must have a constant and unimpaired contact for maximum enzymatic activity. This occurs when the enzyme and substrate are present in dilute aqueous solutions. The performance of the enzyme is reduced or impeded when the substrate is insoluble. Dry solids are enzymatically inert.

2.1.6   Enzymes Work at a Constant Rate

Generally, as long as the reaction conditions do not change, twice the yield of product will be generated in twice the time (Fig. 2.3).

The curve in Figure 2.4 shows an idealized curve of pH dependence found for a purified enzyme acting on a defined substrate. The shoulder of this curve suggests that a second enzyme with a different pH dependence is also active (side activity).

The conversion rate is reduced when there is insufficient substrate available to saturate the enzyme or the enzyme is denatured because of an increase in temperature (inactivation).

2.1.7   pH Dependence of Enzymatic Activity

Every enzyme requires a specific pH value or pH range for optimal activity. The pH activity curve shown in Figure 2.4 is characteristic for an enzyme. This factor is of

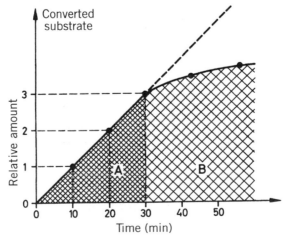

Figure 2.3   Product formation as a function of time. A = enzyme–substrate complex according to Emil Fischer; B = induced-fit theory according to Kosland.

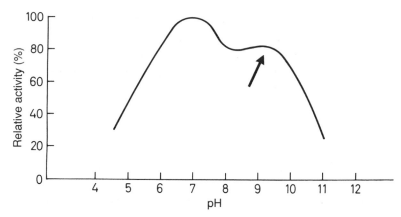

Figure 2.4   Activity as a function of the pH. Enzyme preparation: neutral bacterial protease containing some alkaline protease (→).

prime importance when choosing an enzyme for an industrial process. For example, when clarifying an acidic fruit juice with pH <3, an enzyme with optimum activity in the pH range of 4–5 would show only slight activity at pH 3. In order to operate at maximum enzyme efficiency, another enzyme that has optimum activity at pH 3 must be chosen.

2.1.8   Temperature Dependence of Enzymatic Activity

Enzyme performance usually improves with increasing temperature. Generally, the enzymatic conversion rate doubles on a 10°C temperature increase. This holds true from about 10–40°C. Enzymes may also be active at relatively low temperatures, as is evident in the development of rancidity in butter on long-term refrigeration. Beyond the optimal temperatures, enzymes may be denatured. The temperature at which inactivation begins is characteristic for every enzyme. In industry the optimum temperature range for a given enzyme reaction is that at which the enzyme is still just sufficiently stable.

2.1.9   Enzyme Stability at High Temperatures

The temperatures at which enzymes are stable or labile is of great significance for many technical processes. In some processes enzymes should be sufficiently labile so as to be completely inactivated at temperatures of 70–80°C. On the other hand, enzymes that are active above 100°C are needed for the modern production of glucose from starch. Here, high enzyme stability is of significant economic value. Therefore, the search for temperature stable enzymes has a high priority in enzyme research. It should also be mentioned that high substrate concentrations help stabilize enzymes in some industrial processes. For example, starch hydrolyzing amylases in the presence of 30–40% starch cannot be inactivated by boiling.

2.1.10   Enzyme Sensitivity and Susceptibility to Inactivation

An example of enzyme sensitivity was mentioned in Section 1.4 during discussion of the use of enzymes in modern detergents. Bacterial proteases are insensitive to the action of perborate or anionic detergents, while detergents inactivated the pancreatic proteases employed earlier. In addition to heavy metals, which generally destroy enzymes, there are a number of chemicals and many natural substances that inhibit or completely inactivate enzymes.

2.2   ENZYME STRUCTURE AND ACTION MECHANISMS

The isolation of crystalline enzymes was a deciding factor in resolving the long-ongoing debate on the nature of enzymes. Sumner in 1926 was the first person to produce an enzyme in crystallized form. Because this enzyme catalyzed the hydrolysis of urea to carbon dioxide and ammonia, he called it *urease*. In subsequent years, additional crystalline enzymes were isolated and hydrolyzed into their basic amino acid building blocks.

Many enzymes, such as the hydrolases, consist entirely of protein. Most (industrial) enzymes also belong to this category. Others contain, besides the protein component, a nonprotein prosthetic group that is essential for their activity.

Proteins are amino acid polymers, joined by peptide linkages (polypeptide chains). In contrast to starch or cellulose, where the polymer chains are composed of linked building blocks of the same type, proteins can contain up to 21 different amino acids. In proteins, there are usually hundreds of amino acids bound together. Their sequence is random, and a great variety of different proteins exist. In addition, the polypeptide chains (primary structure) adopt a more or less rigid, helical configuration (secondary structure). A three-dimensional shape develops on the folding of these helixes; this is designated as the tertiary structure.

The specific way in which the protein is folded is dictated by the primary sequence of the amino acids. Most proteins fold spontaneously into their correct form. Heat or other environmental changes can lead to the loss of this conformation and cause the proteins to become denatured.

All proteins in living cells have specific functions. These functions are governed by their interactions with other molecules or by the given amino acid sequences of the polypeptide chain. These are, for example, the regions that bind to other molecules. The location of catalytic activity in enzymes is called the *active site*. Other functions of proteins in living cells include the transport of metabolites and the formation of cell structures. All enzymes are proteins, but not all proteins are enzymes (Yamamoto, 1978).

Most technical enzymes have molecular weights between 20,000 and 70,000 daltons. The active center, however, is composed of only a few amino acids. At least three amino acid residues of trypsin (Ser, His, and Asp) participate in the proteolytic activity of this enzyme (Fig. 2.5a). Initially the protein substrate becomes associated with the enzyme to form an enzyme–substrate complex (Fig. 2.5b). Interaction between the serine residue and the histidine residue within the active center leads to activation of the serine residue. The activated serine residue then reacts with a peptide linkage in the substrate protein. A serine ester between the peptide chain of the substrate and the enzyme is formed (Fig. 2.5c), whereby, in one step, the peptide linkage in the peptide chain is broken. Peptide 2 is released (Fig. 2.5d). Finally, the ester linkage in the active serine is hydrolyzed by the addition of water and peptide 1 is released. The enzyme has thus returned to its original state (Fig. 2.5e).

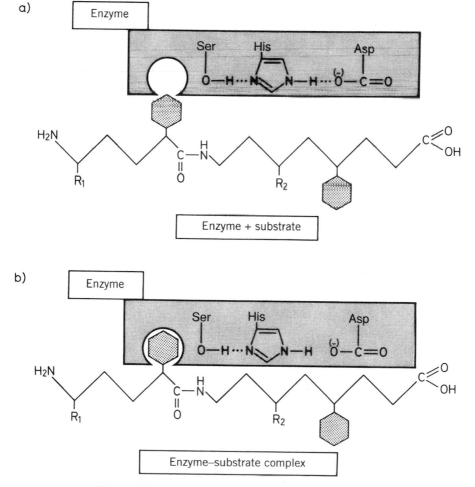

Figure 2.5   Proteolysis catalyzed by trypsin (*continued*).

c)

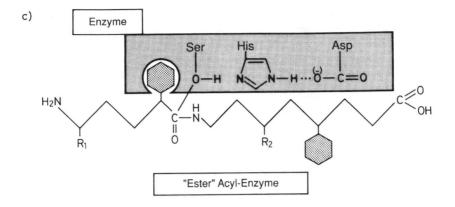

"Ester" Acyl-Enzyme

d)

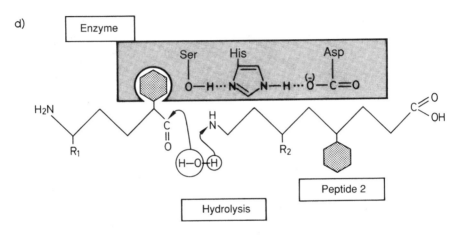

Hydrolysis

Peptide 2

e)

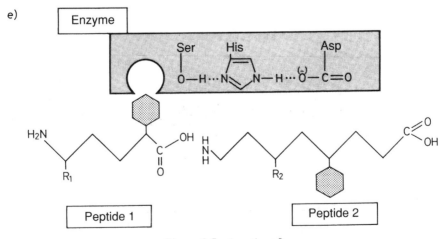

Peptide 1

Peptide 2

Figure 2.5   (*continued*)

2.3 ENZYME ANALYSIS AND ENZYME UNITS

Several methods are available for determining the hydrolytic activity of enzymes. The substrates are primarily naturally occurring polymers. The same polymers that are used as industrial substrates are employed. These polymers are usually diverse with respect to their chemical composition and molecular weight. Defining enzyme activity with these substrates is more difficult than when low molecular weight substrates are used. In industry the latter are seldom used; therefore, allowances are generally made for problems associated in defining activity. Thus, there are very few internationally recognized methods for determining hydrolase activities. Most methods are specific for each manufacturer, and it is difficult to directly compare various activity specifications from different sources.

2.3.1 Automated Enzyme Analysis

Enzyme analyses are time-consuming manual procedures that have demanding personnel requirements. The individual steps must be precisely timed. For processing several samples simultaneously, the various steps must be coordinated. If numerous samples need to be analyzed, it is worthwhile to employ autoanalyzers. Two different systems are commonly used:

- *Fixed-volume analyzers* work with fixed controllable volumes of each reaction compound, similar to manual procedures. Automation allows the processing of more samples per unit time. In a thermostated reaction chamber, the sample, substrate, and reagents are added from dosing stations for the wet chemical analysis of the reaction products. The timer-regulated sample carousel rotates, passing each sample by the various dosing stations. A detection system at the end of the reaction sequence measures the reaction and transfers the signals to a central control unit, which records the data and computes the results (Fig. 2.6).
- Alternative systems are continuously operating analyzers known as *autoanalyzers*, or the flow-injection processors such as those used in clinical analysis.

The importance of analytic methodology for enzyme research demands continuous development in microbial selection and enzyme production and application. Future innovations will be primarily technical. Large numbers of equal samples make programmable robots and pipetting stations economical. Another important consideration is to put existing methods on line to continuously monitor and control individual process steps. This is of particular interest in the production of enzymes from live microorganisms.

2.3.2 Enzyme Units

Some of the methods used internationally for industrial enzymes are listed below. Because of the variation in reaction conditions, different results are often obtained; therefore, it is necessary to be cautious when making comparisons.

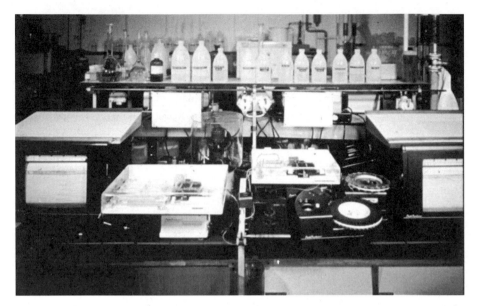

Figure 2.6   Automated amylase analyzer (developed by Röhm GmbH).

2.3.2.1 *Proteases*

- *AU:* The Anson unit (AU) defines protease activity. The test uses hemoglobin at pH 4.7 (original method). There are modifications of this method at pH 5, 7.5, and so on. Crystalline subtilisin has 25–30 AU/g. Technical enzyme preparations contain 2.5–3 AU/g (Anson, 1938; Kuntiz 1947; Sommer, 1967).

- *BAPA:* One BAPA *Na*-benzoyl-L-arginine-*p*-nitranilide) unit corresponds to the amount of protease required to convert 1 μmol of substrate in 1 min under standard reaction conditions.

- *NU:* One Northrop unit (NU) is the amount of protease required to hydrolyze 40% of one liter of casein substrate in 60 min under standard reaction conditions.

- *PU:* One protease unit (PU) is the amount of enzyme that liberates trichloroacetic acid–soluble fragments from casein, equivalent to 1 μg of tyrosine in 1 min, at 30°C under standard reaction conditions.

- *HbU:* One hemoglobin unit (HbU) is the amount of protease that releases 0.0447 mg of nitrogen from amino acids and peptides (determined in a spectrometer at 275 nm) in 30 min under standard reaction conditions. The original method calls for pH 4.7. This method is of particular interest for a protease used in beer stabilization. The Anson method is more appropriate for the neutral pH range. Purified acid fungal protease has an activity of ~1 mil HbU/g protein.

- *LVU:* One Löhlein–Volhard unit (LVU) corresponds to the amount of protease that increases casein fragments in 20 ml of filtrate of a 4% casein solution, equivalent to the action of 5.75×10^{-3} ml 0.1 N NaOH (Künzel, 1955).

2.3.2.2 Carbohydrases

- *DP°:* The unit is defined as "degrees of diastatic power" (DP°) and is the amount of enzyme present in 0.1 ml of a 5% enzyme solution yielding the amount of reducing sugar (in 100 ml of a 1% starch solution at 20°C for 1 h), which reduces 5 ml of Fehling's solution.
- *DU:* The dextrinogen unit (DU) is used to measure α-amylase activity, particularly bacterial α-amylase. One DU is the amount of enzyme that catalyzes the conversion of 1 mg of starch to dextrins under defined reaction conditions. Conditions (Roehm): 1.2% soluble starch (amylum solubile, Merck AG); acetate buffer pH 5.7; reaction time 10 min, 30°C.

According to this method, crystalline bacterial α-amylase has an activity of ~30,000 DU/mg protein. Technical enzyme powders show activities between 100 and 1000 DU/mg; liquid preparations, about 100–300 DU/ml.

- *GAU:* One glucoamylase [amyloglucosidase (AMG)] unit (GAU) is the amount of enzyme that liberates 1 g of glucose from soluble starch in 1 h (Miles, diazyme method).
- *SKB-U:* One SKB unit (SKB) is the amount of amylase that hydrolyzes 1 g of β-limit dextrin to the iodine-negative point in 1 h under defined reaction conditions. In more recent modifications of the method, the β-limit dextrin is replaced by soluble starch (Sandstedt et al., 1939). Crystalline fungal α-amylase has an activity of ~250,000 SKB/g; the activity of technical preparations ranges from 50,000 to 120,000 SKB/g.
- *NF-U:* According to NF (The National Formulary, 1960, USP XIV), one unit (NF) is defined as the time required for 1 g of enzyme preparation to hydrolyze 100 g of starch in a 3.75% starch solution at 40°C. The endpoint is reached when the iodine color disappears.
- *Liquefon-U:* The liquefon method by Sandstedt et al. (1947) determines the time required to reduce the relative viscosity of a defined amount of starch paste by 50%.
- *MWU:* One modified Wohlgemuth unit (MWU) is the amount of enzyme needed to hydrolyze 1 mg of soluble starch to specific dextrins under standard reaction conditions in 30 min.

Pectinases Numerous methods are based on either the decrease in viscosity of a pectin solution or the complete depectinization of apple juice.

- *PG-U:* One polygalacturonase unit (PG) is defined as the enzymatic activity required to lower the viscosity of a standard pectin solution by $1/\eta_{\text{specific}} = 0.000015$ under standard reaction conditions.

- *AJD-U:* One apple juice depectinization (AJD) unit is the amount of enzyme that completely depectinizes 1 liter of a standard apple juice under standard reaction conditions.
- *PA-U:* Mixtures of pectinases have a maximum of ~1500 PA/g protein. This means that 1 g of such a preparation can depectinize 1500 liters of apple juice under standard reaction conditions (after Röhm GmbH).
- *PE-U:* One pectinesterase unit (PE) is the amount of enzyme needed to liberate 1 μmol of titratable carboxyl groups per minute under standard reaction conditions.
- *PTE-U:* Pectin transeliminase (PTE) units are determined by hydrolysis of pectin and measurement of the fragments at 235 nm (Albersheim and Killias, 1962).

Cellulases

- *CU:* One cellulase unit (CU) is the amount of enzyme that releases 1 μmol of glucose from a 1.5% solution of carboxymethylcellulose (CMC) at 30°C and pH 4.5. Determination of the reducing sugars can be measured calorimetrically using, for example, *p*-hydroxybenzoic acid.
- *Xyl-U:* The activity of a xylanase unit (Xyl-U) corresponds to the amount of enzyme that can release reducing sugar equivalent to 1 mmol of xylose per minute from a xylan solution at 30°C and standard reaction conditions. The degradation products are determined according to Nelson (1944) or Somogy (1952).

2.3.2.3 Oxidoreductases

Oxidases   The glucose oxidase unit definition varies with the enzyme source and supplier.

GLUCOSE OXIDASE

- *GO-U:* One glucose oxidase (GO) unit will liberate 1 μmol of H_2O_2 per min at 25°C and pH 7 (ICN Biomedicals, Inc., 1996).
- *GO-U:* One glucose oxidase unit (GO) is defined as the amount of enzyme that will oxidize 1 μmol of β-D-glucose per min at 25°C and pH 4.1 (Calbiochem-Novabiochem, 1996).
- *GO-U:* One glucose oxidase unit (GO) will oxidize 1 μmol of β-D-glucose to D-gluconic acid and H_2O_2 per min at 35°C and pH 5.1 (Sigma Chemical Co., 1996).

CATALASE

- *BU:* One Baker unit (BU) (Baker, 1953), is the amount of catalase (fungal catalase) that will decompose 264 mg of hydrogen peroxide under the reaction conditions defined by Scott and Hammer (1960).

- *CaU:* One catalase unit (CaU; Japanese method [UA], [AM]) is the amount of catalase that degrades 1 mmol of hydrogen peroxide in 1 min at 30°C (iodine thiosulfate titration according to Stellmach, 1988).
- *KU:* One Keil unit (KU) is the amount of catalase (liver catalase) necessary to decompose 1 g of 100% hydrogen peroxide in 10 min at 25°C and pH 7 in an inert atmosphere of CO_2 or N_2 (Keil, 1954; Scott and Hammer, 1960).

Lipases   References to other methods of determining enzymatic activity can be found in the chapters on individual enzymes.

Generally, in commercial usage, the unit of lipase activity (LU) is measured as the amount of enzyme needed to produce a certain amount of acid that is determined by a pH-stat. Commercial manufacturers employ two kinds of assay: one using soluble esters and another one with a substrate such as olive oil in a standardized emulsion.

A more sensitive assay has been developed for clinical applications in which fluorescence is used to follow the hydrolysis of substances such as umbelliferyl heptanoate. Apparently, this assay is also useful in revealing lipases on gel electrophoresis (Tombs, 1995).

2.4 ENZYME KINETICS

Dealing with enzyme kinetics often seems abstract to industrial practitioners and therefore an uninteresting task that should be reserved for basic scientific research, but this should not be the case. The regular use of enzymes often leads to situations that require knowledge of enzyme kinetics.

An enzyme-catalyzed conversion is a dynamic process that can be influenced by a number of factors, which should be considered. These include the type and concentration of the participating substances, temperature, pH value and ionic strength, and natural properties of the enzyme such as its catalytic specificity or its stability under the given conditions.

Enzyme kinetics (*kinetein* = to move) describe the interplay of these factors. The kinetics do not a priori assume complex mathematical correlations to formally describe the dynamic processes during an enzymatic reaction. Enzyme kinetics are primarily an attempt to analyze the data obtained from an enzymatic reaction and to use the data to optimize the reaction. This, of course, provides a detailed description and characterization of the enzyme as well.

2.4.1 Enzymatic Activity

Enzymes catalyze the conversion of a substance S into a product P. The activation energy necessary to obtain a reaction equilibrium is adjusted by the participating enzyme to physiological conditions (normal pressure and room temperature).

Because of the nature of this process, most data that characterize an enzyme can be obtained only indirectly by following the reaction course that is continually influenced by the enzyme. Data on the proteinaceous nature of an enzyme are of little interest to the practitioner.

A useful value is *enzymatic activity,* which describes its productivity under standardized conditions. It describes how much of a given substance S is converted within a given time and under defined conditions into a corresponding product. Enzymatic activity, which is a kinetic quantity, is always dependent on the time and conditions under which the reaction measurements were made.

Enzymatic activity has the dimensions of velocity: dP/dt or $-dS/dt$.

2.4.2 Enzymatic Reaction

An enzymatic reaction, in the simplest case, is the unimolecular reaction developed by Michaelis and Menton (1913). The enzyme–substrate interaction is considered to occur as shown in the following equation:

$$S + E \underset{k_2}{\overset{k_1}{\rightleftharpoons}} E \cdot S \overset{k_3}{\longrightarrow} P + E$$

where E = free enzyme; S = substrate; ES = enzyme–substrate complex; and k_1, k_2, k_3 = rate constants for the formation of ES, release of S, or release of P, respectively. This is a multistep process (Fig. 2.7), usually with a short initial phase.

A physically intimate contact between substrate and enzyme, or, more specifically, with the catalytic center, is necessary for the conversion of a given substrate.

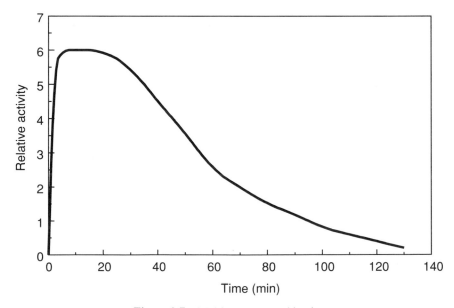

Figure 2.7   Multistep enzyme kinetics.

During this initial phase, a quasi-stationary substrate–enzyme pool accumulates; this plays a major role in kinetic model studies. Increasingly less enzyme remains inactive in this phase because it is not sterically close to the substrate.

The initial phase is normally very short and can be measured only by special methods (short-term kinetics); this initial phase is not significant in practical applications.

Next is the phase of highest efficiency, the *steady state,* in which the enzyme–substrate concentration [ES] reaches a dynamic equilibrium. The formation and disassociation of [ES] are in balance.

$$k_1 \, [E] \, [A] = k_2 \, [ES] + k_3 \, [ES] \tag{1}$$

The concentration of free enzyme in this state is minimal under normal conditions ([S] >> [E]), because nearly everything is bound within the complex. The velocity reaches a maximum under these conditions.

Since both enzyme and substrate are bound, only the given enzyme concentration $[E_{total}]$, or [ES], will limit the reaction velocity and thus product formation.

$$v = V_{max} = \frac{dP}{dt} = k' \cdot [ES] = k' \cdot [E_{total}] \tag{2}$$

Generally, the transition from the enzyme–substrate complex to product is the rate-limiting step; therefore, it can be simply stated that

$$k' \approx k_3 \tag{3}$$

This is the Michaelis–Menten relationship, based on Briggs and Haldane, which can be derived as

$$v = \frac{k_3 \cdot [E_{total}] \cdot [A]}{K_m + [A]} \quad \text{or if} \; V_{max} = k_3 \cdot [E_{total}], \; \text{then} \quad v = \frac{V_{max} \cdot [A]}{K_m + [A]} \tag{4}$$

The constant K_m, which is generally referred to as the *Michaelis–Menten* constant, represents the expression $k_2 + k_3/k_1$; this constant is, together with k_3, the catalytic or turnover constant, are the enzyme-specific units that determine an enzyme's efficiency with a given substrate. The constant K_m corresponds to the substrate concentration at which half of the maximum reaction rate is achieved. The smaller this constant, the less substrate is required for a high reaction velocity.

Again, k_3 is called the *turnover* or *catalytic constant* as mentioned above. This designation implies that it deals with the inherent activity of the enzyme, namely, the conversion of a given substrate into a corresponding product, while K_m approximates (when $k_3 << k_1$) the affinity of a substrate for the given enzyme.

The expression k_3/K_m is often used to express the productivity of an enzyme. The efficiency of an enzyme is greater at smaller K_m; this means, more simply, that when the affinity between the enzyme and substrate is greatest and the turnover rate

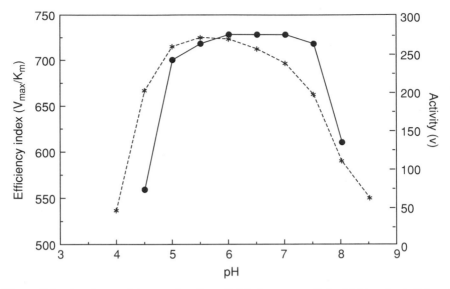

Figure 2.8 Amylase activity as a function of pH, shown as activity (right ordinate *v) and as efficiency index [left ordinate ● (V_{max}/K_m)].

is highest, then the maximum possible reaction velocity has been reached at the given enzyme concentration.

Figure 2.8 shows that there is a close relationship between the efficiency index (V_{max} or k_3/K_m) and enzymatic activity because in both cases substrate saturation is measured under comparable kinetic conditions.

For an industrial process the complete conversion of a substance is frequently desired, as, for example, in starch hydrolysis. Inevitably, the reaction will deviate from the optimum steady-state limiting conditions with time, and parameters other than enzyme dose will noticeably influence the reaction.

The substrate concentration will finally become the limiting factor and must be taken into account. A zero-order reaction ($dS/dt = k \cdot [E]$) will become a reaction of a higher order, in the simplest case, a first-order reaction ($dS/dt = k \cdot [S]$). Therefore, an evaluation of the reaction course with Equation (4) is unrealistic because the critical rate-limiting end phase of an enzymatic reaction is not taken into account.

The completion of the desired reaction is approached only slowly (asymptotically). In such cases, estimates of the time required and of the attainable reaction efficiency are helpful. Incorporating the Briggs–Haldane relationship makes this approximation possible. The time t required to reduce a given substrate concentration [S_0] to the desired value [S] is

$$t = K_m \cdot \ln \frac{S_0}{S} + \frac{S_0 - S}{k_3 \cdot E_{total}} \qquad (5)$$

This is an approximation, because $k_3 \cdot E_{total}$ does not necessarily describe the actual rate for each of the reaction phases. Furthermore, there are other factors; for example, a loss in enzymatic activity during the reaction or retrogradation of reaction metabolites can substantially affect the result.

2.4.3   Procedures for Determining Kinetic Units

This requires measurement of reaction rates at different substrate concentrations. Because K_m and V_{max} cannot be derived exactly from a plot of $v/[S]$ directly (Fig. 2.9), different ways of linearizing the data used which will permit reliable extrapolation. The linearization according to Lineweaver and Burk is the one most commonly used (Fig. 2.10). The Briggs–Haldane relationship can be rearranged to

$$\frac{1}{v} = \frac{K_m}{V_{max} \cdot 1/[S_0]} + \frac{1}{V_{max}}$$

A plot of $1/v$ over $1/[S_0]$ yields a straight line. The y-axis intercept corresponds to $1/V_{max}$, the x-axis intercept corresponds to $-1/K_m$. The use of this graph for a reliable determination of the units in question is limited because various substrate concentrations may be weighted unequally. The result obtained with lower substrate concentrations has a greater influence on the slope of the straight line than that achieved with higher concentrations.

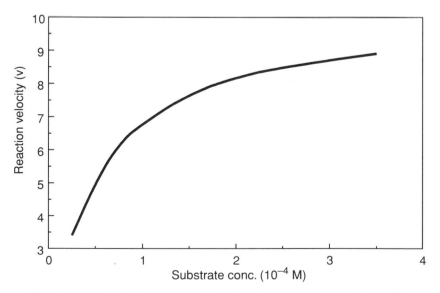

Figure 2.9   Reaction velocity (v) as a function of substrate concentration [S].

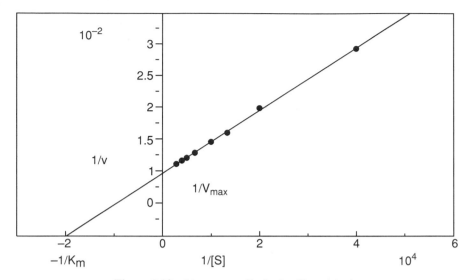

Figure 2.10   Lineweaver–Burk plot (linearizing).

A better, more balanced method is the analysis according to Eadie and Hofstee, which employs the following relationship, derived from equation (4);

$$v = V_{max} - \frac{K_m \cdot v}{[S_0]}$$

When plotting v versus $v/[S_0]$, the intercept of the straight line with the y axis gives V_{max} and the slope of the straight line gives $-K_m$ (Fig. 2.11).

The data processing is greatly simplified and done more accurately with the software especially developed for personal computers (PCs), such as "ENZFITTER" (Elsevier-Biosoft) or "EZ-FIT" (Perrella, 1988). This type of data processing does not involve common linearization procedures, which are known to weight individual measured points unequally, but rather subjects the original data from the Michaelis–Menten kinetics to a nonlinear regression analysis. The various classic linearizations can also be called up and expressed graphically.

2.4.4   Closing Remarks

Enzyme kinetics, in the conventional sense, is essentially an attempt to derive, via skillful experimentation, enzyme-specific parameters that can make a protein an efficient tool. These parameters can be measured only when the enzyme "works," that is, achieves output and can thereby be characterized.

The presence of an enzyme can be detected only indirectly by its catalytic function, which causes changes in the reaction mixture, as well as by the amount of active enzyme that was added. Whether an experiment can in fact yield relevant data

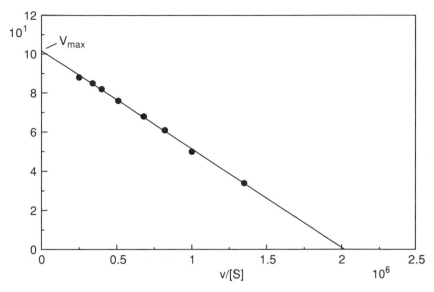

Figure 2.11   Eadie–Hofstee plot (linearizing).

will ultimately depend on the experimenter's comprehensive understanding of enzyme kinetics because different factors such as enzyme concentration can influence the reaction. In order to study enzyme characteristics, the experimenter must be able to recognize and elucidate such factors from the way the reaction proceeds.

These factors or parameters, often thought to be not directly associated with the concept of enzyme kinetics, are, however, instrumental in comparing and quantifying enzymatic activity and performance:

- Enzyme concentration
- Dependence of reaction rate on pH, ionic strength, temperature, and moisture
- Enzyme stability under given conditions
- Substrate specificity of the enzyme
- Presence of activators, inhibitors, and cofactors
- Antagonistic and synergistic effects between various enzymes

The reaction velocity v of an enzyme-catalyzed reaction is the key for such a parameter analysis, and, similar to the performance of a complex system, can in fact be tested only by operating under defined conditions.

2.5   NOMENCLATURE OF INDUSTRIALLY IMPORTANT ENZYMES

See Table 2.1.

TABLE 2.1   Nomenclature of Industrially Important Enzymes

EC No.	CAS No.	Systematic Name[a]	Other Names
		Oxidoreductases[b]	
1.1.3.4	[9001-37-0]	β-D-Glucose:O_2 1-oxidoreductase	Glucose oxidase, Notatin
1.1.3.22	[9002-17-9]	Xanthine:O_2 oxidoreductase	
1.8.3.2	None	Thiol:O_2 oxidoreductase	
1.10.3.3	[9029-44-1	L-Ascorbate:O_2 oxidoreductase	Ascorbic acid oxidase
1.11.1.6	[9001-05-2]	H_2O_2:H_2O_2 oxidoreductase	Catalase
1.11.1.7	[9003-99-0]	Donor:H_2O_2 oxidoreductase	Peroxidase, lactoperoxidase
1.13.11.12	[9029-60-1]	Linoate:O_2 13-oxidoreductase	Lipoxygenase, lipoxidase, carotene oxidase
1.14.18.1	[9002-10-2]	o-Diphenol:O_2 oxidoreductase	Polyphenoloxidase, laccase, tryosinase
		Transferases[c]	
2.4.1.2	—	1,4-α-D-glucan:1,6-α-D-glucan 6-α-D-glucosyltransferase	Dextrin dextranase, dextrin 6-glucosyltransferase
2.4.1.5	[9032-14-8]	Sucrose:1,6-α-D-glucan 6-α-D-glucosyltransferase	Dextran sucrase
2.4.1.18	[9001-97-2]	1,4-α-D-Glucan:1,4-α-D-glucan 6-α-D-(1,4-α-D-glucano)-transferase	1,4-α-D-Glucan branching enzyme, Q-enzyme, amylo-(1,4→1,6)-transglycosilase
2.4.1.19	[9030-09-5]	1,4-α-D-Glucan:4-α-D-(1,4-α-D-glucano)-transferase, cyclizing	Cyclodextrinase, cyclodextrin glucanotransferase, cyclomaltodextrin glucanotransferase
		Hydrolases[d]	
3.1.1.1	[9016-18-6]	Carboxylic ester hydrolase	Carboxyl esterase, aliesterase
3.1.1.2	[9032-73-9]	Arylester hydrolase	Arylesterase, paraoxonase
3.1.1.3	[9001-62-1]	Triacylglycerol acylhydrolase	Lipase, steapsin, tributyrase
3.1.1.4	[9001-84-7]	Phosphatidylcholine 2-acylhydrolase	Lecithinase A, phospholipase A_2
3.1.1.5	[9001-85-8]	2-Lysophosphatidylcholine acylhydrolase, lysolecithin acylhydrolase	Lecithinase B, phospholipase B, lysophospholipase, lysolecithinase
3.1.1.11	[9025-98-3]	Pectin pectylhydrolase	Pectinesterase, pectase
3.1.1.20	[9025-71-2]	Tannin acylhydrolase	Tannase (digallate + H_2O = 2 gallate)
3.1.4.3	[9001-86-9]	Phosphatidylcholine cholinephosphohydrolase (belongs to the phosphodiester hydrolases)	Lecithinase C, phospholipase C, lipophosphodiesterase I

TABLE 2.1 *(Continued)*

EC No.	CAS No.	Systematic Name[a]	Other Names
3.1.4.4	[9001-97-0]	Phosphatidylcholine phosphatidohydrolase II	Lecithinase D, phospholipase D, lipophosphodiesterase II
3.2.1.1	[9000-90-2]	1,4-α-D-Glucan glucanohydrolase	α-Amylase, diastase, ptyalin, glycogenase
3.2.1.2	[9000-91-3]	1,4-α-D-Glucan maltohydrolase	β-Amylase, saccharogen amylase, glycogenase
3.2.1.3	[9032-08-0]	1,4-α-D-Glucan glucohydrolase	Amyloglucosidase, glucoamylase, exo-1,4-α-glucosidase
3.2.1.4	[9012-54-8]	1,4-(1,3;1,4)-β-D-Glucan 4-glycanohydrolase	Cellulase, endo-1,4-β-glucanase
3.2.1.6	[62213-14-3]	1,3-(1,3;1,4)-β-D-Glucan 3(4)-glucanohydrolase[a], endo-1,3(4)-β-D-glucan glucanohydrolase	Laminarinase (full action on 1,3;1,4-glucan linkages) (different from EC 3.2.1.39) (McCleary and Codd, 1989)
3.2.1.7	[9025-67-6]	2,1-β-D-Fructan fructanohydrolase	Inulinase, inulase
3.2.1.8	[9025-57-4]	1,4-β-D-Xylan xylanohydrolase	Xylanase, endo-1,4-β-xylanase (see EC 3.2.1.32)
3.2.1.10	[9032-15-9]	Dextrin 1,6-α-D-glucanohydrolase, oligo-1,6-glucosidase[a], dextrin 6-α-D-glucohydrolase[a]	Isomaltase
3.2.1.11	[9025-70-1]	1,6-α-D-Glucan 6-glucanohydrolase	Dextranase
3.2.1.14	[9001-06-3]	Poly[1,4-(N-acetyl-β-D-glucosaminide)] glycanohydrolase; poly[1,4-β-(2-acetamido-2-deoxy-D-glucoside)] glycanohydrolase	Chitinase, chitodextrinase, poly-β-glucosaminidase
3.2.1.15	[9032-75-1]	Poly-(1,4-α-D-galacturonide) glycanohydrolase	Pectinase, endopolygalacturonase, pectin depolymerase, polygalacturonase
3.2.1.17	[9001-63-2]	Mucopeptide N-acetylmuramoylhydrolase	Lysozyme, muramidase
3.2.1.20	[9001-42-7]	α-D-Glucoside glucohydrolase	Maltase, α-glucosidase
3.2.1.21	[9001-22-3]	β-D-Glucoside glucohydrolase	β-glucosidase
3.2.1.22	[9025-35-8]	α-D-Galactoside galactohydrolase	Melibiase, α-D-galactosidase
3.2.1.23	[9031-11-2]	β-D-Galactoside galactohydrolase	Lactase, β-D-galactosidase
3.2.1.25	[9025-43-8]	β-D-Mannoside mannohydrolase	Mannanase, mannase
3.2.1.26	[9001-57-4]	β-D-Fructofuranoside fructohydrolase	Saccharase, invertase, invertin, β-fructosidase

continued

TABLE 2.1   *(Continued)*

EC No.	CAS No.	Systematic Name[a]	Other Names
3.2.1.32	[9025-55-3]	1,3-β-D-Xylanxylanohydrolase	Xylanase, endo-1,3-β-xylanase (see EC 3.2.1.8)
3.2.1.37	[9025-53-0]	1,4-β-D-Xylanxylanohydrolase	Xylobiase, exo-1,4-β-D-xylosidase, β-xylosidase
3.2.1.39	[9025-37-0]	1,3-β-D-Glucan glucanohydrolase, glucan endo-1,3-β-D-glucosidase[a],	Laminarinase, endo-1,3-β-glucanase (limited action on 1,3; 1,4-glucan linkages) [different from EC 3.2.1.6] (McCleary and Codd, 1989)
3.2.1.41	[9075-68-7]	Pullulan 6-glucanohydrolase	Pullulanase, R-enzyme, debranching enzyme, limit dextrinase
3.2.1.55	[9067-74-7]	α-L-Arabinofuranoside arabinohydrolase	Arabinosidase
3.2.1.68	[9067-73-6]	Glycogen 6-glucanohydrolase	Isoamylase, debranching enzyme
3.2.1.73	[37288-51-0]	1,3;1,4-β-D-Glucan 4-glucanohydrolase, endo-1,4-β-D-glucan glucanohydrolase[a]	Lichenase
3.4.17.2	[9025-24-5]	Petidyl-L-lysin (L-arginine) hydrolase	Carboxypeptidase B, protaminase
3.4.21.1	[9004-07-3]	Chymotrypsin (A and B)	—
3.4.21.4	[9002-07-7]	Trypsin	α-Trypsin, β-trypsin
3.4.21.14	[9014-01-1]	Subtilisin[a], *Aspergillus*, alkaline proteinase	Subtilopeptidase A, *Aspergillus* proteinase B
3.4.21.36	[9004-06-2]	Elastase, pancreatic	Pancreatopeptidase E, pancreatic elastase I
3.4.22.2	[9001-73-4]	Papain	Papainase, papaya peptidase I
3.4.22.3	[9001-33-6]	Ficin	—
3.4.22.4	[37189-34-7]	Bromelain (from pineapple stem)	Bromelin, peptide hydrolase
3.4.22.5	[9001-00-7]	Bromelain (from pineapple juice)	—
3.4.22.6	[9001-09-6]	Chymopapain	Chymopapain A, chymopapain B
3.4.23.1	[9001-75-6]	Pepsin A	Pepsin (various types)
3.4.23.4	[9001-98-3]	Chymosin	Renin, rennet, rennase, chymase, pexin
3.4.23.6	[9025-49-4]	Microbial aspartic proteinases	*Aspergillus* acid proteinase, aspergillopeptidase A, *Aspergillus saitoi* aspartic proteinases
3.4.24.4	[9068-59-1] [9073-78-3] [9036-06-0]	Microbial metalloproteases	Thermolysin *S. griseus* protease

TABLE 2.1   *(Continued)*

EC No.	CAS No.	Systematic Name[a]	Other Names
	[9001-92-7]		*A. oryzae* neutral protease
	[9014-01-1]		*B. subtilis* neutral protease
		Isomerases[e]	
5.3.1.9	[9001-41-6]	D-Glucose-6-phosphate ketolisomerase	Phosphoglucose isomerase, phosphohexoisomerase
		Lyases[f]	
4.2.2.10	[9033-35-6]	Poly(methoxygalacturonide) lyase	Pectin lyase, pectin transeliminase, endopectin lyase
4.2.99.3	[9015-75-2]	Polygalacturonide lyase	Pectate lyase, pectate transeliminase

[a]Recommended name.
[b]Type of reaction: $A'H_2' + B \rightarrow A + B'H_2'$.
[c]Type of reaction: $A'B' + C \rightarrow A + C'B'$.
[d]Type of reaction: $AB + H_2O \rightarrow A'OH' + B'H'$.
[e]Type of reaction: $ABC \rightarrow BAC$.
[f]Type of reaction: $AB \rightarrow A + B$.

Source: Enzyme Nomenclature (recommendations of the Nomenclature Committee of the International Union Biochem. & Mol. Biol. on the Nomenclature and Classification of Enzymes) Academic Press, San Diego, New York, London, 1992.

REFERENCES

Albersheim, P., and Killias, U., *Biochem. Biophys.* **97,** 107 (1962).

Anson, M. L., *J. Gen. Physiol.* **22,** 79 (1938).

Baker, D. L., "Production of catalase from mold," U.S. Patent 2,635,069 (1953).

Calbiochem-Novabiochem International, Inc., *Biochemicals, Immunochemicals, Enzymes, Reagents,* Catalogue, 1996.

Cornish-Bowden, A., and Wharon, K., *Enzyme Kinetics,* IRL Press, Oxford, 1988.

Dixon, M., and Webb, E. C., *Enzymes,* 3rd ed., Academic Press, New York, 1979.

Fischer, E., *Ber. Dtsch. Chem. Ges.* **27,** 2985 (1884).

ICN Biochemicals, Inc., *Biochemicals, Organic Chemicals, Enzymes, Diagnostics,* Catalogue, 1996.

Keil, B., *Armour & Company Bulletin* No. E.C. 2, Chicago, 1954.

Kunitz, M., *J. Gen. Physiol.* **30,** 291 (1947).

Künzel, A., in *Gerbereichemisches Taschenbuch,* Steinkopff, Dresden, Leipzig, 1955, p. 86.

McCleary, B. V., and Codd, R., *J. Cereal Sci.* **9,** 17 (1989).

Michaelis, L., and Menton, M. L., *Biochem. Zeitschr.* **49,** 333 (1913).

Nelson, N., *J. Biol. Chem.* **153,** 375 (1944).

Perrella, F. W., *Ann. Biochem.* **172,** 437 (1988).

Sandstedt, R. M., Kneen, E., and Blish, M. J., *Cereal Chem.* **16,** 712 (1939).

Sandstedt, R. M., Kneen, E., and Blish, M. J., *Cereal Chem.* **24,** 157 (1947).

Scott, D., and Hammer, F. E., *Enzymologia,* Vol. XXII (3), Dr. W. Junk N. V., The Hague, 1960, p. 194.

Segel, H. L., *Enzyme Kinetics,* Wiley, New York, 1975.

Sigma Chemical Co., *Biochemicals, Organic Compounds, Enzymes, Diagnostic Reagents,* Catalogue, 1996.

Sommer, H., in Diemair, W. (ed.), *Handbuch der Lebensmittelchemie. Analytik der Lebensmittel,* Vol 2, Springer, New York, 1967, p. 232.

Somogy, M., *J. Biol. Chem.* **195,** 19 (1952).

Sprössler, B., *Röhm Spektrum* **21,** 12 (1978).

Stellmach, B., *Bestimmungsmethoden Enzyme für die Pharmazie, Lebensmitteltechnick, Biochemie und Medizin,* Dr. F. Steinkopff, Darmstadt, 1988.

Sumner, J. B., *J. Biol. Chem.* **70,** 97 (1926).

The National Formulary, *U.S. Pharmacopoeia,* 11th ed., Washington, DC, 1960.

Tombs, M. P., in Tucker G. A., and Woods, L. F. J. (eds.), *Enzymes in Food Processing,* 2nd ed., Blackie Academic & Professional (imprint of Chapman & Hall), New York, 1995, p. 268.

Yamamoto, H., "Industrial application of enzymes," *Hakkokogaku Kaishi* **56,** 656 (1978).

Zech, R., and Domagk, G., *Enzyme-Biochemie, Pathobiochemie, Klinik, Therapie,* Edition Medizin, VCH, Weinheim, 1986.

3 Description of Enzymes

3.1 CARBOHYDRATE HYDROLYZING ENZYMES

3.1.1 Brief Overview

3.1.1.1 Substrates Carbohydrates are widely distributed in nature. Fifty to eighty percent of the dry weight of plants consists of carbohydrates, which include the important structural components cellulose, pectin, and starch. The term *carbohydrate* is taken from the French *hydrate de carbone*. Carbohydrates are composed principally of carbon and water, as shown in the chemical formula for glucose: $C_6(H_2O)_6$. Polymeric carbohydrates are built from simple carbohydrate monomers such as glucose. Oligosaccharides, such as sucrose (cane sugar) or maltose, are composed of only a few of the monomers linked together. Additional linkage of oligosaccharides leads to long-chain polysaccharides, which play an essential role in human and animal nutrition and are the basis of many natural fermentation products, such as beer, wine, and vinegar. They also provide the raw materials for the production of many industrial products, such as antibiotics and enzymes. Carbohydrates give cotton cloth, writing paper, and a valuable source of fuel. Enzymes can efficiently degrade carbohydrates without combustion for many practical applications.

3.1.1.2 Enzymes The action of enzymes can be experienced at breakfast time. The physician tells clients "chew your bread well," and when they do, they experience a sweet taste. The enzymes in human saliva, the salivic amylases, convert tasteless starch in bread into sweet-tasting sugar. Cooked starch forms rigid gels, which are called *pastes*. A trace amount of salivic amylase converts such a paste into a liquid in seconds. The high-molecular-weight, hydrophilic starch is hydrolyzed into low-molecular-weight sugars and dextrins. This hydrolysis is essential for our nutrition because humans, animals, and even microorganisms absorb carbohydrates only in the hydrolyzed form as sugar.

Scientifically, the carbohydrate cleaving enzymes are called *glycosidases* or more commonly, *carbohydrases*. They hydrolyze polysaccharides such as starch, cellulose, and pectin. Their action is apparent in lowering the viscosity of watery solutions of carbohydrates such as pectin and carboxymethylcellulose solutions. The activity of the enzyme is determined by the amount of free reducing sugars produced by its action.

3.1.1.3 Glycosidic Linkage The monomeric building blocks of carbohydrates and their polymerization to oligomers and polymers has been mentioned earlier.

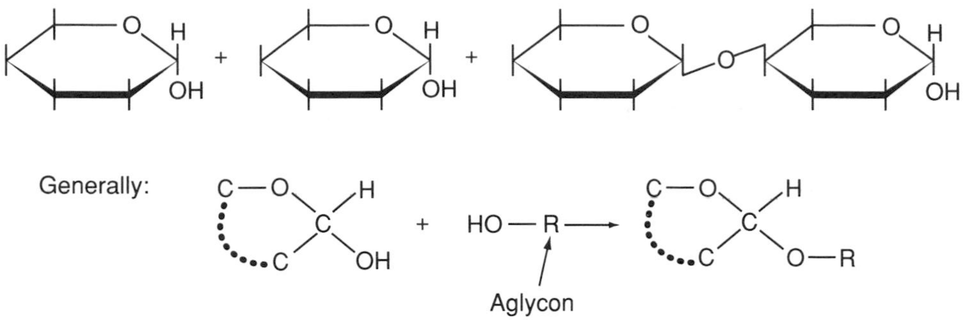

Generally:

Aglycon

Figure 3.1   The glycosidic linkage.

The linkage between monomeric sugars, such as glucose, to disaccharides, is termed a *glycosidic linkage* (Fig. 3.1).

A *glycoside* is scientifically defined as a full acetal of a sugar, where the free acetal hydroxy group of the sugar is replaced by an "R" residue. If "R" is another sugar residue, then the molecule is a disaccharide. If both the sugar and "R" are glucose, then maltose is formed. "R" can also be a different molecule containing a hydroxyl group, such as methanol or phenol. These are, in general, termed *aglycons*. Such glycosides occur widely in nature. They are found in flavor substances and in natural dyes.

3.1.1.4 Enzyme Specificity   Carbohydrases hydrolyze the specific glycosidic linkages of certain monosaccharide residues. They are able to cleave short-chain oligosaccharides as well as polysaccharides with various structures. Table 3.1 shows some of the products of enzymatic hydrolysis of some oligosaccharides. Table 3.2 depicts some naturally occurring polysaccharides, their monomeric building blocks, their linkages, and the enzymes used for their scission.

The mechanism of scission is consistent with a transfer of a substrate glycosyl residue to water (Fig. 3.2).

The position of transfer can be located using ^{18}O-labeled water. An α-glycosidase cleaves at the α-position, whereas a β-fructofuranoside cleaves at the β-position. Free reducing sugars are formed by hydrolysis.

TABLE 3.1   Oligosaccharides

Substrate	Hydrolysis Product	Linkage	Enzyme
Maltose	2 Glucose molecules	α-1,4	α-Glycosidase
Sucrose	Glucose, fructose	α-1,2	—
α-Methylglycoside	Glucose, methanol	—	—
Cellobiose	2 Glucose molecules	β-1,4	β-Glycosidase
Raffinose	Galactose, sucrose	α-1,6	α-Galactosidase

TABLE 3.2   Polysaccharides

Polysaccharide	Monomer Unit	Linkage	Enzyme
Starch	Glucose	α-1,4	α-Amylases
—	—	—	β-Amylases
—	—	α-1,6	Isoamylases
Cellulose	Glucose	β-1,4	Cellulases
Dextran	Glucose	α-1,6	—
Xylan	Xylose	β-1,3; β-1,4	Xylanases
Pectin	Galacturonic acid	α-1,4	Pectinases
Araban	Arabinose	α-1,5; α-1,3	Arabanases
Inulin	Fructose	β-1,2	Inulinase

The specificity of carbohydrases depends on

1. The configuration of the glycosidic linkage. For example, maltose can be hydrolyzed only by α-glucosidase and not by β-glucosidase.
2. The chemical structure of the monomeric sugars joined by the linkage.
3. The molecular weight of the substrate. α-Amylases are able to hydrolyze high-molecular-weight amylose very rapidly, but maltose or maltotriose are hydrolyzed only very slowly, if at all. Enzymatic hydrolysis is, by definition, reversible. The aqueous reaction equilibrium favors the hydrolysis products.

In nonaqueous systems, glycosyl residues can be transferred to other sugars instead of water. This reaction, termed *transglycosylation,* yields new oligoglycosides.

By volume and value, carbohydrases are the most important of the industrially produced enzymes. Their major application is in starch processing, which encompasses the production of dextrins, glucose, maltose, and different types of fructose syrup. These products are needed in all areas of the food industry. Carbohydrases are used in distilleries, breweries, and the production of baked goods. The production of fruit juices or fruit juice concentrates without these enzymes is unimaginable.

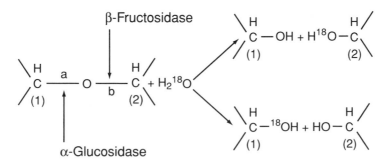

Figure 3.2   Sucrose hydrolysis with enzymes of different specificities.

3.1.2 Amylases

3.1.2.1 Starch As a reminder, a few facts about starch are given here. Starch is the major component of most foods and the most important energy source in our food supply. Starch products are obtained from grains, roots, and tubers. Glycogen is stored as a reserve carbohydrate in the liver of animals and humans. Starch granules contain amylose and amylopectin, which are bound to each other by hydrogen bonding. The types of cross-linking and the degree of polymerization determine the various characteristics of the different types of starch. These factors dictate, among other properties, the different swelling power and gelatinization temperature of various starches, properties that are of great importance for the food processing industry. Amylose consists of a linear chain of 500 or more α-1,4-bound glycosidic linkages. By contrast, amylopectin is highly branched. The branches occur in intervals of 20 glucose residues and are formed via α-1,6 linkages. So-called waxy starch in corn consists of almost 100% amylopectin. If amylopectin is treated with pure β-amylase, the hydrolysis stops at the 1,6 linkages, giving so-called limit dextrins. When treated with iodine solution, starch forms a deep-blue-colored iodine complex, while amylopectin gives a red colorization. The point at which the characteristic blue color disappears during starch degradation is called the *iodine-negative value* or the *achroic point.*

3.1.2.2 Gelatinization of Starch On heating a suspension of starch in water, the starch swells when the specific gelatinization temperature is reached. The sudden water uptake leads to a strong increase in viscosity and to a change in the chemical and enzymatic vulnerability of the starch. Gelatinization plays an important role in the process of bread baking, in the production of thickening agents based on starch, and in the production of textiles where starch and dextrins are used as glue or sizing. As the gelatinized starch cools, the amylose fraction crystallizes rapidly by forming intermolecular hydrogen bonds. This process, known as *retrogradation,* is one of the reasons why daily bread stales so rapidly.

3.1.2.3 Stepwise Starch Degradation The amylases split starch into dextrins and sugars by hydrolysis of the a-1,4 glycosidic linkages. They are common in nature, occurring in saliva, pancreas, and many plants such as cereals. In industry, amylases are produced from cultures of bacteria and fungi. There are α- and β-amylases. α-Amylases split the glycosidic linkage in the interior of the starch chain. On the other hand, β-amylases split maltose from the nonreducing end of the chains. In addition, some amyloglucosidases liberate glucose residues stepwise from the end of the chain (Fig. 3.3).

3.1.2.4 Endoamylases (EC 3.2.11)
Reference: Fogarty and Kelly (1983), MacGregor (1988)

The endoamylases hydrolyze the α-1,4 glycosidic linkages in starch, glycogen, and derivatized speech. Depending on the source, the properties of different amylases vary widely. Many α-amylases can be produced in pure crystalline form, such as

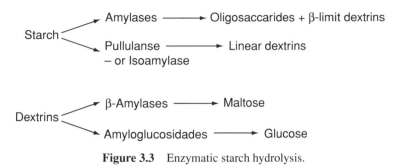

Figure 3.3   Enzymatic starch hydrolysis.

those from malt, pancreas, *Aspergillus oryzae,* and *Bacillus subtilis.* Their properties have been well investigated. They differ in their molecular weights, temperature stability, optimum-activity pH, and hydrolytic specificity. Because of this specificity, different α-amylases produce oligosaccharides of various chain lengths and yields. The fast and complete degradation of starch requires pregelatinization of the substrate. Bacterial α-amylase degrades gelatinized starch 300 times faster and fungal α-amylase 100,000 times faster than native starch (Walker and Hope, 1963). The rate of hydrolysis depends primarily on the degree of polymerization of the starch, dropping markedly with a lower degree of polymerization. Amylose is hydrolyzed faster than amylopectin. Amylase-induced starch hydrolysis is shown in Figure 3.4.

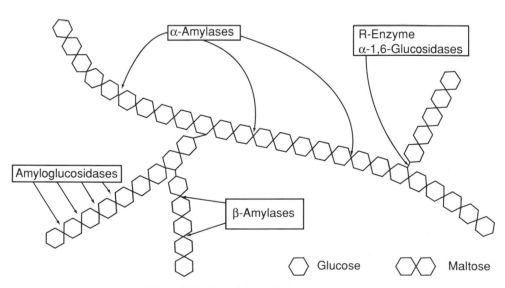

Figure 3.4   Starch hydrolysis by amylases.

Bacterial α-Amylase (EC 3.2.1.1, CAS 9000-90-2)

SOURCE   In industry, bacterial α-amylases are produced mainly from cultures of *B. subtilis* var. *amyloliquefaciens*. In addition, industrial preparations from *B. subtilis* or *B. subtilis* var. *amylosacchariticus* are available. The latter produces more glucose from starch than the liquid α-amylase from *B. amyloliquefaciens* (Pazur and Okada, 1967). A comprehensive overview of the properties of bacterial amylases can be found in Fogarty and Kelly (1983).

PROPERTIES

Molecular Weight: 49,000 daltons (Fischer and Stein, 1960); 60,000 daltons (Granum, 1979) from *B. amyloliquefaciens;* 48,000 daltons from *B. subtilis* (Kakiuchi, 1965). In the presence of zinc, a dimer is formed with a molecular weight of 100,000 daltons.

Cofactors: α-Amylase contains no coenzymes, but as a calcium metalloenzyme, it requires at least one gram-atom of calcium per mole of enzyme in order to maintain its activity. This calcium also stabilizes the enzyme against heat denaturation and attack from proteases. Complexing calcium with EDTA leads to denaturation of the free amylase by heat, urea, or acid and rapid degradation by proteases (Fischer and Stein, 1960).

Activity: The K_m (starch) is between 1 and 3.4×10^{-3} for amylases of different *B. subtilis* strains. The specific activity is about 4×10^5 SKB units per gram of enzyme protein (Figs. 3.5 and 3.6).

Temperature Stability: Ninety percent of the activity is maintained on incubation with 150 ppm Ca^{2+}, at pH 6.5–7.5 for one hour at 65–70°C (Fig. 3.7). High substrate concentration, such as 25–40% starch, also stabilizes the enzyme, so that total inactivation, even at boiling temperatures, requires a long treatment time.

Inactivation: The enzyme, depending on temperature, is rapidly inactivated at pH values lower than 4.5–4.7. In the presence of high substrate concentrations, the pH must be adjusted to below 4.0. Treatment for 10–15 min at 120°C is required for heat denaturation of pH 6–8.

Inhibitors: Chelating agents such as polyphosphate, EDTA, oxalate, free chlorine, and oxidizing agents.

TECHNICAL PREPARATIONS

Activity: Industrial α-amylase preparations are offered in liquid and powder forms. The liquid preparations are mainly just centrifuged, filtered culture liquids, stabilized with sodium chloride. Their activity is about 500–3000 SKB units per gram (for definition of SKB units, refer to "Enzyme Units," Section 2.3.2). Powdered products, prepared by precipitation or spray drying, are generally standardized with sodium chloride, calcium sulfate, starch, dextrin, or sugar. Their activity is about 5000–50,000 SKB/g.

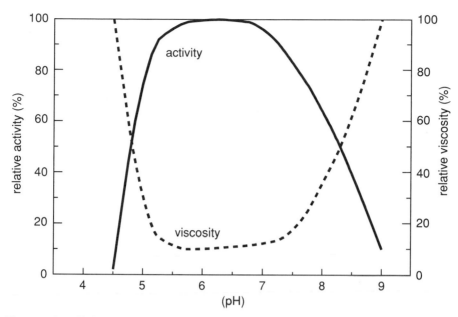

Figure 3.5 pH dependence of bacterial amylase from *B. amyloliquefaciens;* measured by loss of viscosity and increase in reducing sugars.

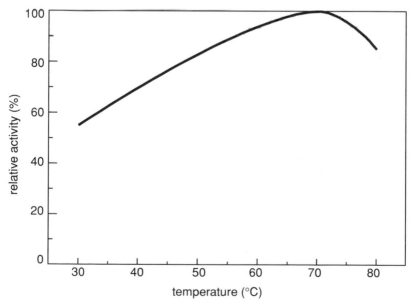

Figure 3.6 Temperature dependence of bacterial amylase (2% soluble starch, pH 6.0).

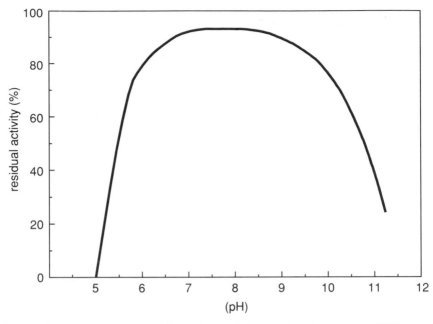

Figure 3.7 Effect of pH on stability of bacterial α-amylase (temperature 70°C; reaction time 1 h, 0.5 g CaCl$_2$ and 6.0 g NaCl per liter).

Stability: Activity measurements of stored samples show that the enzymes are highly stable under dry storage. Loss in activity is about 5–10% per year at 20°C. Liquid products lose about 10% of their activity in 6 months at 15–20°C.

Secondary Activities: Technical enzyme preparations contain, in addition to α-amylases, neutral and alkaline bacterial proteases, β-glucanases, and other hemicellulases.

Hydrolysis Pattern: The reaction rate of bacterial α-amylase increases with soluble starch and decreases with chain length. The closer the cleavage site is to the α-1,6 branching, the slower the reaction. Maltosyl-α-1,6-maltotriose cannot be degraded to lower-molecular-weight compounds. The initial products of amylose are G$_2$, G$_3$, G$_6$, and G$_7$ oligomers (Fig. 3.11).

Products and Manufacturers: Canalpha [BN], Nervanase [AM], Bactamyl [GB], Optamyl [SO], Aquazym [NO], Rohalase A [RM]; for further products, see the products list in Appendix B.

INTERMEDIATE STABLE BACTERIAL AMYLASE   A bacterial amylase with a stability intermediate between that of the fungal α-amylase and that of the bacterial amylase produced by *B. megaterium* was described by Hebeda et al. (1991). The optimum pH is 6.0, with 80% activity at pH 5.5–6.5 and an optimum temperature of ~70°C. The enzyme is inactivated between 85 and 90°C and serves as an antistaling enzyme in baking.

Product: Megafresh [EB].

Thermostable Bacterial α-Amylases (EC 3.2.1.1; CAS 9000-91-2)   The properties of thermostable α-amylases, and foremost the degree of heat stability, depend on the source of the enzyme. The data on heat stability in the literature can be misleading if exact details of the experimental conditions are not given. It is particularly important to know whether stability was determined in aqueous solution with calcium ions and sodium chloride or in the presence of substrate. Currently, the α-amylase used most frequently as a technical enzyme is the heat-stable α-amylase from cultures of *B. licheniformis.* It is described in greater detail later in this book. The enzyme from *B. stearothermophilus* shows a temperature optimum at about 80°C in the presence of substrate. The optimum pH is about 3.0 (Manning and Campbell, 1961). Tamura, Kanno, and Ishii (1973) compared different α-amylases with that from *B. stearothermophilus,* under identical reaction conditions in the presence of 5 mM Ca^{2+} ions (Table 3.3).

Srivastava (1987) isolated a strain from *B. stearothermophilus,* which produces both α- and β-amylases. Molecular weights were 48,000 daltons for α-amylase, containing 13.5% carbohydrate; 57,000 daltons for the β-amylase, which contained 0.8% carbohydrate.

Optimal activity for both enzymes is at pH 7.0. Calcium improves the heat stability, so that for incubation at 70°C in the presence of 15 mM Ca^{2+}, the preparation still showed 80% activity after 40 minutes. This stability value is higher than that for the amylase from *B. amyloliquefaciens* and lower than that for the enzyme for *B. licheniformis.*

α-Amylase from B. licheniformis (EC 3.2.1.1)   This enzyme, which is presently used in starch liquefaction, was first used industrially in 1973 (Madsen et al., 1973).

PROPERTIES

Molecular Weight: 28,000 daltons [SDS–gel electrophoresis; Krishnan and Chandra (1983)].

Activation Energy: 5.1×10^5 J/mol.

Cofactors: Na^+, Ca^{2+}, Mg^{2+} activate.

Activity: Eighty percent of maximum activity is shown between pH 4.2 and 8.0 (4.3 mM Ca^{2+}, 60°C, 10 min, 46% soluble starch) (Fig. 3.8). In excess of

TABLE 3.3   Comparison of Various Bacterial α-Amylases

Enzyme Source	pH Range	Molecular Weight (Daltons)	Temperature Optimum (°C)
B. stearothermophilus	4.0–5.2	96,000	80
B. stubtilis	4.5–6.5	49,000	60
B. licheniformis	5.0–9.0	22,500	78
B. stearothermophilus (Ogasakawa)	5.0–6.0	48,000	70

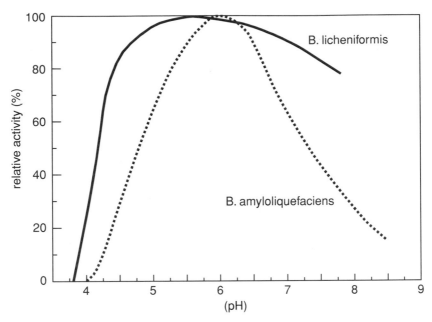

Figure 3.8 pH dependence of bacterial α-amylases from *B. amyloliquefaciens* and *B. licheniformis* at 60°C. Substrate: 0.4% soluble starch in 0.05 M acetate buffer, pH 4–5.6; 0.05 M tris-maleate buffer, pH 6–8.5; and 4.3 mM Ca^{2+} [NO].

90°C, the optimum activity under these conditions is at pH 5.7 for a reaction time of 20 min. In a 30–40% starch slurry, typical for industrial processes, short reaction times at temperatures of up to 110°C can be used (Fig. 3.9).

Stability: Figure 3.10 illustrates the effect of calcium ion concentration on enzyme stability. Addition of 3.4 ppm Ca^{2+} is sufficient to stabilize the enzyme. In contrast, for the α-amylase from *B. amyloliquefaciens,* addition of 150 ppm Ca^{2+} is necessary. The enzyme is stable in the presence of sodium tripolyphosphate.

Inhibitors: NO_3^-, F^-, SO_3^{2-}, $S_2O_3^{2-}$, MoO_4^-, *p*-chloromercuribenzoate, sodium iodoacetate.

Stabilizers: Cysteine, thiourea, mercaptoethanol, sodium glycerophosphate (Krishnan and Chandra, 1983).

Inactivation: The enzyme is totally inactivated by adjusting the pH to 3.5–4.0 and incubating for 5–30 min at 90°C.

Hydrolysis Pattern: Although it was earlier assumed that starch degradation followed strict statistical laws, it was found later that during the first hydrolysis phase, G_3 (maltotriose) was preferentially released from the nonreducing end. *B. amyloliquefaciens* amylase preferentially yields G_6-oligosaccharides, which subsequently are further hydrolyzed extremely slowly. *B. licheniformis* amylase produces mainly G_5 oligosaccharides (Fig. 3.11); these are also hydrolyzed further.

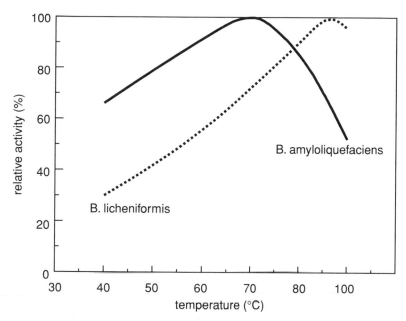

Figure 3.9   Temperature dependence of bacterial α-amylases. Substrate: 0.46% soluble starch, pH 5.7 and 4.3 mM Ca^{2+}; reaction time 7–20 min [RM].

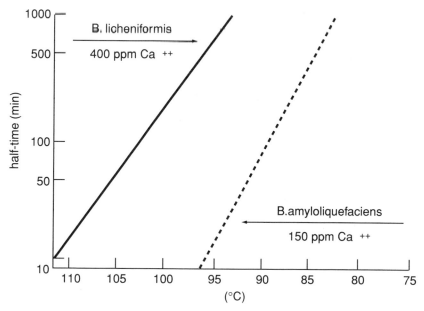

Figure 3.10   Effect of temperature on half-life of bacterial α-amylases. Substrate: 33% starch (Norman, 1981).

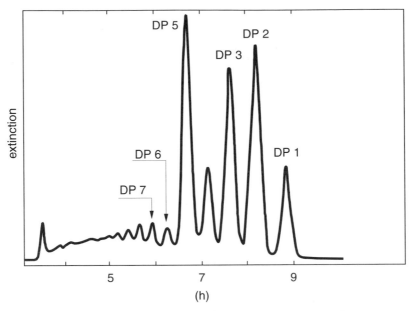

Figure 3.11   Oligosaccharides produced from cornstarch on hydrolysis by bacterial α-amylases (Norman, 1981). (DP = degree of polymerization.)

TECHNICAL PREPARATIONS

Activity: Most preparations are liquids with activities ranging from 2000 to 6000 SKB/g.

Manufacturers: [NO], [GB], [RM], [SO], [DA], [BN], [SE], and others.

Other α-Amylases   In recent years, innumerable α-amylases have been described, some of which have found technical applications. These include special oligosaccharide-producing α-amylases. Certain specific oligosaccharides are of interest for the manufacture of glycosides with pharmaceutical applications and for the preparation of dietetic products.

Maltotetraose Producing α-Amylase   An amylase isolated from Pseudomonas stutzeri preferentially produces maltotetraose from high-molecular-weight dextrins (Kobayashi et al., 1986a and b).

PROPERTIES

Molecular Weight: 62,000 daltons.

Isoelectric Point: pH 4.7.

Activity: Optimum pH 6.7–7.0; maximum temperature 45°C. Potato starch was liquefied with bacterial amylase and reacted with the *P. stutzeri* enzyme for 17 h at pH 7.0 and 45°C. The potato starch used yielded 70% maltotetraose.

Maltopentaose-Producing α-Amylase   Fermentation of *B. cereus* yields an amylase that produces maltopentaose from soluble starch with a 65% yield. (Jpn. Patent, Sapporo Breweries, 1985).

PROPERTIES

Molecular Weight: ~90,000 daltons.

pH Optimum: pH 6–7.

Temperature Optimum: 50–55°C.

Hydrolysis Pattern: Amylose, amylopectin, and maltohexaose are cleaved. β-Cyclodextrin, pullulan, dextran, and maltopentaose are not cleaved.

Maltohexaose Producing β-Amylases   *Aerobacter aerogenes* is known as a producer of pullulanase. Kainuma et al. (1972) obtained an amylase from cultures of this organism, which produced maltohexaose from amylose and amylopectin. The yield, based on the amylopectin, was about 36%. During the first phase, high-molecular-weight limit dextrins were formed; thereafter, only maltohexaose was released from the nonreducing end of G_7–G_9 oligosaccharides. This mechanism of action categorizes this enzyme as a β-amylase.

Maltogenic Amylases (See Also "β-Amylases")   Koaze et al. (1974, 1975) found an amylase from *Streptomyces* sp. that cleaved starch to maltose (71%), glucose (4%), and oligosaccharides by an endo mechanism. The mixture can be hydrogenated to sugar alcohols, predominantly maltitol. Pure maltose, which can be obtained using ion exchangers, is used to replace glucose in infusion solutions, where the same osmotic pressure can be obtained by using twice the weight of glucose. Another, mainly maltose-forming, α-amylase is obtained industrially from *Streptomyces hygroscopicus*. Optimum activity is at pH 5–6. Above 60°C the enzyme is inactivated (Meiji Information 1).

Manufacturer: [MJ].

Fungal α-Amylases (EC 3.2.1.1; CAS 9000-90-2)   Fungal cultures were cultivated on rice in China and Japan during the nineteenth century. Those cultures, known as *koji,* were extracted with water and were used as Takadiastase for the production of fermentable sugars, which, in turn, were used to make alcohol. Dried preparations were used as digestive aids in medicine (Takamine, 1884).

Fungal α-amylases were prepared from various strains of *Aspergillus. A. oryzae* is the most commonly used organism, but *A. niger, A. awamori,* and *A. usamii* are also described as being good producers. Technically, the enzyme is mainly prepared using *A. oryzae.* In Europe and in the United States, the enzyme is manufactured by submersion fermentation (Kvesitadze et al., 1974). In East Asia, some processes still use fixed-bed cultures (koji) with wheat or rice bran as substrate.

Fungal α-Amylase from A. oryzae

PROPERTIES

Molecular Weight: 51,000 daltons.

Molecular Structure: The fungal amylase is a glycoprotein containing mannose residues.

Cofactors: The *A. oryzae* amylase binds 10 g-atoms of calcium per mole of enzyme; 9 g-atoms of these are bound only loosely and can be removed without loss of catalytic activity (Oikawa and Maeda, 1957). Activity is lost with the removal of the remaining calcium with EDTA.

Specific Activity: 50,000 SKB/mg of enzyme protein.

pH Optimum: Fungal amylases from different sources yield different pH-activity curves. The optimum pH for the enzyme from *A. oryzae* is 5.0, that from *A. usamii* is 4 (Fig. 3.12).

Temperature Optimum: 50–60°C at pH 5.0 *(A. oryzae)* (Fig. 3.13).

Stability: Fungal α-amylases are less heat-stable than bacterial amylases and are inactivated before the gelatinization temperature of starch is reached (68–70°C). The enzyme from *A. oryzae* is stable for about 30 min at pH 5.7

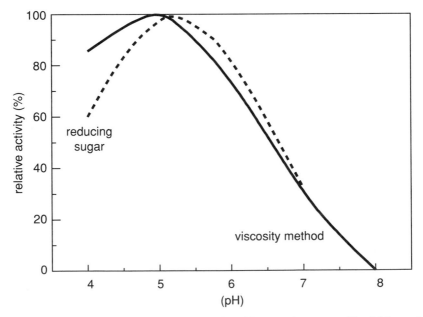

Figure 3.12   pH dependence of fungal α-amylase (*A. oryzae*); measured by (*a*) loss of viscosity and (*b*) increase in reducing sugars.

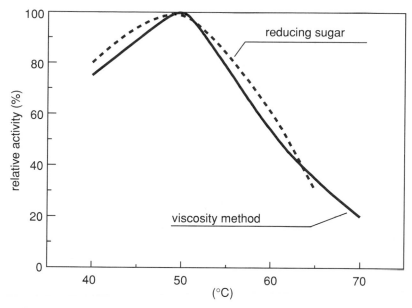

Figure 3.13   Temperature dependence of fungal α-amylase; measured by (*a*) loss of viscosity and (*b*) increase in reducing sugars. Substrate: 4% soluble starch, pH 5.0; reaction time 10 min.

and 50°C. Below pH 3.5 and at higher temperatures (30–40°C), the enzyme is rapidly inactivated. This must be taken into consideration when the enzyme is used in fruit juices. Ca^{2+} ions stabilize the enzyme. The enzyme from *A. usamii* shows somewhat better stability in the acidic pH range.

Hydrolysis Pattern: During hydrolysis of starch with fungal amylases, no glucose is formed from dextrin polymers larger than G_8 (Fig. 3.14).

TECHNICAL PREPARATIONS

Activity: Powdered preparations, obtained by spray drying or precipitation with alcohol, are standardized according to the type of application with salts, starch, maltodextrin, lactose, or glucose. The activity of the preparations used in flour milling is about 500–5000 SKB/g. Concentrates for the starch or baking industry are set to 50,000–120,000 SKB/g. Liquid concentrates for the starch industry are offered with activities from 6000 to 40,000 SKB/g.

Stability: The drop in activity using dry storage at about 20°C is approximately 5–10% per year.

Secondary Activities: Of particular practical relevance are companion activities due to acidic proteases and cellulases. In certain applications, many of these

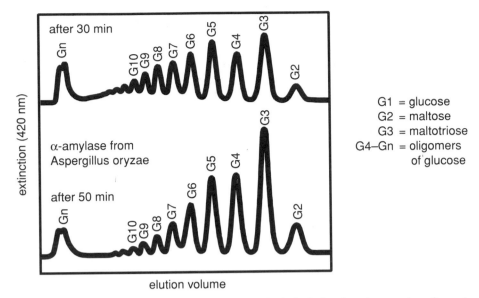

Figure 3.14 Products formed during amylose hydrolysis by fungal α-amylase from *A. oryzae* (Dellweg and Jahn, 1972).

additional activities are desirable, whereas for others they may be less desirable or even harmful. Protease activities between 100 and 30,000 HUT units can be present, and this must be taken into account for applications in flour processing. In addition, preparations from fixed-bed cultures, such as the well-known Takadiastase, are a valuable source of many enzymes. Secondary enzymes found included α- and β-glycosidases, nucleases, hemicellulases, various proteases, and peptidases.

Manufacturers: [RM], [NO], [GB], [AM], [UA], [SN], [SA], [SO], [SOE], and others.

Fungal β-Amylase from A. fumigatus K 27   Abe et al. (1988) described an *Aspergillus* α-amylase that is particularly effective in degrading raw starch. The fungal strain, which presumably belongs to the *A. fumigatus* group, produces a mixture of about 70% amyloglucosidase and 30% α-amylase. The pH optimum for raw-starch degradation is between 4.6 and 5.5 at 55°C. Cornstarch can be completely degraded at 55°C in 7–24 h, depending on the amount of enzyme used. The amount of enzyme required, however, is relatively high. About 1 ml of culture solution of a 5–6-day-old culture growth is required to degrade 250 mg of cornstarch in 24 h. By comparison, 1 ml of amyloglucosidase culture medium from 5-day-old culture of *A. niger* (a 5-day culture) can cleave about 1 kg of gelatinized starch in 24 h at 60°C.

α-Amylase from Chalara paradoxa   An enzyme that is particularly effective in the degradation of raw starch is obtained from cultures of *Chalara paradoxa* (Kainuma et al., 1985). The optimum activity range lies between pH 4.5 and 5.5 at 45–55°C.

PROPERTIES AND MANUFACTURER

> *Activity:* Five milligrams of the only commercial preparation produces 2000 mM of reducing sugar from raw cornstarch at pH 5.0 and 40°C. One enzyme unit is defined as the amount of enzyme required to produce 1 mM reducing sugar (in form of glucose) from raw cornstarch at pH 5.0 and 40°C (Meiji Information 2). (See Fig. 3.15.)
> *Manufacturer:* [MJ].

Malt α-Amylase   Those α-amylases important in breweries and distilleries coexist with β-amylases in malted grains. The terms used for the grain amylases are *diastase* and *diastatic activity,* which describe the cooperative action of both enzymes. The presence of the two enzymes was first unequivocally shown by Sumner and Somers (1947). While β-amylase is already present in large amounts in unmalted grains, α-amylase is generated only during the malting process. A good overview on the enzyme formation during this process has been published by Schwimmer (1981).

During malting, α-amylase activity increases several-hundred-fold in 4–5 days. "Green" grain α-amylase differs from that of unmalted grain in its properties in that it is able to attack intact wheat starch (Kruger and Marchylo, 1985).

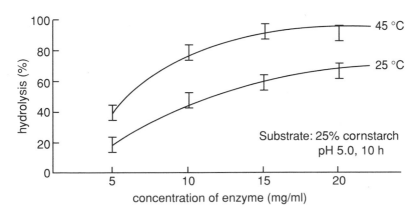

Figure 3.15   Hydrolysis of raw starch by amylase from *Chalara paradoxa* [MJ]. Substrate: 25% cornstarch, pH 5.0, 10 h.

PROPERTIES

Molecular Weight: 52,000 daltons (malted barley). Different types of grain contain multiple forms of the enzyme, with a molecular-weight distribution of 42,000–54,000 daltons.

pH Optimum: pH 5.5 (malted barley).

Temperature: Optimum is 70°C; inactivation is rapid at higher temperatures. At high substrate concentrations, this α-amylase is effective beyond the gelatinization point of starch.

Stability: The enzyme has a good stability between pH 4.5 and 9.5 at 25°C. Inactivation is rapid below pH 4.5.

Inhibitors: Potassium periodate, oxidizing agents.

Hydrolysis Pattern: β-Limit dextrins are formed during the first phase; on more extensive hydrolysis, oligosaccharides from G_2 to G_6.

Pancreatic α-Amylase   Pancreatic α-amylase was formerly prepared by fractional precipitation of aqueous extracts from hog or bovine pancreas. The technical enzyme preparation was used principally for textile sizing. Because of their low temperature stability, the preparations are not suitable for today's rapid high-temperature sizing processes for which heat-stable bacterial α-amylases are now used. Pancreatic amylase is used as a digestive substitute in a pharmaceutical preparation of pancreatic enzymes called *pancreatin.*

See Figure 3.16.

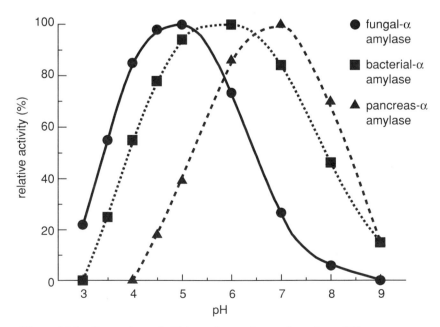

Figure 3.16   Comparison of pH dependence of α-amylases from different sources.

PROPERTIES, PRODUCTS, AND MANUFACTURERS

Activity: Activity optimum is between pH 6.0 and 7.0. The enzyme is activated by NaCl and stabilized by Ca^{2+}.

Inactivation: The enzyme is inactivated in aqueous solutions if accompanied by pancreatic protease. It is also inhibited by chelating agents and by SH-group blocking substances, such as iodoacetamide or heavy-metal ions.

Products and Manufacturers: [RM], soluble preparations; [HÖ], [KC], [BF], and others. Product: pancreatin.

Amylases Cleaving α-1,6 Linkages   The important raw starches contain about 80% amylopectin. As was mentioned previously, α-amylases cleave the α-1,4 linkages, whereas the α-1,6 linkages resist attack. Intensive hydrolysis of amylopectin with α-amylases leads to the "α-limit dextrins." In similar fashion, treatment with β-amylases leads to the "β-limit dextrins," as cleavage here also terminates at the α-1,6 branches (Robyt and Whelan, 1968).

The amylases that cleave α-1,6 linkages can be divided into three groups:

1. Amylo-1,6-glycosidases (in higher organisms)
2. Pullulanases, also R-enzyme or debranching amylases
3. Isoamylases

Pullulanase (EC 3.2.1.41; CAS 9075-68-7)   Pullulan (Fig. 3.17), which is produced by the fungus *Aureobasidium pullulans,* consists of maltosyl-(G_2) and maltotriosyl-(G_3) residues, which are joined by α-1,6 linkages. Tetraosyl residues are also present.

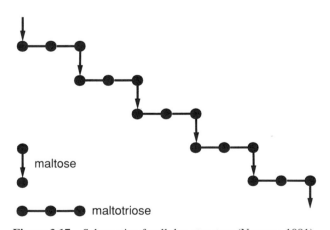

Figure 3.17   Schematic of pullulan structure (Norman, 1981).

Pullulan degrading enzymes are produced by plants and microorganisms. Even purified enzymes cleave not only pullulan but also amylopectin, β-dextrins, and glycogen.

The first isolation of pullanase was from *Aerobacter aerogenes* by Bender and Wallenfels (1961). The enzymes from cultures of *Klebsiella pneumoniae* (L) and *B. cereus* (Takasaki, 1976) are of technical importance.

Pullulanase from K. pneumoniae (Norman, 1981)

PROPERTIES

Activity: pH optimum at 5.5–6.0 (50°C).

Temperature: Maximum 50–55°C at pH 5.0.

See Figures 3.18 and 3.19.

A technical pullulanase preparation from *Bacillus* sp. shows 80% of maximum activity at pH 5–7 (see Fig. 3.20).

Low-molecular-weight α-1,6-oligosaccharides such as maltotriosyl-α-1,6-maltotriose are cleaved 10 times faster than amylopectin.

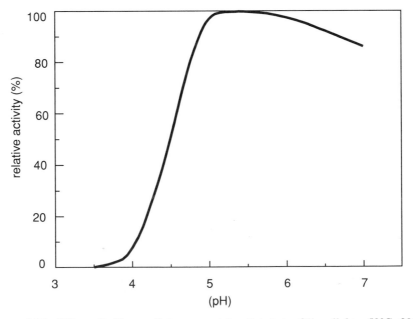

Figure 3.18 Effect of pH on pullulanase activity. Substrate: 2% pullulan, 50°C, 30 min (Norman, 1981).

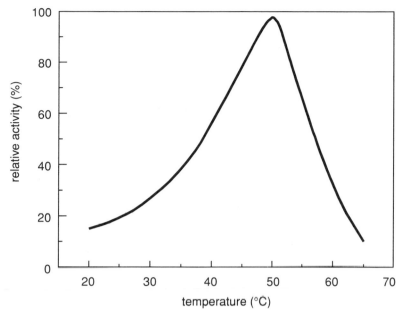

Figure 3.19   Effect of temperature on pullulanase activity. Substrate: 2% pullulan, pH 5.0, 30 min (Norman, 1981).

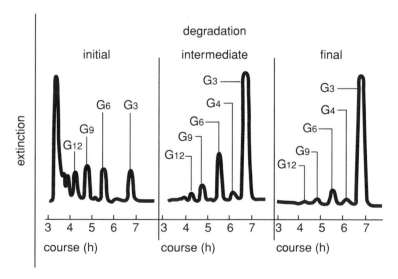

Figure 3.20   Products formed during pullulan hydrolysis by an enzyme from *Bacillus* sp. (50°C, pH 4.5).

Pullulanase from Aerobacter aerogenes

PROPERTIES

Molecular Weight: 145,000 daltons.

Molecular Structure: Single-stranded polypeptide chain.

Isoelectric point: pH 3.88; 4.46; 7.6. The crystalline enzyme is separated by electrofocusing into three protein bands.

Optimal Activity: At pH 6.0 and 50°C.

Activators: Ca^{2+}, ~5 × 10^{-4} M.

Inhibitors: Ca^{2+}, >10^{-2} M; Co^{2+}, Ni^{2+}, Cd^{2+}, Zn^{2+}, Hg^{2+}, EDTA.

Stability: pH 4.0–8.0 at ≤ 50°C.

Pullulanase from Bacillus deramificans   According to Amory and Konieczyny-Janda (1994), this enzyme has a better thermostability in the acidic pH range than do other pullulanases from *Bacillus*. The gene coding for the pullulanase was isolated from *B. deramificans* and expressed in *B. licheniformis*.

PROPERTIES

Molecular Weight: 105,000 daltons.

Molecular Structure: Single polypeptide chain.

Isoelectric Point: pH 4.5.

Optimal Activity: pH 4.0–4.5 at 55–60°C.

In the saccharification process the formation of disaccharides can be minimized by reduction of the amyloglucosidase dosage, which is used together with the pullulanase.

Product: Optimax (SO).

Isoamylases (EC 3.2.1.68; CAS 9067-73-6)   These debranching enzymes, of which there are many different types, cleave pullulan slowly and amylopectin rapidly, as shown in Table 3.4 from Yokobayashi et al. (1970).

TABLE 3.4   Effect of Isoamylase and Pullulanase on Various Substrates

	Relative Rate of Hydrolysis	
Substrate	Isoamylase	Pullulanase
Pullulan	—	100
Amylopectin	100	15
Glycogen	124	1
6^3-0-α-Maltosylmaltotriose	2.8	22
6^3-0-α-Maltotriosylmaltotriose	9.7	162
6^3-0-α-Maltotriosylmaltotetraose	33	146

Isoamylase from Pseudomonas sp.

PROPERTIES

pH Activity: 80% maximum activity in the pH range of 3.5–5.5.

Temperature: Maximum 50°C at pH 4–4.5.

Inactivation: In 10 min at pH 2.0 and 50°C.

Isoamylase from B. cereus var mycoides This organism, which produces isoamylase as well as β-amylase and pullulanase, was first described by Y. Takasaki (1976). The characteristics of the enzyme mixture are very similar regarding optimum pH, optimum temperature, and stability. This mixture is thus ideally suited for the production of maltose by the simultaneous action of different enzymes, and 88–90% maltose can be obtained from liquefied starch (DE = 1) with only 0.1% glucose as a by-product.

Manufacturers: [NO], [GB], and others.

REFERENCES

Abe, J., Bergmann, W., Obata, K., and Hizukuri, S., *Appl. Microbiol. Biotechnol.* **27,** 447 (1988).

Amory, A., and Konieczny-Janda, G. "A novel debranching enzyme for the production of dextrose syrup," lecture presented at the 45th Starch Convention, Detmold, Germany, April 1994.

Bender, H., and Wallenfels, K., *Biochem. Z.* **334,** 79 (1961).

Dellweg, H., and Jahn, M., *Monatsschr. Brauerei* **25,** 271 (1972).

Fischer, E. A., and Stein, E. A., in Boyer, P. D., Lardy, H. A., and Myrback, K. (eds.), *The Enzymes,* 2nd ed., Vol. 4, Academic Press, New York, 1960, pp. 313–342.

Fogarty, W. M., and Kelly, T., in Rose, A. H. (ed.), *Economic Microbiology,* Vol. 5, Academic Press, New York, 1983, pp. 116-158.

Granum, P. E., *J. Food Biochem.* **3,** 1 (1979).

Hebeda, R. E., Bowles, L. K., and Teague, W. M., *Cereal Foods World* **36,** 619–624 (1991).

Kainuma, K., Kobayashi, S., Ito, T., and Suzuki, S., *FEBS Lett.* **26,** 281 (1972).

Kainuma, K., Ishigami, H., and Kobayashi, S., *Denpum Kagaku* **32,** 136 (1985).

Kakiuchi, J., *J. Phys. Chem.* **69,** 1829 (1965).

Koaze, Y., Nakajiama, Y., Hidaka, H., Niwa, T., Adachi, T., Yoshida, K., Ito, J., Niida, T., Shomura, T., and Ueda, M., "Production of new amylases by cultivation of *Streptomyces* and uses of these new amylases," U.S. Patent 3,804,717 (1974).

Koaze, Y., Nakajiama, Y., Hidaka, H., Niwa, T., Adachi, T., Yoshida, K., Ito, J., Niida, T., Shomura, T., and Ueda, M., "Production of maltose with amylases produced by *Streptomyces*," U.S. Patent 3,868,464 (1975).

Kobayashi, S., *Appl. Microb. Biotechnol.* **25,** 137 (1986).

Kobayashi, S., Okemoto, T., Nemoto, Y., Hashimoto, H., and Hara, K., "Novel maltotetraose-producing enzyme and production thereof," Jpn. Patent Appl. 85/43568 (March 7, 1985); Jpn. Patent 61/202687 (1986a).

Kobayashi, S., Okemoto, T., Nemoto, Y., Hashimoto, H., and Hara, K., "Production of maltotetraose," Jpn. Patent Appl. 85/43569 (March 7, 1985); Jpn. Patent 61/202700 (1986b).

Krishnan, T., and Chandra, A. K., *Appl. Environ. Microbiol.* **46,** 430 (1983).

Kruger, J. E., and Marchylo, B. A., *Cereal Chem.* **62,** 11 (1985).

Kvesitadze, G. I., Kokonashvili, G. N., and Fenixova, R. V., "Method for preparing α-amylase," U.S. Patent 3,826,716 (1974).

MacGregor, E. A., *J. Prot. Chem.* **7,** 399 (1988).

Madsen, G. B., Norman, B. E., and Scott, S., *Die Stärke* **25,** 304 (1973).

Manning, G. B., and Campbell, L. L., *J. Biol. Chem.* **236,** 2952 (1961).

Norman, B. E., in Birch, G. G., Blakebrough, N., and Parker, K. J. (eds.), *Enzymes and Food Processing,* Applied Science Publishers, London, 1981.

Oikawa, A., and Maeda, A., *J. Biochem. Tokyo* **44,** 745 (1957).

Pazur, J. H., and Okada, S., *Carbohydr. Res.* **4,** 371 (1967).

Robyt, J. E., and Whelan, W. J., in Radley, J. A. (ed.), *Starch and Its Derivatives,* 4th ed., Chapman & Hall, London, 1968, p. 477.

Schwimmer, S., *Source Book of Food Enzymology,* Avi, Westport, CT, 1981, p. 552.

Srivastava, R. A. K., *Enzyme Microbiol. Technol.* **9,** 749 (1987).

Sumner, J. B., and Somers, G. F., *Chemistry and Methods of Enzymes,* 2nd ed., Academic Press, New York, 1947.

Takamine, J., "Preparing and making taka-koji," U.S. Patent 525,820 (1894).

Takasaki, Y., *Agric. Biol. Chem.* **40,** 1515, 1523 (1976).

Tamura, K., Kanno, M., and Ishii, J., German Discl. DE 2 717 333 (Nov. 10, 1973).

Walker, G. J., and Hope, P. M., *Biochem. J.* **86,** 255 (1963).

Yokobayashi, K., Misaka, A., and Harada, T., *Biochim. Biophys. Acta* **212,** 458 (1970).

3.1.2.5 Exoamylases The β-amylases (EC 3.2.1.2; CAS 9000-91-3), also known as *saccharogenic* or *maltogenic amylases,* are found in cereals, malted cereals, sweet potatoes, and other plants. β-Amylases hydrolyze maltose residues at the α-1,4 linkage from the nonreducing end of the starch chain. The designation does not relate to the configuration of the glycosidic linkage that is hydrolyzed, but rather to the free hydroxyl of the cleavage product: β-maltose. β-Amylase does not split α-1,6 linkages. The reaction products from amylopectin are maltose and β-limit dextrin (40–45%). The β-amylases occur in different plants as isozymes with various characteristics. Molecular weights vary from 57,000 to 64,000 daltons in cereals and 152,000 daltons for the enzyme in sweet potatoes. The heat stability of β-amylases depends on their origin. When malt, which contains a mixture of α- and β-amylases, is heated to 70°C in the presence of Ca^{2+}, the β-amylase is inactivated. The β-amylase from soybeans is inactivated in 30 min at 70°C.

β-Amylase from Malted Wheat Flour

PROPERTIES

Structure: Plant β-amylases are sulfhydryl enzymes that do not need metal ions for activation.

Molecular Weight: 57,000 daltons.

Activity: One mole of the enzyme cleaves 252,000 glycosidic linkages per min at 30°C and pH 4.8 (Englard et al., 1951).

pH Activity: Optimum pH 5.0; 90% activity in the pH range of 4.0–6.0.

pH Stability: Optimum pH 4.5–9.0.

Temperature: Maximum 55–57°C, inactivation above 60°C.

Inhibitors: Heavy metals and SH-group blocking agents.

TECHNICAL PREPARATIONS   Industrial preparations are isolated from barley; in East Asia, from soybeans. They differ from typical malt extracts. Barley β-amylase is used in the saccharification of liquid starch to maltose syrup. Depending on whether the starting material is obtained by acid liquefaction or enzymatic degradation, a sugar mixture with the following composition is obtained: 2–8% glucose, 45–60% maltose, and 10–25% maltotriose. The mixture has a dextrose equivalent (DE) of 35–50.

Activity: The "diastatic power" is defined according to FFC (*Food Chem. Codex III*, p. 484). The unit is called "degrees of diastatic power" of DP°. (See Section 2.3.2, "Enzyme Units.") This method of determination cannot be used for malt extracts because 20–40% of the malt extract activity can be attributed to α-amylase. Traditionally, Windisch–Kohlbach units are still being used in the brewing industry.

Liquid products with 1500 DP°/ml are commercially available.

pH Optimum: 5.3 for saccharification (Fig. 3.21).

Optimum Stability: At pH 6–7.

Storage Stability: ∼6 months at 4°C.

β-Amylases from Microorganisms   It is known that β-amylases are produced by a number of microorganisms. Examples are *B. cereus*, *B. megaterium*, *Pseudomonas* sp., and *Streptomyces* sp.

The amylase from *Bacillus polymyxa* can hydrolyze amylopectin to maltose at a 94% yield (Robyt and French, 1964). Originally, it was assumed that the enzyme possessed α-amylase as well as β-amylase properties. Fogarty and Griffin (1975) were able to show that the high conversion yield of starch to maltose was due to

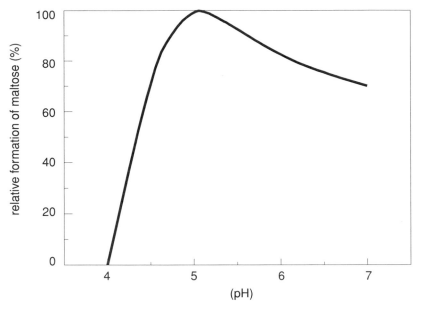

Figure 3.21   pH dependence of maltose formation [FR]. Substrate: 34% enzymatically hydrolyzed starch (14 DE); temperature 60°C. Enzyme: 0.2 liter of Spezyme BBA [FR] per ton of dry solids.

two types of activity, one from a α-1,6-linkage debranching enzyme and the other from a β-amylase, with properties similar to those of plant β-amylases (Fogarty, 1983a).

β-Amylase from B. megaterium   Okada and Higashibara (1974) produced β-amylases from various strains of *B. megaterium*. These differed slightly in their pH optima and stability. The pH optima were in the pH range of 6.0–6.5. The enzymes were inhibited by SH-group blocking agents such as *p*-chloromercuribenzoate and were reactivated by cysteine. The molecular weight of the pure β-amylase is about 50,000 daltons; the isoelectric point is at pH 9.14.

β-Amylase from Streptomyces sp.   Koaze et al. (1974) investigated various *Streptomyces* strains, which produced maltose from starch. The glucose:maltose ratio averaged 0.055:1, the pH optimum was 4.5–5.0, and the optimum temperature was at 50–60°C.

Maltogenic Amylase from B. stearothermophilus   Outtrup and Norman (1984) described a thermostable maltogenic amylase from a gen of *B. stearothermophilus* expressed in *B. subtilis*. The enzyme is produced by Novo Nordisk (Novo company information, 1994). The optimum pH is 5.0 (80% of activity between pH 4.5 and

6.0). The optimum of temperature is 75°C. Qi Si et al. (1994) investigated the functional mechanism of the antistaling effect of the enzyme in baking. For further information on this effect, see Section 5.4.

Product: Novamyl [NO].

Amyloglucosidases (EC 3.2.1.3; CAS 9032-08-0) Amyloglucosidase (AMG) has for some time been one of the most important technical enzymes for the manufacture of glucose and for conversion of carbohydrates to fermentable sugars. It is used in distilleries, the manufacture of baked goods, and the production of fruit juice. In 1957, Ueda, in Japan, discovered different active forms of amyloglucosidase in culture filtrates from *Aspergillus awamori* (Ueda, 1957). Beginning with the manufacture of large quantities of glucose syrup in the 1970s in the United States, the importance of this enzyme grew significantly. Presently, several thousand tons of enzyme are produced annually.

The enzyme, also called *glucoamylase, α-1,4-glucan-glucohydrolase,* or *α-amylase,* cleaves β-D-glucose via an exo mechanism for nonreducing chain ends of amylopectin, amylose, and glycogen.

For the technical production, two types of microorganisms are used. These microorganisms produce different enzymes. The majority of glucoamylases are produced today from *Aspergillus* sp. by a submersion process. A smaller quantity is produced in fixed-bed processes with *Rhizopus,* mainly in Japan. A commercial product from cultures of *Endomycopsis capsularis* (Ebertova, 1966) is also in use.

Amyloglucosidase from *A. niger* is the most important enzyme of this type for technical manufacture and applications. Besides α-1,4 linkages, this enzyme also cleaves α-1,6 and α-1,3 linkages (Fogarty, 1983b) (Table 3.5).

The oligosaccharides maltose and maltotriose are hydrolyzed much more slowly than amylose. This is the case for the enzyme from *Rhizopus* as well as from *Aspergillus,* as shown by Fukumoto (1968) (Table 3.6).

Most strains of *Aspergillus niger* produce at least two, and sometimes three, isozymes. Isozymes AMG 1 and AMG 2 were described by Hayashida and Yoshino (1978). According to Svensson et al. (1982), all commercially available products contain at least two isozymes, the distribution of which changes during fermentation and from batch to batch.

Both enzymes are glycoproteins. The carbohydrates are *O*-glycosidically bonded to the peptide chain. The major component is mannose, which is linked at about 44 sites to seryl, threonyl, or asparagyl residues in the form of di-, tri-, or tetrasaccha-

TABLE 3.5 Relative Efficiency of Amyloglucosidase in Disaccharides

Disaccharide	Linkage	Relative Rate of Hydrolysis
Maltose	α-1,4	100
Isomaltose	α-1,6	1.0
Nigerose	α-1,3	0.2

TABLE 3.6   Relative Efficiency of Fungal Amyloglucosidase in Oligosaccharides

	Relative Rate of Hydrolysis	
Substrate	R. delemar	A. niger
Amylose	312	300
Amylopectin	1222	1260
β-Limit dextrin	937	800
Maltotriose	182	142
Maltose	100	100

rides. In the literature, there is occasional confusion about the classification of the two isozymes. In Table 3.7, the AMG with the higher molecular weight is labeled as AMG-HM; the one with the lower molecular weight, as AMG-LM. According to Svensson et al. (1982), the isozymes can be described as shown in Table 3.7.

AMG-HM has strong debranching activity, and is absorbed on raw starch, which it efficiently degrades. AMG 1 from *Rhizopus* sp. shows a similar action. Highly branched waxy cornstarch is more readily hydrolyzed than amylose (Ueda and Kano, 1975).

According to Hayashida and Yoshino (1978), a gradual degradation by protease and glycosidases (fungal α-mannosidases) occurs during cultivation of the microorganisms, which results in the formation of multiple forms of AMG.

Partial hydrolysis of AMG-HM with subtilisin results in AMG-LM and an enzymatically inactive glycopeptide with a molecular weight of 13,200 daltons (Hayashida et al., 1988).

AMG-LM shows only weak debranching activity and degrades raw starch only slightly. Similar properties were found for AMG from *Rhizopus*.

According to Konieczny-Janda [SO], the molecular structures of the two enzymes can be described as shown in Figure 3.22.

TABLE 3.7   Characteristics of Two Amyloglucosidases

	Isozyme	
Characteristics	AMG[a]-HM	AMG[a]-LM
Molecular weight (daltons), from AA[b] composition	82,000	70,000
Molecular weight (daltons), from sedimentation	110,000	81,000
Molecular weight (daltons), per Pazur and Ando (1959)	112,000 (AMG 11)	99,000
pI	3.25	3.58
pI per Pazur and Ando (1959)	4.0	3.4
Monosaccharide equivalents per mole of enzyme mannose, glucose, galactose	99–100	99–102

[a]Amyloglucosidase; HM, high molecular weight; LM, low molecular weight.
[b]Amino acid.

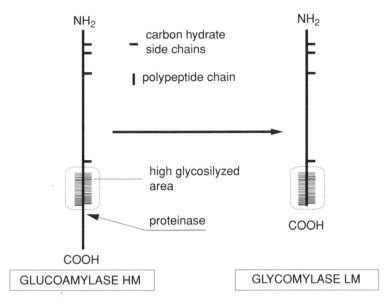

Figure 3.22 Schematic representation of AMG-HM being converted to AMG-LM by protease.

Both enzymes show a slightly different temperature dependence. At temperatures in excess of 65°C, the loss of activity for AMG-HM is somewhat less than that for AMG-LM. Also, under acidic conditions (i.e., pH < 3.5), the larger enzyme molecule loses its activity more slowly. An AMG that is particularly alkali and heat-stable is produced by the thermophilic fungus *Humicola lanuginosa* (Taylor et al., 1978). The pH optimum is around 6, the enzyme is stable up to pH 11, and the temperature optimum is 65–70°C.

Degradation of Raw Starch   While gelatinized starch is rapidly degraded by endo- and exoamylases, nonswollen starch grains are difficult to degrade. According to Walker and Hope (1963), crystallized α-amylases from human saliva, hog pancreas, *B. subtilis,* and *A. oryzae* all, to a greater or lesser extent, degrade various raw starches. For example, hog pancreatic enzyme, at 30°C and pH 6.5, hydrolyzes 5% of potato starch and 55% of cassava starch in 24 h. The size of the starch granules is important (Kulp, 1973) because small granules are more rapidly attacked.

It is important for the manufacture of glucose or the production of alcohol to save the energy that is used to gelatinize starch. These costs represent 50% of the total energy costs in alcohol production.

The degradation of raw starch with amyloglucosidase from *Rhizopus* is the subject of many publications and patents. Takaya et al. (1982) found that amyloglucosidases from *R. niveus, R. amagasakiense,* and *Endomycopsis* hydrolyze corn-starch in similar fashion. Enzymatic activity takes place over the surface of the starch grain, thereby producing holes that run radially from the surface inward.

A fungal amylase that, besides other activities, also contains amyloglucosidase (isoelectric point 8.0), seems to be especially suitable for the degradation of raw starch; it is described by Dwiggins et al. (1984). In the manufacture of bread, en-zymatic hydrolysis of raw starch also plays a role in dough preparation. Dur-ing milling, about 6–8% of the starch granules are damaged. They subsequently swell in cold water and are hydrolyzed by amylases (see Section 5.4 on flour and baking).

Amyloglucosidase from A. niger
Reference: Fogarty (1983c)

PROPERTIES

> *Molecular Weight:* See Table 3.7.
>
> *Structure:* Various glycosylated isozymes; *A. saitoi* has a content of 18% neutral sugars.
>
> *Isoelectric Point:* See Table 3.7.
>
> *pH Activity:* Depending on the strains used and culture conditions, slightly shifted pH optima are found. Substrate concentration and temperature influ-ence the optimum reaction conditions. At 33% substrate concentration (dex-trin 2 DE), 90% of maximum activity is obtained between pH 3.3 and 4.7 at 60°C.
>
> *Stability:* Optimum stability is at pH 4.0 and 60°C. Most preparations are stable in solution for several hours with substrate at pH 2.0 and 30°C. AMG-LM is more stable than the isozyme AMG-HM. (See Figs. 3.23–3.25.)
>
> *Inactivation:* In solutions with less than 10% starch, total inactivation occurs in 10 min at pH 3–5 and 80°C, with higher starch concentrations, 90–95°C is necessary.

TECHNICAL PREPARATIONS

> *Activity:* The technical preparations used for the production of glucose from starch are mostly stabilized liquid products with an activity between 200 and 400 AMG units/g (see section on analytic method). Salt and benzoate are used as preservatives, and glucose, maltodextrins, or sorbitol in concentra-tions of ≤30% is used to stabilize activity. Powdered products with 500–2000 AMG units/g are produced by spray drying.
>
> *Side Activities:* Technical enzymes contain, in addition to α-amylase, a transgly-cosidase activity that lowers the yield of glucose in starch saccharification (see discussion on transglycosidases).
>
> *Manufacturers:* [MC], [GB], [RM], [NO], [AM], [SO], [SOEG], and others.

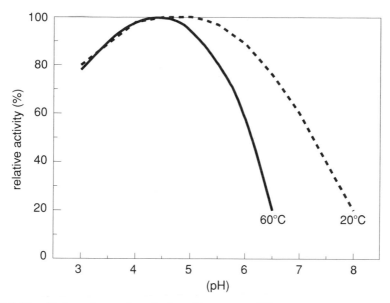

Figure 3.23 pH dependence of amyloglucosidase at two different temperatures [RM].

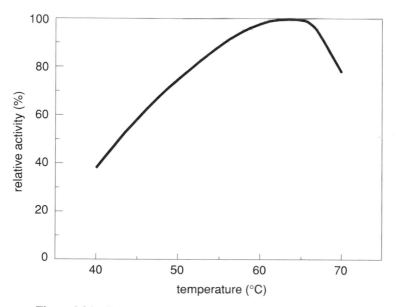

Figure 3.24 Temperature dependence of amyloglucosidase [RM].

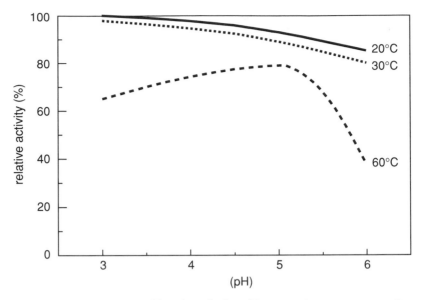

Figure 3.25   pH-dependent stability of amyloglucosidase at various temperatures (in aqueous solution) [RM].

Amyloglucosidase from Rhizopus sp.

PROPERTIES

   Molecular Weight: 100,000 daltons *(R. delemar)*.
   Structure: Glycoprotein, innumerable isozymes.
   Isoelectric Point: pH 7.2 *(R. delemar)*.
   Activity: Optimum pH 4.5; optimum temperature 40°C *(R. delemar)*.
   Stability: Optimum stability at 55°C is pH 3.5–4.5.

TECHNICAL PREPARATIONS   *Rhizopus* AMG preparations are generally produced in powder form by precipitation with alcohol. Liquid preparations are unstable, presumably because of their protease content. The protease is largely denatured by heating to 60°C in the presence of glycerol, whereas the AMG remains intact. In this way, Plainer et al. (1987) were able to impart good stability to liquid products.
   Powdered products with about 1000 AMG units are stable when kept dry.

   Manufacturers: [SN], [AM].
   Side Activities: Amylases, transglycosidases, proteases, esterases, lactase, and
      cellulases.

Enzyme Affinity: Rhizopus delemar (Ono et al. 1964): K. maltose, 1.16 mM; K. maltotriose, 0.2 mM; K. maltopentaose, 0.12 mM; K amylose, 0.00384 mM.

REFERENCES

Dwiggins, B. L., Pickens, C. E., and Niekamp, C. W., "Raw starch saccharification," U.S. Patent 4,618,579 (1984).

Ebertova, H., *Folia Microbiol.* **11,** 14 (1966).

Englard, S., Sorof, S., and Singer, T. P., *J. Appl. Chem. Biotechnol.* **25,** 229 (1951).

Fogarty, W. M., *Microbial Enzymes and Biotechnology,* Applied Science Publishers, London, 1983: (a) p. 35; (b) pp. 23–25; (c) pp. 22–34.

Fogarty, W. M., and Griffin, P. J., *J. Appl. Chem. Biotechnol.* **25,** 229 (1975).

Fukumoto, J., in Radley, J. A. (ed.), *Starch and Its Derivatives,* 4th ed., Chapman & Hall, London, 1968, p. 507.

Hayashida, S., Nakahara, K., Kuroda, K., Miyata, T., and Iwanaga, S., *Agric. Biol. Chem.* **53,** 135 (1988).

Hayashida, S., and Yoshino, E., *Agric. Biol. Chem.* **42,** 927 (1978).

Hebeda, R. E., Bowles, L. K., and Teague, W. M., *Cereal Foods World* **36,** 619–624 (1991).

Koaze, Y., Nakajiama, Y., Hidaka, H., Niwa, T., Adachi, T., Yoshida, K., Ito, J., Niida, T., Shomura, T., and Ueda, M., "Production of new amylases by cultivation of *Streptomyces* and uses of these new amylases," U.S. Patent 3,804,717 (1974).

Konieczny-Janda, G., technical information [MKC], now [SO].

Kulp, K., *Cereal Chem.* **50,** 697 (1973).

Okada, S., and Higashibara, M., "Method for producing β-amylase by bacterial fermentation," U.S. Patent 3,804,718 (1974).

Ono, S., Hiromi, K., and Jimbo, M., *J. Biochem. Tokyo* **55,** 315 (1964).

Outtrup, H., and Norman, B. E., *Die Stärke* **36,** 405 (1984).

Pazur, J. H., and Ando, T., *J. Biochem.* **234** (1959).

Plainer, H., Sprössler, B., and Uhlig, H., "Method for making beer," U.S. Patent 4,684,525 (1987).

Qi Si, J., and Simonsen, R., "Novo Nordisk Report EF 9414394," in *Proc. Internatl. Symp. AACC/ICC/CCOA,* Beijing, Nov. 1994.

Robyt, J. E., and French, D., *Arch. Biochem. Biophys.* **194,** 338 (1964).

Svensson, B., Pederson, T. G., Svendsen, I. S., Sakai, T., and Ottesen, M., *Carlsberg Res. Commun.* **47,** 55 (1982).

Takaya, T., Glover, D. V., Sugimoto, Y., Tanaka, M., and Fuwa, H., *Denpum Kagaku* **29,** 287 (1982).

Taylor, P. M., Napier, E. J., and Fleming, I. D., *Carbohydr. Res.* **61,** 301 (1978).

Ueda, S., *Proc. Internatl. Symp. Enzyme Chem.,* 1975, p. 491.

Ueda, S., and Kano, S., *Die Stärke* **27,** 123 (1975).

Walker, G. J., and Hope, P. M., *Biochem. J.* **86,** 255 (1963).

3.1.2.6   Transglucosidase (EC 2.4.1.24; CAS 9030-12-0) α-1,4-D-glucan-6-α-glucotransferase   Transglucosylation reactions have been known for some time. The reaction is essentially similar to enzymatic hydrolysis. The enzymatic hydrolysis of a glycoside can be conceived of as a transfer of a glucose residue (donor) to water (acceptor). Other molecules such as sugar can be acceptors instead of water.

The reverse reaction of amyloglucosidase at higher substrate concentrations and longer incubation time has already been mentioned. The reversion products are maltose and isomaltose. Disaccharides can also be synthesized through the mechanism of glucose transfer. The enzyme that catalyzes this reaction is a transglucosidase, formed as a by-product in the production of amyloglucosidase. The extent of transglycosylation activity can be estimated by adding the enzyme to a 30–50% glucose solution and determining the amount of isomaltose formed.

The transfer of a glucose residue from a α-1,4 position on maltose or maltotriose to a α-1,6 position leads to panose and isomaltose (Pazur and Ando, 1961). An overview of the properties of the transglucosidase from *A. niger* can be found in Fogarty and Benson (1982).

Transglucosidase from A. niger

PROPERTIES

> *Molecular Weight:*   150,000 daltons.
> *pH Optimum:*   4.5–5.0.
> *Temperature optimum:*   60°C.

Because transglucosidase reduces the yield of glucose during the saccharification of starch, many researchers have worked on the removal or inactivation of this enzyme (Kathrein, 1963). Bentonite is an effective absorbent over a range of pH values and is commonly employed. According to U.S. Patent 3,108,928 (1963), the pH is raised to 9.9 with magnesium oxide, and then after incubation at 30°C for 40 min, the pH is readjusted to 5.0. Recently developed genetic methods have led to the production of transglucosidase-free *Aspergillus* strains.

Fructosyltransferase (EC 2.4.1.9; CAS 9030-16-4)   This enzyme, generated by *A. niger,* is used by Adachi and Hidaka at Meiji (Japan) for the production of fructooligosaccharides under the tradename Neosugar. The product is a mixture of β-1,2-linked fructose residues (F) with a terminal glucose (G), as, for example, F–F–G (1-kestose) or F–F–F–G (nystose). The product has special dietetic properties and is used as a low-calorie sweetener. These fructooligosaccharides are not hydrolyzed by the human digestive system, but are absorbed by the intestinal *Bifidobacterium* sp., the *Bacterium fragilis* group, and *Peptostreptococcus.* The growth of the bacteria is strongly stimulated (Mitsuoka et al., 1987), thereby eliminating digestive problems caused by lack of fermentation.

Cyclodextrin glucosyltransferase (CGT) (EC 2.4.1.19; CAS 9030-09-5), cyclodextrinase
References: Cramer and Hettler (1967), Fogarty (1983)

Cyclodextrins are circular oligomers with 6-, 7-, or 8-glucose molecules, known as α-, β-, and γ-cyclodextrin, respectively. They are formed by the action of the enzyme (CGT) on starch or dextrins. They are named *Schardinger dextrins* after the scientist who discovered them.

Cyclodextrins have gained importance in recent years because of their ability to form inclusion compounds with other molecules. In this way, volatile concentrated aroma molecules that are sensitive to the action of oxygen, light, or heat can be stabilized by complexing with cyclodextrins. In pharmaceuticals, this encapsulation has many advantages, such as improving solubility or dispersability. New applications for cyclodextrins are continually being developed in the area of cosmetics and perfumes.

Cyclodextrins have been known for some time. The nature of their formation by an enzyme from *B. macerans* was first closely investigated by Cramer and Steinle (1955). The reaction rate was monitored by the disappearance of the iodine color during the enzyme's reaction with starch. It was believed, at one time, that the formation of cyclodextrin depended on the complex action of three enzymes: a dextrinizing amylase, a cyclodextrin synthetase, and a glucosyltransferase. However, after completely extracting and purifying the hydrolytic activity from a crude enzyme preparation of *B. macerans,* Kobayashi et al. (1978) found that the pure crystalline enzyme contained all three activities (Table 3.8).

A cyclodextrinase (CGT) from *B. megaterium* with high transglucosidase activity was described by Kitahata and Okada (1975). Alkalophilic enzymes are also sources for the CGT. Nakamura and Horikoshi (1976) found that unpurified enzymes from such organisms showed two pH optima, one at pH 4.5–5.0 and the

TABLE 3.8   Properties of Cyclodextrin Glucosyltransferase from *B. macerans* (Fogarty, 1983)

Property	Depinto and Campbell (1968)	Kobayashi et al. (1978)	Stavn and Granum (1979)
Molecular weight (daltons)	139,300[a]	145,000[a]	67,000[b]
pI	—	—	5.4
pH optimum	6.1–6.2	6.0	5.4–5.8
Temperature optimum	—	60°C	60°C
K_m (starch)	3.33 mg/ml	—	5.7 mg/ml

[a]Two subunits with identical amino acid compositions.
[b]Single chain.

other at pH 7.5–8.5. A purified enzyme isolated from the organism had a molecular weight between 85,000 and 88,000 daltons and showed optimum activity at pH 7 and 50°C.

On binding to a carrier, the enzyme with the two pH optima lost its alkaline optimum, and the temperature stability was not improved by immobilization.

REFERENCES

Adachi, T., and Hidaka, H., technical information [MJ].

Cramer, F., and Hettler, H., *Naturwissenschaften* **54,** 625 (1967).

Cramer, F., and Steinle, H., *Liebigs Ann. Chem.* **595,** 81 (1955).

Depinto, J. A., and Campbell, L. L., *Biochemistry* **7,** 114 (1968).

Fogarty, W. M. (ed.), *Microbial Enzymes and Biotechnology,* Applied Science Publishers, London, 1983, pp. 32–33.

Fogarty, W. M., and Benson, C. P., *Biotech. Lett.* **4,** 61 (1982).

Kathrein, H. R., "Treatment and use of enzymes for the hydrolysis of starch," U.S. Patent 3,108,928 (1963).

Kitahata, S., and Okada, S., *Agric. Biol. Chem.* **39,** 2185 (1975).

Kobayashi, S., Kainuma, K., and Susuki, S., *Carbohydr. Res.* **61,** 229 (1978).

Mitsuoka, T., Hidaka, H., and Eida, T., *Die Nahrung* **31,** 427 (1987).

Nakamura, N., and Horikoshi, K., *Agric. Biol. Chem.* **40,** 1785 (1976).

Pazur, J. H., and Ando, T., *Arch. Biochem.* **93,** 43 (1961).

Stavin, A. and Granum, P. E., *Carbohydr. Res.* **75,** 243 (1979).

3.1.3   Carbohydrases Other Than Amylases

3.1.3.1   Inulinase (EC 3.2.1.7; CAS 9025-67-6)

2,1-(1,3;1,4)-β-D-Glucan-3(4)-glucanohydrolase   Inulin (see structure in Fig. 3.26) is a linear, β-2,1-linked polymer of fructose. A sucrose residue is bound to the end of the chain. The molecular weight is about 6000 daltons. Inulin serves as an energy reserve in many plants, especially the composites, such as chicory, Jerusalem artichoke, and dahlia. Pure inulin is almost insoluble in cold water, but can be extracted from plant material with hot water. Inulin can be hydrolyzed using inorganic acids at pH 1–2 in 1–2 h at 80–100°C (Zittan, 1981). Because fructose is unstable at low pH values, such treatment readily leads to by-products that are deeply colored. Therefore, it is desirable to completely hydrolyze inulin to fructose under gentle conditions. Fructose, with its high degree of sweetness, is important in the dietetic and beverage industries. The enzymatic hydrolysis of inulin offers an alternative method of fructose production beside the standard procedure with starch as the source material.

Inulinase is produced by several microorganisms; those from yeasts (*Candida* sp. and *Kluyveromyces fragilis*) and *Aspergillus* are of particular interest for the

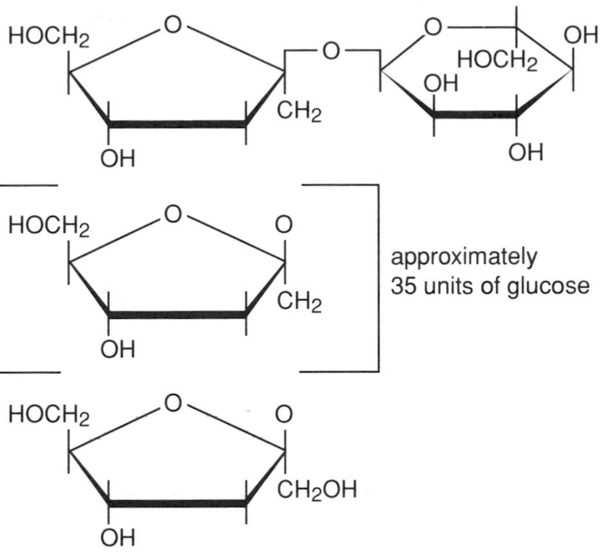

Figure 3.26 Inulin structure.

manufacture of technical enzyme products. The yeast inulinase from *Candida* sp. (Groot, Wassink, and Fleming, 1980) has an optimum activity at pH 4.5 and exhibits 80% of maximum activity in the pH range of 3.5–6.0. Optimum temperature is 45°C at pH 4.5, the enzyme is inactivated above 55°C.

Inulinase from *Aspergillus* was isolated, crystallized, and characterized by Nakamura (1978). According to Zittan (1981), there are two enzymes in *A. ficuum*—an endoinulase (EC 3.2.1.7) and an exoenzyme (EC 3.2.1.80)—a fact known for other carbohydrases as well.

The enzyme described by Zittan (1981) has the following properties: Optimal activity is at about pH 4.5 and 60°C. Figure 3.27 shows the dependence of *Aspergillus* and yeast inulinase activity on pH.

Products and Manufacturers: Inulinase from *Aspergillus* [NO], [MJ], and others.

Dextranase (EC 3.2.1.11; CAS 9025-70-1) In the manufacture of sugar (sucrose) from sugarcane or sugarbeet, dextran can be formed by *Leuconostoc mesenteroides*. This causes not only a loss of sugar content but also significant problems in filtration and crystallization during the production process. The enzyme dextran sucrase produced by *Leuconostoc* synthesizes α-1,6 dextran from sucrose (nF-G) according to

$$\text{Sucrose (nF-G)} \rightarrow \text{dextran (Gn)} + \text{fructose (nF)}$$

This enzyme was isolated by Tilburg (1972) from *Penicillium funiculosum*, who cocultivated *Penicillium* with the dextran producer *Leuconostoc* in a sugar medium.

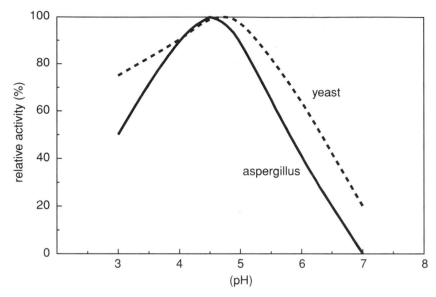

Figure 3.27   pH dependence of *Aspergillus* inulinase.

Dextran forms chains of approximately 50,000 glucose residues, has extremely high viscosity, and interferes with filtration and clarification of sugar juices. Dextranase cleaves the α-1,6 linkages of dextran in a stepwise fashion. The reaction products are isomaltose and isomaltotriose.

PROPERTIES

Activity: One unit of enzyme is the amount that produces the reducing sugar equivalent of 1 g of maltose in one hour from dextran at 40°C [NO].

Temperature Optimum: 60°C.

Stability Optimum: In buffer, pH 5.4, the enzyme is stable in sugar solution for about one hour at 50°C. At 70°C, 50% of activity is lost in about 15 min. See Figure 3.28.

Penicillium funiculosum forms two isozymes (pI 3.98 and pI 4.19). The molecular weight is about 44,000 daltons. The enzymes are activated by Co^{2+} and Mn^{2+} and inactivated by Hg^{2+} and Cu^{2+}.

Dextranase from Arthrobacter sp. This enzyme cleaves insoluble glucans more rapidly than the enzyme from *Penicillium* sp. It is active between pH 5.5 and 7.5, with an optimum at pH 6.5–7.0. The optimum temperature is between 37 and 45°C. Molecular-weight determinations: 40,000 (gel filtration) and 62,000 daltons [sodium dodecylsulfate (SDS)–electrophoresis].

Manufacturer: [NO], [GO], and others.

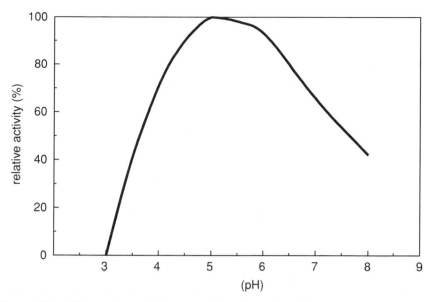

Figure 3.28   pH dependence of dextranase from *Penicillium liliacinum* (McIlvaine buffer, 40°C [NO]).

3.1.3.2   *Glycosidases*   The scission of the glycosidic linkage for polysaccharide-cleaving carbohydrases and the endo- and exoamylases was discussed earlier. In this section, the enzymes cleaving oligosaccharides will be described.

The enzymes in this group show definite specificities:

1. Either α- and β-specificity, according to which the enzymes are actually classified.
2. Sugar or structure specificity, which means that the enzymes recognize the configuration of the glycosidic hydroxyl group.
3. Specificity can depend on the degree of polymerization of the oligomers.
4. Specificity relative to the aglycon that is the nonsugar moiety of a glycoside. A large number of important natural products that can be cleaved by specific glycosidases belong to this class of substances. An example is amygdalin, the hydrolysis of which has been studied for a number of years (Fig. 3.29). See also Table 3.9.

α-Glucosidases (EC 3.2.1.20; CAS 9001-42-7) α-D-Glucoside glucohydrolase
The best studied enzyme of this group is maltase. This enzyme splits one molecule of maltose into two molecules of glucose and always accompanies the amylases in the animal digestive tract. α-1,4-Linked glucose residues are cleaved from the nonreducing end of the oligosaccharides and α-D-glucosides. Polysaccharides are hydrolyzed only very slowly. α-Glucosidases are very specific with respect to the glucose residues of the glucoside, but less specific with respect to the aglycons.

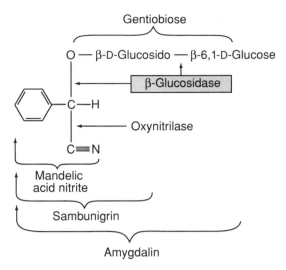

Figure 3.29   Enzymatic hydrolysis of amygdalin.

Maltase from animal organs is able to cleave sucrose; however, raffinose is not cleaved by the intestinal mucous membrane.

Yeast is a good source of α-glucosidase. This enzyme hydrolyzes maltose and maltose derivatives, aryl- and alkyl-α-D-glucoside and sucrose.

The pH optimum for the maltases from *A. niger* and malted barley is 4.5.

TABLE 3.9   Glycosidases, Substrates, Products

Enzyme	Substrate	Hydrolysis Product
α-Glucosidases	Maltose	2 Glucose molecules
	Sucrose	D-Glucose + D-fructose
	Melezitose	2 Glucose molecules + D-fructose
β-Glucosidases	Cellobiose	2 Glucose molecules
	Prunin	Glucose + naringenin
	Gentobiose	2 Glucose molecules
α-Galactosidases	Raffinose	D-Galactose + saccharose
	Stachyose	D-Galactose + raffinose, galactose branches in galactomannan chains,
	—	D-Galactose, and polymers
β-Galactosidases	Lactose	D-Galactose and D-glucose
β-Rhamnosidase	Naringin	D-Rhamnose + prunin
β-Fructosidase	Sucrose	D-Fructose + D-glucose
β-Xylosidase	Xylobiose	D-Xylose
α-L-Arabinofurnanosidase	D-Nitrophenyl-α-L-arabinofuranoside	D-Nitrophenyl and L-arabinose

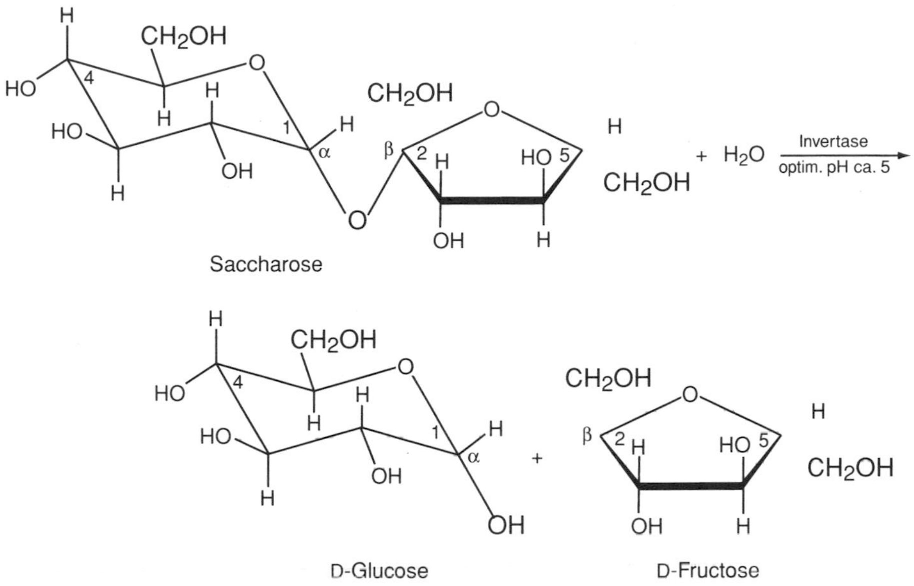

Figure 3.30   The action of invertase.

Invertase (EC 3.2.1.26; CAS 9001-57-4) β-fructosidase, β-D-fructofuranoside Fructohydrolase   The disaccharide sucrose (cane sugar) consists of glucose and fructose, coupled by a glycosidic linkage between the C-1 atom of glucose and the C-2 atom of fructose. Hydrolysis of this bond leads to a mixture of reducing sugars that is sweeter than cane sugar because of the high sweetening power of the liberated fructose (Fig. 3.30).

The term for this reaction is *inversion,* and the resultant monosaccharide mixture is called *invert sugar.* The term *inversion* relates to the observation that an aqueous sucrose solution rotates the polarized plane of light to the right whereas invert sugar rotates the plane to the left because of the fructose, which shows a high degree of rotation to the left. During hydrolysis, an inversion in the direction of rotation of the polarized light is observed.

In industry, hydrolysis is usually accomplished with inorganic acids. The drawback of this traditional method is that hydrolytic by-products are produced that have to be removed by purification. As invertase is easily immobilized and highly efficient with regard to product formation, the enzymatic process can be considered superior to the chemical method.

The candy industry in particular has shown interest in this enzyme. It is used in the production of marzipan and soft chocolate fillings, as well as for artificial honey and sugar invert syrup. The latter is used in many branches of the food industry. One special application for invert syrup is as winter food for bees.

Berthelot (1860) was the first to isolate the enzyme from yeast and it was subsequently used in extensive studies on enzyme kinetics. Invertase is an intracellular

enzyme. This means that the enzyme remains within the cell during the yeast cultivation. The free enzyme is obtained by lysis of the cell wall.

Today, yeast is used almost exclusively for the production of commercial invertase.

PROPERTIES, PRODUCTS, AND MANUFACTURERS

Molecular Weight: For the pure intracellular invertase, 135,000 daltons; for the
extracellular enzyme, 270,000 daltons, of which ~50% consists of bound
mannan from yeast.

pH Activity: The optimum activity ranges between pH 4 and 5.5.

Temperature: Because various commercial preparations show different stabili-
ties, exact data for optimum reaction temperatures are not available. In most
cases 55°C is the optimum temperature for dilute, and 65–70°C for concentrated sugar solutions. (See Fig. 3.31.)

Stability: There are significant differences between the intracellular and the extracellular enzymes. Technical preparations are stable at 30°C between pH 3
and 7.5.

Immobilized Invertase: See Chapter 5.

Products and Manufacturers: Invertin [MK], [SE]; Maxinvert [GB]; Rohalase I
10 [RM].

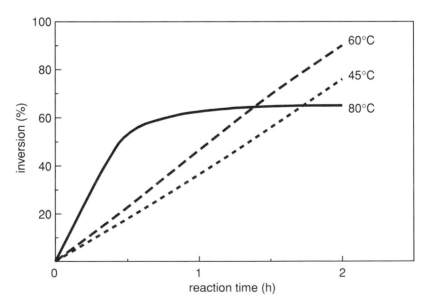

Figure 3.31   Temperature-dependent inversion of a 70% sucrose syrup [GB]. K_m values:
26 mM for sucrose, 150 mM for raffinose (Gascon et al., 1968).

α-D-Galactosidase (EC 3.2.1.22; CAS 9025-35-8), α-D-galactoside galactohydrolase
Reference: McCleary (1988)

This enzyme is present in plants and animals and is also produced by various microorganisms. The enzyme has been isolated from alfalfa, guar beans (McCleary, 1988), and green coffee beans. Several strains of *A. niger* produce the enzyme together with several hemicellulases.

3.1.3.3 β-Glucosidases (EC 3.2.1.21; CAS 9001-22-3): β-D-glucoside glucohydrolase

Naringinase (α-L-rhamnosidase)
Reference: Neubeck (1975)

Citrus fruits contain two different types of bitter substances known as *flavonoids* (naringin) and the limonoids (limonin).

Naringin, the 7-(2-rhamnosido-β-glucoside) of 4',5,7'-trihydroxy-flavonone, occurs in large amounts in grapefruit and bitter oranges. For the average taste, a small content of naringin is desirable. However, if grapefruit is reduced to juice, an undesirable level of bitterness may be generated during extraction.

In Japan, Kishi (1955) discovered an enzyme from *A. niger* that degraded naringin.

Today's commercial products contain two different glycosidases: a rhamnosidase (EC 3.2.1.43) that hydrolyzes naringin to prunin by releasing rhamnose, and a β-glucosidase, which cleaves prunin to naringenin and glucose:

$$\text{Naringin} \xrightarrow{\text{rhamnosidase}} \text{prunin} + \text{rhamnose} \xrightarrow{\beta\text{-glucosidase}} \text{naringenin} + \text{glucose}$$

Naringinase preparation must be free of pectin-depolymerizing enzymes; otherwise an undesirable clarification of the citrus juice will result. In a study on the hydrolysis of naringin by Pilnik (1982), it was shown that as naringin disappears, the intermediate products, prunin and the end-product naringenin, were formed.

ENZYME CHARACTERISTICS   The optimum activity of commercial preparations is around pH 3.5; 80% of the maximum activity lies between pH 3 and 4.7. The effect of pH and temperature on activity can be seen in Figure 3.32.

The loss in stability of naringinase with decreasing pH and increasing temperature can be seen in Figure 3.33.

TECHNICAL PREPARATIONS

Fungal Sources: A. niger, Penicillium sp., and others.

Products and Manufacturers: Naringinase [AM]; Kumitanase [SA], and others.

Hesperidinase   Hesperidin is another glycoside that occurs in large quantities in citrus fruits. It is the 7-(2-rhamnosido-β-glucoside) of 4'-methoxy-3',5,7'-trihydroxy-

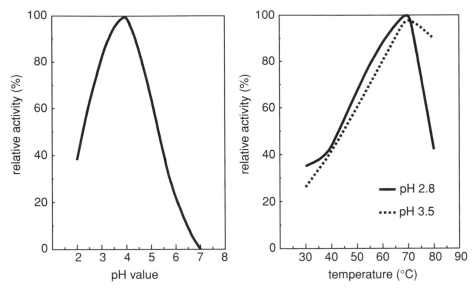

Figure 3.32   pH and temperature dependence of naringinase activity [AM].

flavanone (hesperidin). It sometimes crystallizes in frozen fruits or during the manufacture of orange juice concentrates and is almost tasteless. Hesperidin crystals can
interfere with the stability of turbidity in orange juice (Rothschild and Karsenty,
1974).

The use of a hesperidinase in orange juice is described by Sawayama et al.
(1966). This paper deals with a specific rhamnosidase from *A. niger.*

CLEAVAGE OF LIMONIN When discussing the bitter substances in citrus fruits,
limonin (Fig. 3.34) must also be mentioned. Limonin is an exceptionally bitter
triterpenoid dilactone that is present in navel oranges as a nonbitter precursor molecule, the limonic acid A-ring lactone. Limonin is released from the precursor molecule by acid. The precursor, which is bound to the albedo layer of the fruit, can be
released into the juice on cell rupture during a hard frost or when pressing the
fruits. It is then converted by the acidic juice into the bitter limonin (Joslyn and
Pilnik, 1961).

A number of methods have been sought to eliminate these types of bitterness.

Limonin dehydrogenase (or limonoate dehydrogenase, CAS 37325-58-9), found
in *Arthrobacter globiformis* and *Pseudomonas* (Hasegawa et al., 1972), convert
limonin and limonin-A-ring-lactone to a 17-dehydro derivative. In this way, formation of the bitter D-ring lactone cannot occur. This interesting possibility of solving
a very important problem for the citrus industry still fails because the dehydroge-
nases described depend on expensive coenzyme systems (NAD) and further, the pH
optimum is not in the range of acidic citrus juices.

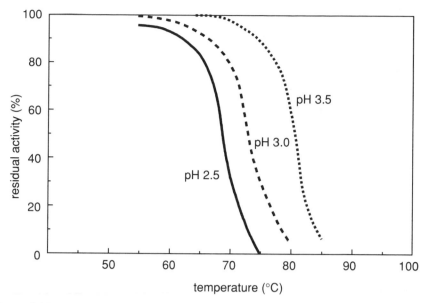

Figure 3.33 Temperature stability of a naringinase preparation [AM] at various pH in 0.1 M buffer, 10 min.

Lysozyme (EC 3.2.1.17; CAS 9001-63-2), Muramidase Lysozyme is widely distributed in animal tissue, plants, and microorganisms. It was first isolated by Fleming (1922). It is also known as an antibiotic because it is able to kill certain bacteria by hydrolyzing their cell walls. Lysozyme is present in milk (especially human milk), tears, and egg albumin. Human milk contains about 3000 times more lysozyme than cow milk. This antibacterial activity is one advantage of breastfeeding over using cow milk. Lysozyme is currently produced on a large scale from egg albumin. Mainland China is a major producer. According to Scott et al. (1987), it is estimated that 100 tons are produced annually worldwide.

Lysozyme is isolated by isoelectric precipitation from albumin and adsorption on ion exchangers, then concentrated by ultrafiltration with repeated recrystallizations.

Unpurified papain is also rich in lysozyme. Twenty percent of the soluble protein in the latex consists of lysozyme (Cayle et al., 1964). There are various applications for the enzyme in food and medicinal technologies. Among these is the use of lysozyme in the production of certain cheeses, where it is important to kill harmful microorganisms such as *Clostridium tyrobutyricum*.

P. Jolles and J. Jolles (1961) have described the structure of lysozyme in human milk. Weidel (1966) found that part of the cell wall of certain gram-positive bacteria consists of a single sacklike macromolecule. This murein sack is very specifically cleaved by lysozyme. It hydrolyzes the β-1,4 linkage between the N-acetyl

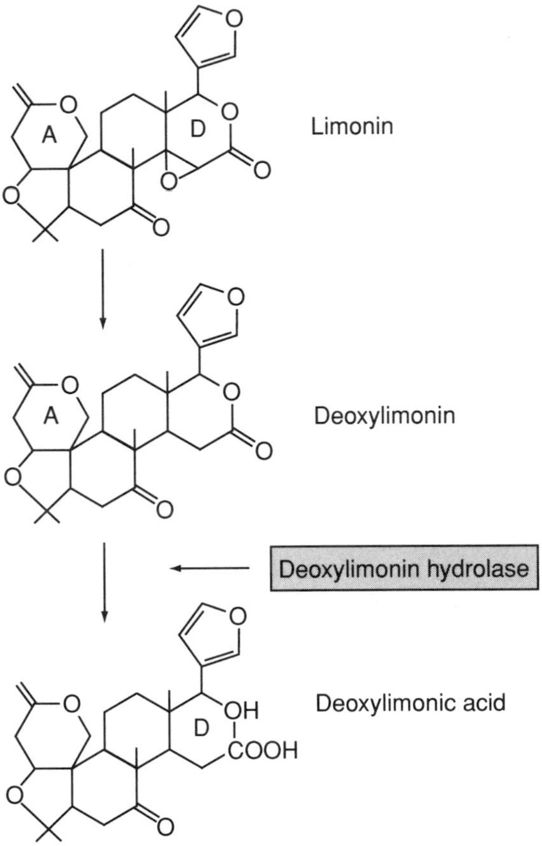

Figure 3.34   Limonin hydrolysis.

muramic acid (Fig. 3.35), and the 2-acetamido-2-deoxyglucose residues in mucopolysaccharides or mucopeptides.

PROPERTIES AND MANUFACTURERS   Lysozyme is marketed as a white crystalline powder. It is readily water-soluble but is insoluble in organic solvents. It is a polypeptide with 129 amino acid residues with a molecular weight of 14,388 daltons. Lysozyme is a strongly basic protein with an isoelectric point between pH 10.5 and 11.7.

K_m: 4.0×10^{-3} with p-nitrophenyl-β-D-chitobioside as substrate.

Activity: The pH optimum is 7–8 with *Micrococcus* cells as substrate; the temperature optimum is 35°C.

Stability: Optimum is pH 4–5. Aqueous solutions are stable under aseptic conditions at 2–4°C.

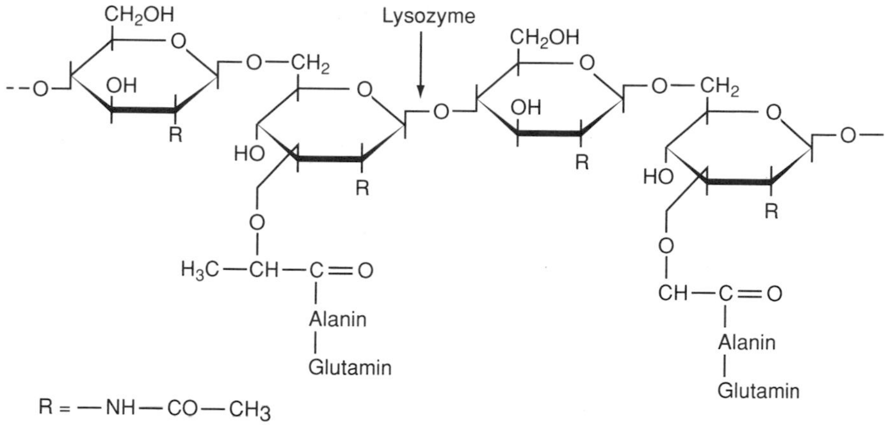

Figure 3.35   Muramic acid hydrolysis with lysozyme.

Inhibitors: Surfactants such as dodecylsulfate, oxidizing agents, and heavy metals.

Analysis: Determination of lysozyme by the Federation International Pharmaceutique (FIP) (Ruyssen and Lauwers, 1978), or by the Calbiochem lysozyme assay (Shugar, 1952).

Substrate: Micrococcus luteus cells, determined by the decreased absorption at 450 nm. Crystalline lysozyme shows about 25,000 FIP units per milligram.

Manufacturers: [GB], [NG], [SOE], [WB].

Lysozyme from Microorganisms
Reference: Jolles (1969), Osserman (1972)

A number of microorganisms have been cited as producers of lysozyme, or lytic enzymes with the specificity of lysozyme; examples are *B. subtilis* (Okada and Kitahara, 1972), *B. cereus* (Csuzi, 1968), and *Lactobacillus casei* (Shirota et al., 1974).

According to Woehner et al. (1986), a bacteria lysing enzyme showing lysozyme-type activity can be produced by fermentation of *Streptomyces coelicolor.*

3.1.3.4   *Lactases, β-Galactosidases (EC 3.2.1.23; CAS 9001-22-3)*

Introduction   Lactase catalyzes the hydrolysis of lactose (milk sugar), a disaccharide present in milk, to glucose and galactose.

The enzyme often occurs together with β-glucosidases in yeast, fungi, and bacterial cultures. It is formed in the pancreatic and lactal glands of mammals. Lactase is immobilized on the brush-border of the intestinal wall (epithelium of the jejunum). Lactase deficiency, due to biochemical or genetic aberrations leads to flatulence, diarrhea, and bleeding. Various populations, particularly those of East Asia and Africa, suffer from lactase deficiency (Kretschmer, 1972). Only adults are affected. Because of the minor nutritional importance of milk products in these regions, lactose intolerance presents no significant problem.

Milk contains about 4.5–5% lactose. This sugar is relatively insoluble, having only about half the sweetness of sucrose. Increasing interest in the enzymatic scission of lactose into a more valuable mixture of monosaccharides is due to the fact that large amounts of lactose containing whey are produced annually. Until recently whey was used either as animal feed or simply discharged, where it becomes a source of high-cost wastewater treatment. Even today, where much of the whey is ultrafiltered for the extraction of milk protein, there is still the problem of how to utilize the lactose in the permeate. Lactose is used, albeit to a limited extent, by the pharmaceutical industry. The annual global yield of whey is estimated at about 50 million tons, which corresponds to 2.5 million metric tons of lactose.

Enzymatic cleavage of lactose can be achieved with free or carrier-bound lactases. Yeast lactase is used to make milk tolerable for people with lactase deficiency. The enzyme can also be used for the manufacture of yogurt and for the hydrolysis of lactose in whey. The advantages of whey hydrolysates to that of lactose are

1. The hydrolysate is sweeter (~70% of sucrose sweetness).
2. The glucose–galactose mixture is significantly more soluble than lactose, a property important in food processing.

The lactose hydrolysate can be used as a syrup for the manufacture of baked goods, in the production of ice cream and desserts, and in dressings and lemonades. Lactase hydrolyzes lactose according to the scheme in Figure 3.36.

Innumerable microorganisms produce lactases of different properties. The lactase produced by *Escherichia coli* has been used by many researchers involved in research on the genetics of *E. coli*. *E. coli* lactase is produced in industrial quantities for diagnostic purposes [BM]. For applications in food processing, the enzyme is extracted from yeast and *Aspergillus* sp. These two enzymes differ in their activity and stability optima as a function of pH (Table 3.10).

Because the reaction parameters are strongly dependent on the enzyme, comprehensive pH and temperature dependences are presented in detail.

Optimal activity of the *A. oryzae* enzyme is at pH 4.5, which corresponds to the pH value of acid whey (pH 4–5). Enzymatic activity is only about 40% of that of

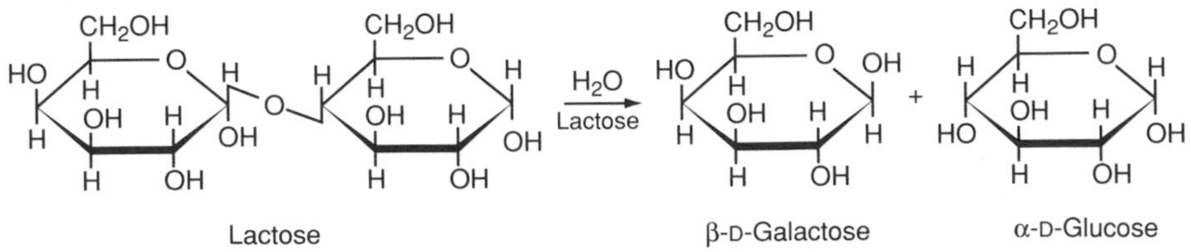

Figure 3.36   Lactose hydrolysis.

TABLE 3.10   Enzyme Characteristics (Günther and Bürger, 1982)

Characteristics	Yeast Lactase	*Aspergillus* Lactase
Organism	*K. fragilis*	*A. oryzae*
Activity	5000 EU/g (pH 6.2/30°C)	8600 EU/g (pH 4.5/30°C)
pH optimum	6.0–6.2	4.0–5.0
pH stability	5.5–8.5	4.0–8.5
Temperature optimum	43°C	50–60°C
Temperature stability	43°C (max)	60°C (max)
pI	4.0/6.2	3.5/4.4
Molecular weight (daltons)	130,000	130,000
K_m	20 mM	60 mM

the optimum in the hydrolysis of lactose in milk or sweet whey (pH 6.2). The enzyme from *S. lactis* shows optimal activity in the pH range of milk (Fig. 3.37).

Yeast Lactase   The enzymes from *K. lactis* (Fig. 3.38) or *K. fragilis*, together with the fungal lactase from *A. oryzae* (Fig. 3.39), cover the whole spectrum of lactase applications. The use of yeast lactase is difficult in some cases because of its sensitivity to pH, temperature, and ions. Calcium ions inhibit—whereas manganese, magnesium, and potassium ions activate—yeast lactase. Fungal lactase is more sta-

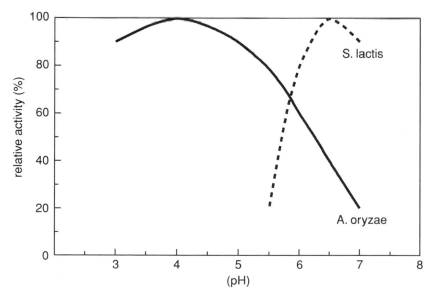

Figure 3.37   pH dependence of lactase from *A. oryzae* and *S. lactis* [5% lactose solution, 30°C; Plainer et al. (1982) [RM]].

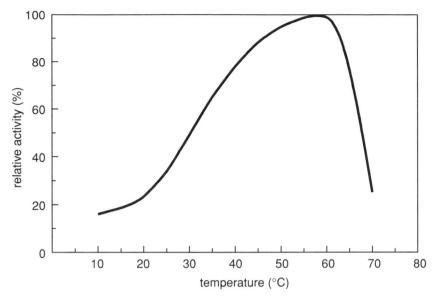

Figure 3.38 Temperature dependence of lactase from *A. oryzae* [5% lactose solution, reaction time 15 min; Plainer, 1986 [RM]].

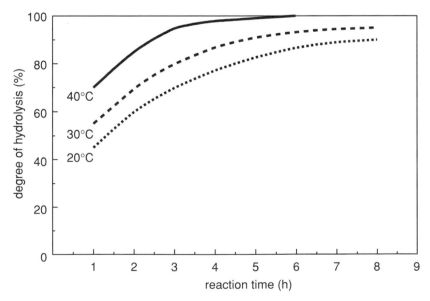

Figure 3.39 Temperature dependence of lactase from *K. lactis* [AM].

ble but has the disadvantage in that the galactose produced by the hydrolysis inhibits product formation. It is recommended that hydrolysis be undertaken in dilute ~10% substrate solutions.

Lactase from A. niger Technical lactase from *A. niger* shows optimum activity at pH 3.5–4.5 and can tolerate temperatures up to 50°C. Product inhibition with this enzyme is also high.

Lactase from E. coli

> *Molecular Weight:* 540,000 daltons.
> *Optimum Activity:* pH 7.0; K_m (lactose), 3.8×10^{-3}.
> *Stability:* After incubation at 30°C for 30 min, at pH 6.6–7.6, a residual activity of 80% remains.

Lactase from B. stearothermophilus The enzyme described by Goodman and Pederson (1976) has a molecular weight of 215,000 daltons and an optimum activity at pH 6–6.4. Activity drops off rapidly at acidic pH. The optimum temperature is 65°C; the K_m value for lactose (at 55°C) is 0.11 mM, and that for O-nitrophenyl-β-D-galactopyranoside is 9.8 mM.

Lactase from B. circulans The enzyme is temperature stabile, but is unsuitable for practical use because of extremely strong product inhibition.

Because lactose is currently a waste product, efforts are being made to make lactose hydrolysis more cost-effective; a lactose hydrolysate product needs to be competitive with sugar or glucose–fructose syrup. One possibility would be to use an immobilized enzyme for the hydrolysis.

Immobilized Lactase Some typical data from an immobilized lactase, Plexazym LA 1 [RM], is given here. Plexazym LA 1 consists of macroporous methacrylate beads to which lactase from *A. oryzae* is bound. The hydrolysis of lactose, either in the form of whey or from filtrates, occurs in a fixed bed at pH 4–5 and 30–35°C. The flow rate amounts to 35–50 times the fixed-bed volume per hour. The capacity of the immobilized lactase is in the order of 200 tons sour whey per kilogram of Plexazym LA 1.

Analysis of the Lactases

1. *Determination with o-Nitrophenyl Galactopyranoside (ONPG) as a Substrate:* With the method according to Morisi et al. (1973), the release of *o*-nitrophenyl is measured photometrically at 420 nm. The reaction proceeds with the *A. oryzae* enzyme at pH 4.5, with the yeast lactase at pH 6.5 and 30°C.

2. *Determination with Lactose as Substrate:* The enzymatic activity is assayed measuring the glucose formed during hydrolysis. It is determined either by

enzymatic glucose test (offered by [BM], [MK]), or more rapidly with an enzyme electrode. The conditions are 5% lactose solution, 30°C at pH 4.5 or 6.5, depending on the type of enzyme.

Products and Manufacturers: Mostly stabilized liquid products with 1000–4000 U/ml (lactose) or 3000–12000 ONPG units/ml are available. Products: Lactase 7041 C [RM]; Plexazym LA 1 [RM]; Maxilact [GB]; Hydrolact [SE]; Lactase Y-L [AM]; Sumizyme [SN]; Lactozym [NO].

REFERENCES

Berthelot, *Compte Red. Hebdomad. Seances Acad. des Sciences* **51,** 980 (1860).

Cayle, T., Saletan, L. T., and Lopes-Ramos, B., *Americ. Soc. Brew. Chem. Proc.,* 142 (1964).

Csuzi, S., *Biochim. Biophys. Acta* **3**(1), 41 (1968).

Fleming, A., *Proc. Roy. Soc. B* **93,** 306 (1922).

Goodman, R. E., and Pederson, D. M., *Canadian J. Microbiol.* **23,** 798 (1976).

Groot Wassink, J. W. D., and Fleming, S. E., *Enzyme Microb. Technol.* **2,** 3 (1980).

Günther, E., and Bürger, E., "A method of manufacturing lactose-hydrolyzed yoghurt by means of β-galactosidase," in Dupuy, P. (ed.), *Use of Enzymes in Food Technology,* Lavoisier, Paris, 1982.

Hasegawa, S., Bennet, R. D., Maier, V. P., and King, A. D., *Agric. Food Chem.* **20,** 1031 (1972).

Hasegawa, S., and Maier, V. P., "New methods to remove citrus bitterness," Annual Meeting of ACS, Miami Beach, FL, C&E News, May 27, 1985.

Jolles, P., and Jolles, J., *Nature* **192,** 1187 (1961).

Jolles, P., *Angew. Chem.* **81**(7), 244 (1969).

Joslyn, M., and Pilnik, W., "Enzymes and enzyme activity," in Sinclair, W. B. (ed.), *The Orange: Its Biochemistry and Physiology,* Univ. of California Press, Berkeley, 1961.

Kishi, K., *Kagaku Koggyo (Osaka)* **29,** 140 (1955); *Chem. Abstr.* **49,** 14106 i (1955).

Kretschmer, N., *Sci. Am.* **227,** 71 (1972).

McCleary, B. V., "α-D-Galactosidase from lucerne and guarseed," in *Methods in Enzymology,* Vol. 160, *Biomass,* 1988, p. 627.

Morisi, F., Pastore, M., and Viglia, A., *J. Dairy Sci.* **56,** 1123 (1973).

Nakamura, T., Kurokawa, T., Nakatsu, S., and Ueda, S., "Purified inulinase from *A. niger* (Miji)," *Nippon Nogeikagaku Kaishi* **52,** 159 and 581 (1978).

Neubeck, C. E., in Reed, G. (ed.), *Enzymes in Food Processing,* 2nd ed., 1975, p. 431.

Okada, S., and Kitahara, S., *Hakko Kagaku Zasshi* **51**(10), 705 (1973).

Osserman, E. (ed.), *Lysozyme Proc. Conf.,* Academic Press, New York, 1972.

Pilnik, W., in Dupuy, P. (ed.), *Utilisation des Enzymes en Technologie Alimentaire,* Lavoisier, Paris, 1982.

Plainer, H., Sprössler, B., and Uhlig, H., "L'hydrolyse du lactose par des enzymes immobilisées," in Dupuy, P. (ed.), *Utilisation des Enzymes en Technologie Alimentaire,* Lavoisier, Paris, 1982.

Plainer, H. (1986), Company information from Röhm GmbH Darmstadt.

Rothschild, G., and Karsenty, A., *J. Food Sci.* **39**, 1042 (1974).

Ruyssen, R., and Lauwers, A., *Pharmaceutical Enzymes, Properties and Assay Methods,* E. Story-Scientia P.V.B.A., Gent 1978.

Sawayama, Z., Shimoda, Y., Oku, M., and Matsumoto, K., *Toyo Inst. Food Technol.* **7**, 126 (1966).

Scott, D., Hammer, F. D., and Szalkucki, T., "Bioconversions—enzyme technology," in Knorr, D. (ed.), Food Technology, Marcel Dekker, New York, 1987.

Shirota, M., Mutai, M., Sakurai, T., Yokokura, T., and Isawa, H., Jpn. Patent 1,759,601 (1974).

Shugar, D., *Biochim. Biophys. Acta* **8**, 30 (1952).

Weidel, W., in Karlson, P, *Lehrbuch der Biochemie,* fifth edition, Georg Thieme, Stuttgart, 1966, pp. 273, 274.

Woehner, G., Voelskow, H., Präve, P., Lueck, E., and von Rymon-Lipinsky, G., Ger. Offen. DE 3, 440, 735 (1986).

Zittan, L., "Enzymatic hydrolysis of inulin, an alternative way to fructose production," *Die Stärke* **33**(11), 373 (1981).

3.1.4 Cellulose Degradation

3.1.4.1 Substrate The polymers cellulose and hemicellulose constitute the major portion of the organic matter on our planet. They form the cell walls of plants and are consequently the principal components of wood, cotton, and grass. Cellulose, as a renewable raw material, is an ideal resource for the industrial production of paper, textiles, and synthetic products. If economical processes could be developed for its enzymatic hydrolysis, cellulose would also be an ideal source of fermentable carbohydrates from which food and fuel could be produced. Along with cellulose, wood contains large quantities of hemicellulose, of which xylan is a major component. This polysaccharide and its monomeric building block, xylose, are raw materials for industrial fermentation, and xylose has been used as a dietetic sugar for a number of years.

The enzymatic degradation of plant tissue and plant constituents has been investigated for some time by different food processors. To understand the enzymes that make the degradation of high-molecular-weight carbohydrates possible, one should be familiar with their structures and compositions.

Plant Cell Structure The plant cell consists of cytoplasm, which contains soluble and insoluble substances and is surrounded by a cell membrane. The cell wall lies tightly against the outer surface of this cytoplasmic membrane. The thickness of the rigid cell wall varies between $0.1\mu m$ and several micrometers. Neighboring cells have a common wall, similar to the inside walls of a house. The cell walls not only join the single cells to form a tissue but also contain channels for circulating liquids within the tissue. The cell wall is porous and penetratable for practically every type of molecule, while the plasma membrane has selective permeability. Water, which diffuses through the cell membrane by osmotic pressure, increases the intracellular pressure and thereby forces the membrane against the cell wall.

After cell division in the growing region (meristematic area), the cells expand in one dimension. The cell wall in young cells that are still growing is thinner and more pliable than walls in mature cells. This is the primary cell wall (Fig. 3.40). On completion of growth, the cell wall thickens and develops the secondary cell wall. The composition of cell walls differs according to the cell's function. The walls obtain their rigidity through a network of cellulose microfibrils (CMF) that are intrically interwoven with a matrix of other polysaccharides and proteins (Table 3.11).

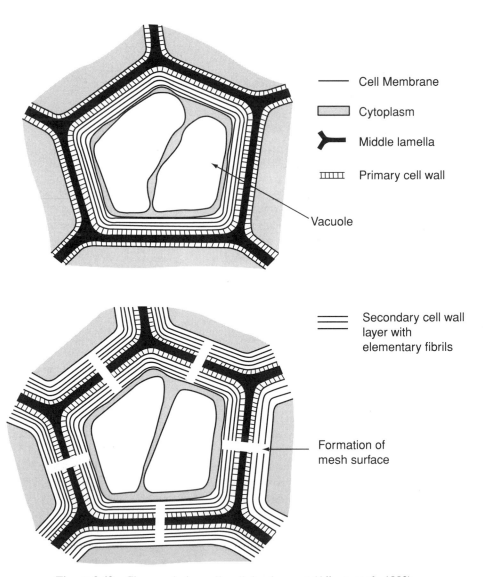

Figure 3.40   Changes during cell wall development (Alberts et al., 1983).

TABLE 3.11   Cell Wall Polysaccharides of Higher Plants (Aspinal, 1980)

Polysaccharide	Structure	Monomer Unit
Cellulose	β-1,4-D-Glucan	Glucose
Pectins	Polygalacturonic acids	Galacturonic acid and its methyl ester
	Rhamnogalacturonic acids	Rhamnose
	Galactans, arabinogalactans	Arabinose
Hemicelluloses	Xylans	Xylose
	Xyloglucans	Xylose, glucose
	β-1,3-; β-1,4-D-Glucans	Glucose
	Galactomannans	Galactose, mannose
Other polysaccharides	Arabinogalactans	Galactose, arabinose
	Glucuronomannans	Glucuronic acid, mannose

Cellulose   Cellulose consists of linear chains of several thousand glucose residues, linked by β-1,4 glycosidic bonds. The chains are stabilized by internal hydrogen bonds. Bundles of about 40–60 chains are aligned in parallel and linked via hydrogen bonds. Such lengthy, crystalline aggregates, known as *microfibrils,* are, in turn, subunits of macrofibrils that have a cross-sectional diameter of about 0.5 nm (Fig. 3.41). Twenty to thirty percent of the primary plant cell wall consists of cellulose.

The cellulose in native cotton has a degree of polymerization (DP) of about 10,000; this means that about 10,000 glucose residues are glycosidically linked in

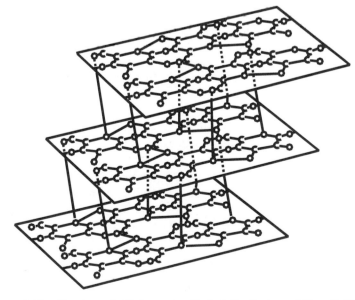

Figure 3.41   Structure of cellulose microfibrils (after Sarko and Muggli, 1975).

each molecule. In contrast, wood pulp has a DP of only 600–1000 and regenerated cellulose, about 200–600. Cotton consists of approximately 90% cellulose; wood, approximately 60%.

All glucose chains run in the same direction; that is, the reducing ends lie at the same end of the microfibrils. These are, in turn, surrounded by an ordered network of cellulose and polysaccharide chains that are tightly, but not covalently, bound to the microfibrils.

Hemicelluloses   Good literature reviews on hemicelluloses and the so-called plant mucins can be found in Pomeranz (1985), Feldman (1987), and Wucherpfennig and Dietrich (1987). The last review refers largely to colloids that occur in fruit juices and wine.

In addition to cellulose, the plant cell wall contains a number of other polysaccharides, some of which are soluble in alkali. The insoluble fraction is also known as *α-cellulose;* the alkali soluble and acid precipitable fraction, as *β-cellulose.* The fraction that remains soluble in the acidic solution contains predominantly hemicellulose.

Hemicelluloses consist of a heterogeneous group of polymers of mannose, galactose, certain *β*-glucans, the heteropolymers from uronic acid, and methyl- or acetyl-substituted hexoses. Also included are the pentosans that are polymers of xylose and arabinose (Fig. 3.42). The cell wall is made up of about 20% hemicellulose.

XYLOGLUCANS   Xyloglucans are bound directly to microfibrils by hydrogen bonds. The main chain consists of *β*-1,4-linked glucose residues, branched primarily with xylose residues (*β*-1,6-linked) as well as some arabinose residues. Additional side chains contain galactose and occasional fucose residues (Fig. 3.43).

ARABINOGALACTANS   The main chain of this polymer consists of alternating sequences of galactose and arabinose oligomers, to which both monomers are bound as side chains. They are bound to the glycoproteins of the cell wall. Serine and hydroxyproline residues form the bridges between the protein and the carbohydrates. There are two types of arabinogalactan, which differ in their linkage patterns (Whitaker, 1984):

1. Arabinogalactan I has *β*-1,4-linked D-galactose residues in the main chain and side chains with *α*-1,5-L-arabinose residues linked to the main chain with *α*-1,3 linkages.
2. Arabinogalactan II has *β*-1,3-linked D-galactose residues with *β*-1,6-D-galactose and *α*-1,3 and *β*-1,3-L-arabinose residues in the side chains (Fig. 3.44).

RHAMNOGALACTURONANS   The rhamnogalacturonans can be categorized as pectic substances (Fig. 3.45).

PECTINS   Pectins are a group of acidic polysaccharides that occur in the middle lamella of the primary plant cell wall. In the middle lamella, the pectins are bound to calcium and are responsible for the cohesion and texture of the plant tissue. The

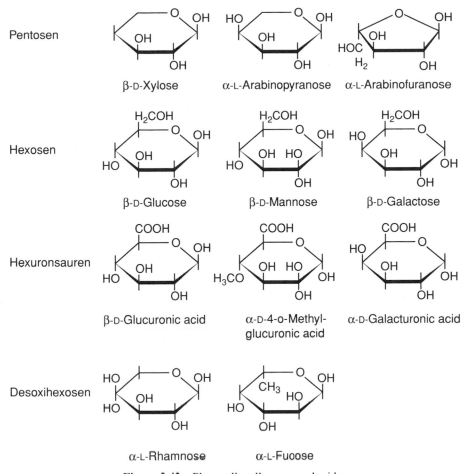

Figure 3.42 Plant cell wall monosaccharides.

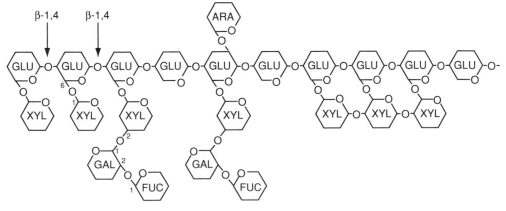

Figure 3.43 Structure of xyloglucan.

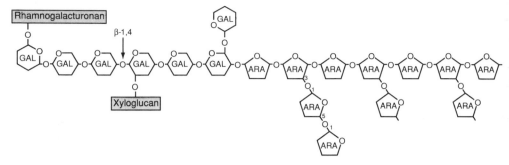

Figure 3.44   Structure of arabinogalactan (Albersheim, 1975).

degradation of the middle lamella by enzymes leads to disintegration of the plant
tissue into single cells. This type of hydrolysis is called *maceration*.

The pectins are not uniform chemical substances, but are very heterogeneous
mixtures of polysaccharides with different molecular weights and degrees of esterification. The major building blocks are nonesterified or (anhydro)galacturonic acid
units esterified with methanol. Galacturonic acid polymers are formed by α-1,4
linkages. Bound to these chains are additional rhamnose-rich polymers with side
chains containing arabinose, galactose, and xylose residues. They are accompanied
by hemicelluloses such as arabinans and galactans. Extracted pectins show molecular weights from 3000 to 300,000 daltons. The solubility of the pectins increases
with increasing degree of esterification and lower molecular weight.

The negatively charged polymers are cross-linked by calcium ions forming rigid
gels. The insoluble pectin that constitutes the middle lamella is, because of its function, known as "cell cement." It plays a special role in fruit ripening, which is related to structural changes in the tissue. In the manufacture of cloud-stable fruit and
vegetable juices, attempts are made to specifically dissolve the middle lamellar

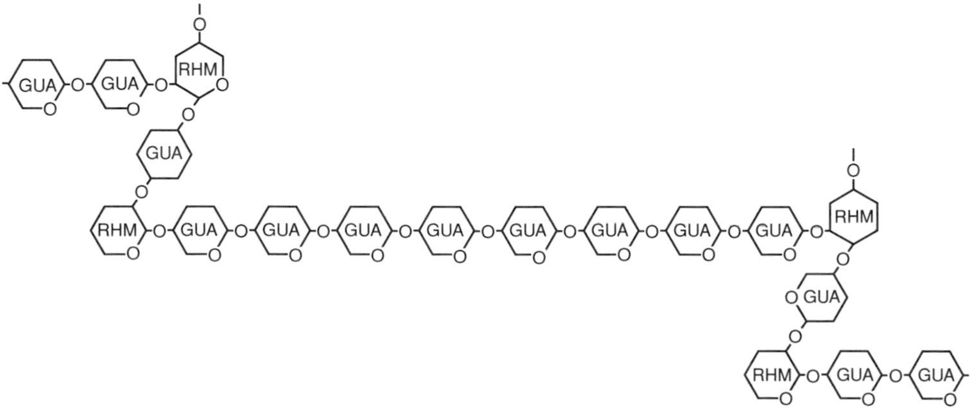

Figure 3.45   Structure of rhamnogalacturonan.

TABLE 3.12   Pectin Content of Plant Material (Whitaker, 1984)

Source	Pectin, Dry Weight (%)
Apple pomace	15–18
Lemons	30–35
Oranges	30–40
Beets	25–30
Carrots	~7

pectin, also called *protopectin*. The water-soluble pectins, formed mainly during fruit ripening, are highly esterified with methanol.

Pectins play a vital role in processing raw plant materials into foods (Girschner, 1981) (Table 3.12).

The plant cell wall components discussed above and their structural composition are illustrated in Figure 3.46 (see also Table 3.13).

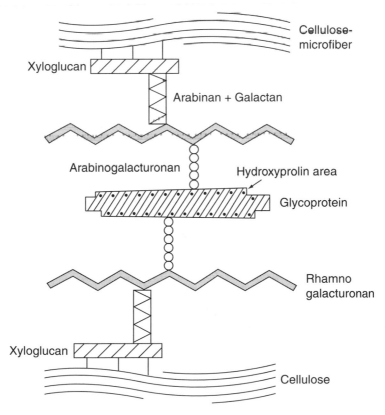

Figure 3.46   Schematic illustration of the primary plant cell wall (Robinson, 1977; modified after Franz, 1985).

TABLE 3.13   Composition of Some Plant Cell Walls

Apples	Galacturonic acid, 8.2%; glucose, 75.9%; galactose, 8.9%; arabinose, 6.0%; xylose, 1.2% (Jermyn and Isherwood, 1956)
Tomatoes	Cellulose, 17%; pectin, 22%; protein, 17%; arabinogalactan, 13–23%; xylose and glucose, 13–23% (Williams and Bevenue, 1954)
Pears	Glucan, 21.4%; galactan, 3.5%; mannan, 1.1%; xylan, 21%; arabinan, 10%; polygalacturonic acid, 11.5%; lignin, 16.1% (Jermyn and Isherwood, 1965)
Wheat endosperm	Arabinoxylan, 2–3%
Rye	Pentosans, ~8% (Wieg, 1984)

3.1.4.2 Fungal Cellulases   In nature, cellulose is degraded by microorganisms that rot wood and by soil bacteria. Bacterial flora present in the intestinal tracts of plant eaters rapidly degrade cellulose. Fungi and bacteria produce cellulases; they are also present in the seeds of higher plants and malt. Further, they are found in worms, caterpillars, and snails. The enzyme complex that has been isolated from snails *(Helix pomatia)* contains a number of very efficient cellulases in addition to β-glucosidases, hemicellulases, and α-glycosidases. While native cellulose in its natural environment is rapidly degraded by certain enzymes, degradation by enzymes isolated from organisms occurs very slowly. Enzymatic activity is significantly improved if cellulose is chemically or physically pretreated. Cellulose that has been preswollen in phosphoric acid or alkali is attacked very rapidly. This applies even more so to chemically modified cellulose derivatives such as carboxymethylcellulose (CMC) or hydroxyethylcellulose (HEC), which are often used to determine cellulase activity. Liquefaction of wallpaper glue that contains CMC can be observed after fungal growth on the wallpaper.

It is beyond the scope of this book to relate the history of cellulase degradation; however, a brief summary of the literature of the last 100 years reveals some interesting studies on the degradation of cellulose during plant seed germination. The plant-cell-degrading enzymes, called *cytases,* are less thermostable than diastases (amylases). De Bary (1886) described the action of a fungus that he had grown on beets, "the mycelium of which poisoned the cells, and softened the beet tissue as if it had been cooked." Karrer and Schubert (1927) and their school intensively studied the action of snail enzymes on regenerated cellulose (artificial silk). Selby and Maitland (1965) reported on the cellulases of the soil fungus *Myrothecium verrucaria*—a group of cellulases that has been studied especially extensively.

The work of Reese et al. (1950) helped elucidate an understanding of the complex action of cellulases. The two-step degradation process suggested requires an activation reaction, followed by hydrolytic cleavage. The activating but not hydrolyzing enzyme was named "C_1 activity." According to this concept, microorganisms that are capable of degrading crystalline cellulose have C_1 activity. This activity is not present in enzymes that attack only substituted cellulose (CMC). These have only CMCase or C_x activity as shown below illustrating the enzymatic hydrolysis of cellulase (after Reese et al., 1950):

$$\text{Cellulose} \xrightarrow{C_1} \text{reactive cellulose} \xrightarrow{C_x} \text{cellobiose} \xrightarrow{\beta\text{-glucosidase}} \text{glucose}$$

According to current knowledge, the degradation of native cellulose is believed to require the synergistic action of three different hydrolytic enzymes (Wood and McCrae, 1979). This means that the attack occurs simultaneously by at least the first two enzymes:

1. Exocellulase, also called *exobiohydrolase*.
2. Endocellulase, also known as *endoglucanase*.
3. β-Glucosidase, also called *cellobiase*.

According to this scheme, the initial reaction at the amorphous area on the surface of cellulose microfibrils is conducted by an endoglucanase (with a special C_x activity?). This leads to new terminal groups, thus providing substrate for the next step (Fig. 3.47).

The terminal groups are formed on cleaving the cellobiose from the nonreducing end of glucan chains by cellobiohydrolase. This enzyme shows the highest affinity for cellulose. According to Wood and McCrae (1979), the isolated and purified cellobiohydrolase is capable of cleaving up to 80% of microcrystalline cellulose. As the reaction with cellobiohydrolase progresses, substrate is produced for the endoglucanase (C_x activity). As this reaction sequence proceeds, the released cellobiose molecules are steadily hydrolyzed by the third enzyme, cellobiase. This

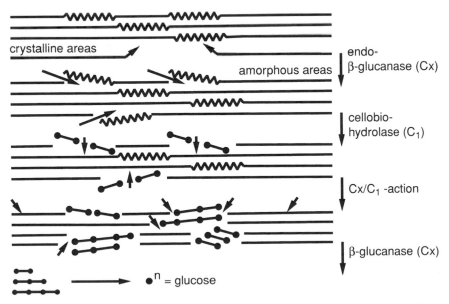

Figure 3.47 Schematic illustration of the stepwise enzymatic hydrolysis of cellulose (Montenecourt and Eveleigh, 1979).

step is very important because the accumulation of cellobiose strongly inhibits the cellulose degradation reaction (step 2).

Lactones such as gluconolactone or cellobionolactone are strong inhibitors of cellobiase. According to Bruchmann et al. (1987), the lactone ring hydrolyzing lactonases that occur in *A. niger* or in *T. reesei* play an important role in cellulose degradation.

A number of other influencing factors as well as other reaction mechanisms have been suggested for the initial attack described above. Eriksen (1981) claims that an oxidative degradation is possible in the first step. A cellobiose oxidase could thus be involved in the reaction.

Further information on the degradation of cellulose can be found in excellent reviews by Enari (1983) and Gosköyr and Eriksen (1980). A description of the three enzymes involved in cellulose degradation follows. The sources of microbial cellulase that are important for technical production are also described. The principal strains are *Trichoderma reesi* and *Aspergillus* sp.

Cellobiohydrolase (EC 3.2.1.91; CAS 37329-65-0), 1,4-β-Glucan Cellobiohydrolase   This exocellulase cleaves cellobiose from the nonreducing end of the glucan chain and from cellodextrins (cellulose fragments with three to six glucose residues). A number of cellobiohydrolases can be isolated by using modern fractionation methods (Table 3.14).

The cellobiohydrolases CBH I and CBH II differ in their immunologic reactivity. CBH I is more active, and there are five isozymes, all of which have similar molecular weights and amino acid profiles. They contain 0.1–10% carbohydrate. The amino acid sequence was determined by Fagerstam et al. (1984).

Wood and McCrae (1972) have shown that the purified C_1 activity is a cellobiohydrolase that degrades phosphoric acid–treated cellulose well but acts on CMC very slowly. The reaction product is almost totally cellobiose (95%).

Endoglucanase (EC 3.2.1.4; CAS 9012-54-8), 1,4-β-D-Glucan-4-glucanohydrolase This enzyme splits the β-1,4 glycosidic linkages randomly in the internal regions of cellulose chains of cellodextrins, phosphoric acid–treated cellulose, and cellulose derivatives (CMC and HEC).

TABLE 3.14   Cellobiohydrolases from *Trichoderma reesei*

Enzyme Source	Molecular Weight (daltons)	pI	Carbohydrate (%)	Reference
Cellulase				
Onozuka	42,000	3.79	9.2	Berghem et al. (1976)
(5 isozymes)	53,000	—	—	—
Meicellase	48,000	—	0.1	Gum and Brown (1976)
T. reesei	65,000	4.2–3.6	—	Nummi et al. (1983)
"CBH I"	60,000	3.9	—	Pettersson et al. (1981)
"CBH II"	60,000	3.6	—	Pettersson et al. (1981)

TABLE 3.15   Fungal Endoglucanases

Enzyme No./ Source	Molecular Weight (daltons)	pI	Carbohydrate (%)	Reference
I/*T. reesei*	20,000	7.5	—	Pettersson et al. (1981)
II/*T. reesei*	40,000	4.6	—	—
	12,500	4.6	21	Berghem et al. (1976)
	50,000	3.9	12	—
	47,000	5.74	—	—
T. reesei	43,000	4.0	—	Berghem and Pettersson (1974)

The amorphous regions of crystalline cellulose are apparently attacked by a specific endoglucanase.

According to Beldman et al. (1988), six specific and non-specific endoglucanases can be distinguished in *T. viride*. The nonspecific types also cleave xylan.

Aspergillus sp. strains, especially *A. niger* (see Table 3.15), form endoglucanases but only very small amounts of exoglucanases (C_1 activity).

The degradation of cellulose by bacteria differs from that by fungi. Bacteria produce extracellular endoglucanases but no exoglucanases and only intracellular β-glucosidase.

β-Glucosidase (EC 3.2.1.21; CAS 9001-22-3), Cellobiase, β-D-Glucoside-glucohydrolase)   Cellobiose and cellodextrins, including cellohexanose, are hydrolyzed to glucose by this enzyme. Cellulose and higher cellodextrins are not attacked. The enzyme is produced by several different *Trichoderma* species, especially by *Aspergillus* species, primarily *A. niger* (Table 3.16).

Cellulases from Humicola sp.   During the last few years several different cellulases, some of which are thermostable, have been found in *Humicola* species. They include *H. insolens, H. grisea,* and *H. grisea thermoidea* (Kikkoman, 1986).

The enzyme from *H. lanuginosa* (Olutiola, 1982) is a CMCase and has a molecular weight of 70,000 daltons and optimum activity at pH 5 and 45°C.

TABLE 3.16   Fungal β-Glucosidases (Enari, 1983)

Enzyme Source	Molecular Weight (daltons)	pI	Reference
T. reesei	47,000	5.74	Berghem and Pettersson (1974)
T. reesei	35,000	—	Enari et al. (1980)
	130,000	—	
A. oryzae	218,000	4.3	Mega and Matsushima (1979)
A. niger	150,000	—	Enari et al. (1980)

A cellulase from *H. insolens* (Celluzyme [NO]) that is used in the textile industry has an optimum activity at pH 7–8.5.

Bacterial Cellulases   Very little is known about the cellulases from bacteria. Hagget et al. (1979) reported on cellulases from a *Cellulomonas* strain. Cellulases from bacteria active in the alkaline range were described by Horikoshi and Akiba (1982) and Kawai et al. (1988) and in Japanese Patent 007511 (1988).

PROPERTIES

Molecular Weight: About 35,000 daltons (by gel filtration).

Activity: Endoglucanase activity ranges within 80–100% at pH 5–10.5 with CMC as substrate. The enzyme shows no activity toward cellulose swollen in phosphoric acid. After incubation at pH 9 for 1 h and 30°C, the activity is reduced by only 10%. The maximum temperature tolerated at pH 9 is 55°C. *p*-Nitrophenyl-cellobioside is cleaved.

Activators: NaCl, KCl, $CaCl_2$, activation at ~10%.

Stability: Complexing agents do not hinder activity, nor do soaps, Softanol 70 H, or Hostapur SAS. The cellulase is completely resistant to attack by proteases from *B. subtilis* or *B. licheniformis*. (Alkalase [NO]; Maxatase [GB]).

After it was discovered that bacterial cellulases can be utilized as active components in detergents, industrial research and production groups focused on this new area.

An alkaline-stable cellulase from *Bacillus* with a molecular weight of ~16,000 daltons and a temperature optimum at 60°C was described by Oshino et al. (1986).

Cellulase from Acidothermus cellulyticus   Poutanen et al. (1987) discovered a thermophilic bacterial strain that was isolated from the hot springs in Yellowstone National Park. The cellulase from this bacterium has a temperature optimum at 80°C and could be of great interest to industry. Here again, three types of cellulose-degrading enzymes were found. The endoglucanase activity seems to be the most stable (83°C) (Fig. 3.48).

3.1.4.3 Analysis of Cellulases   Because of the complex nature of both substrate and enzyme, analysis of cellulases is not simple. Cellulase preparations from different microorganisms show very different enzyme profiles. The composition of an enzyme preparation from the same organism depends strongly on the conditions under which the organism was cultivated. The type and composition of the culture medium are an important factor.

The activity of the individual enzymes is of primary interest to researchers. Very specific analytic methods are necessary to describe their properties. For industry, the complex action of an enzyme preparation and its specific activity in a particular area of application are more relevant. Consequently, different methods have been developed for either biochemical investigations or for industrial applications.

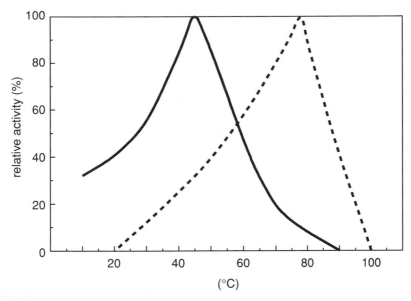

Figure 3.48 Temperature dependence of cellulase from *Acidothermus* (*a*) and *Trichoderma reesei* (wild strain) (*b*) on filter paper.

The substrates for analyzing the complex, synergistic action of endoglucanases and exoglucanases, the so-called C_1 activity, are insoluble crystalline celluloses such as Avicel, Solkafloc, and especially filter paper (see Table 3.17). The standardization of these substrates is difficult because of their macromolecular structure. Degree of polymerization, degree of crystallinity, purity, and substrate source all influence the analytic results.

Small molecule substrates such as *p*-nitrophenyl-β-glucoside, cellobiose, or oligocellulodextrins would be ideal for standardization. However, such substances are not readily cleaved by all cellulase preparations and would also be sensitive to β-glucosidase activity.

Methods for Analyzing C_1 Activity (Filter Paper Activity)   Filter paper activity according to Mandels (1975) is defined as the amount of sugar that is formed in one hour when Whatman No. 1 filter paper (50 mg) is incubated with 0.5 ml of enzyme solution at pH 4.8 (citrate buffer) in a total volume of 1.5 ml.

Because the reaction does not run linearly, it is recommended that the enzyme be so diluted, that ~2.0 mg of reducing sugar is formed per hour. This method yields between 0.2 and 2 FPU per milliliter in the culture liquids.

To circumvent the problems that arise in the analysis of insoluble substrates, water-soluble derivatives such as carboxymethylcellulose (CMC) or hydroxyethylcellulose (HEC) are used extensively. It is important to remember, however, that these methods determine not the desired "real" cellulase activity but mainly endoglucanase activity.

TABLE 3.17   Methods for Determination of Cellulase Activity (Enari, 1983)

Measured Activity	Substrate	Determination Method	Reference
Total cellulase activity (C_1)	Filter paper Avicel Solkafloc Linters	Glucose	—
Solubilizing activity	Filter paper	Weight reduction	Halliwell and Riaz, 1970
	Avicel	Optical density	
	Cellulose azur	Optical density	Leisola and Linko, 1976
	Colored cellulose		
Endoglucanase	CMC	Reducing sugar	Mandels and Weber, 1969
	HEC Phosphorous acid Soaked cellulose	Decrease of viscosity	Almin and Eriksson, 1967
Exoglucanase	Linters	Reducing sugar	Mandels and Weber, 1969
	Avicel Filter paper		
β-Glucosidase	Cellobiose	Glucose	Selby and Maitland, 1967
	p-Nitrophenyl-β-D-Glucosid		

Although neutral HEC is better suited for the analytic determination, CMC with a defined degree of substitution is most commonly employed.

Methods for Determining Endoglucanase Activity

1. Viscometric determination according to FIP (Ruyssen and Lauwers, 1978). HEC and CMC form highly viscous solutions in water. The viscosity drops rapidly on the action of cellulases. It was found that several different endoglucanases show changes in specific viscosity proportional to enzyme concentration and reaction time (Eriksen and Gosköyr, 1976). This is valid only for a certain range of viscosity decrease, namely, around 50% relative viscosity. The FIP standard "endocellulase" is used as reference. Moreover, it is difficult to determine the molecular reaction rate of a cellulase from the change in viscosity. However, Almin and Eriksen (1967) have described a method in which absolute enzyme units can be calculated from the viscosity change of a precisely standardized HEC substrate.

2. Analysis of endoglucanase activity by measuring the reducing sugars. In general, the amount of reducing sugar produced during hydrolysis is measured ei-

ther with dinitrosalicylic acid or by Nelson–Somogy procedures. For the analysis of cellulase, the ferric cyanide method of Wood and McCrae (1972) is preferred.

3.1.4.4 Technical Cellulase Preparations Technical cellulase preparations are usually obtained from *Trichoderma reesei, Aspergillus niger,* and more recently, bacterial strains. In Japan, preparations were initially obtained from *Trichoderma viride* using surface cultures in the well-known koji process.

Steamed wheat bran was used as culture medium, which, after inoculation with the spores or mycelia of the given microorganism, was incubated in trays. Optimal culture conditions were determined by performing an extensive series of tests in which the pH, temperature, moisture, and other parameters were varied. On harvest of the culture (after 3–5 days), the growth medium was extracted with water, the extract concentrated, and the enzyme precipitated with alcohol. Even today, some companies still produce *Trichoderma* cellulases using the koji process. Despite some inconvenience, the process offers advantages such as in the manufacture of certain products with a low production volume. The same can be said for cellulase prepared from *A. niger.* According to Yamada (1977), about 45 tons of cellulases were produced from *T. viride* and *A. niger* in 1976.

The manufacture of cellulases with submerged fermentation was first described in various Japanese patents 29 years ago (Kawaji et al., 1964).

Trichoderma Cellulases In the laboratories of the U.S. Army in Natik and at Rutgers University, the wild strain of *Trichoderma viride* QM 6 was the first source for the production of cellulases. Later, mutations and selections were able to increase the productivity of this strain. Mutant QM 9414 was chosen as the basis for further high-performance mutants (Rehm, 1983).

Use of cellulase preparations from *T. reesei* was limited to the manufacture of digestive aid products only a few years ago. In research, the cellulases are needed in combination with pectinases for the production of protoplasts (Melchers et al., 1978) (cellulase Onuzuka, Rohament CT/Rohament P).

Cellulase prices have been considerably lowered by improving cellulase production yields. Consequently, new areas of application for cellulases in food production and agriculture have become economically feasible. These applications include producing silage from grasses, addition to animal feed, the degradation of hemicelluloses during starch production from corn and wheat, and processing fruit and vegetables.

Commercial products, in both liquid and solid state, differ with respect to composition of cellulase activities and accompanying enzymes. The latter are of importance for applications in specific areas. The following side activities were found in enzyme preparations from *Trichoderma:* various xylanases, proteases, and a whole array of pectinases. In addition, the presence of β-1,3-glucanases, mannases and galactomannases, amylases, and glycosidases has been described.

Especially loaded with side activities are the preparations from *Trichoderma harzianum,* which contain a series of cell wall lysing enzymes. These enzymes can be used for the lysis of yeast cells and for the production of protoplasts from *Aspergillus* and *Penicillium.*

There has been much literature on the relationship between the enzymatic composition of cellulases in technical preparations and the practical applications of these cellulases. Van Belle's group (1986) investigated the cellulase complexes in order to find a relationship between analytic composition and practical activity in the grass silages (Bertin and Hellings, 1986). A systematic analysis revealed the following enzymatic activities: cellulases, mannase, arabinogalactanase, arabanase, and pectinase. These studies were intended to help develop a specific working enzyme preparation for silage treatment.

Liquid cellulase preparations have an activity of 5–10 FPU/ml, the solid preparations about 10–50 FPU/g. The CMCase (C_x activity) differs from the filter paper units (C_1 activity). This activity ranges within 1000–1500 CU/ml in liquid preparations and 200–4000 CU/g in solid ones.

PROPERTIES   The relation of temperature and pH to the activity of some commercial preparations is shown in Figures 3.49–3.54.

PRODUCTS AND MANUFACTURERS   Celluclast and Novozyme 234 [NO], Econase C15 [AO], Rohament CT [RM], Avizyme [FR], Sumizyme C [SN], Meicelase MC [MJ], Cellulase TAP [AM], Cytolase 123 (GR).

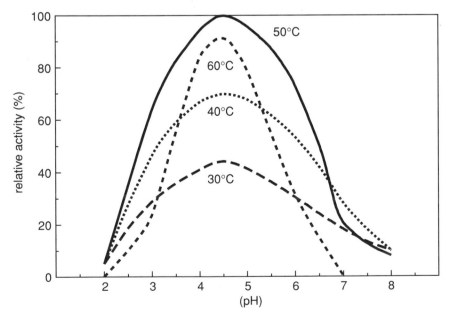

Figure 3.49   pH and temperature dependence of cellulase during Avicel hydrolysis (Meiji Co., 1987).

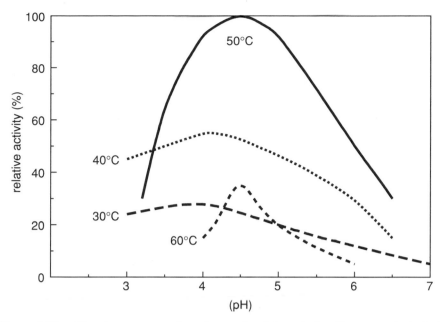

Figure 3.50 pH and temperature dependence of cellulase during filter paper hydrolysis (Meiji Co., 1987).

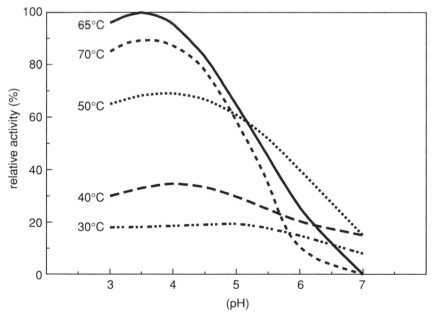

Figure 3.51 pH and temperature dependence of cellulase (CMCase) during carboxymethylcellulose hydrolysis (measured as increase in reducing sugars).

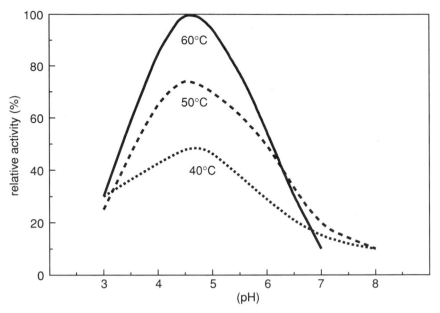

Figure 3.52 pH and temperature dependence of β-glucosidase from *T. viride* (Meiji Co., 1987).

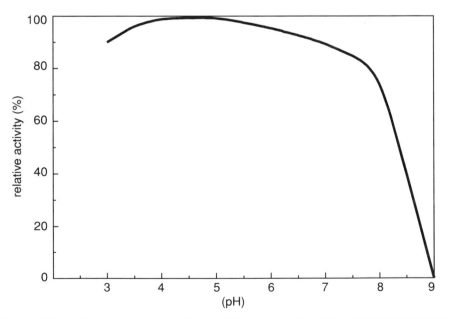

Figure 3.53 pH dependence of cellulase stability; reaction time 12 h; pH 5.0; 25°C ([RM], 1988).

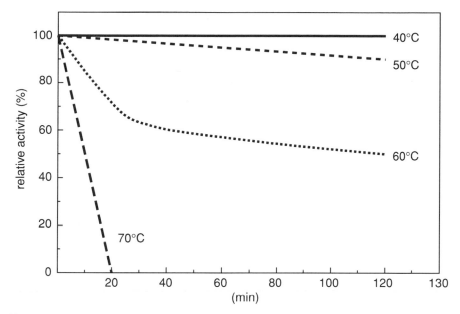

Figure 3.54   Effect of temperature and time on the stability of cellulase from *T. reesei* in aqueous solution ([RM], 1988).

Aspergillus Cellulases   In the scientific literature, only those enzymes that are capable of degrading native cellulose are called *cellulases*. However, in industry, other carbohydrases capable of splitting the β-1,4-glycosidic linkage are also often described as cellulases or CMCases. Such cellulase complexes contain numerous types of hemicellulases, amylases, and proteases in addition to standard cellulase activity. These products are very manufacturer-specific, and their composition depends not only on the producing *Aspergillus* strain but also on the culture conditions used by the manufacturer.

Because of the diverse activities of the various hydrolases, these preparations are used to improve the filtration rates of hydrolysates from barley and corn (breweries and distilleries). In addition, the efficiency of pectinases in fruit and vegetable processing can be improved or modified by *Aspergillus*-derived cellulase preparations.

Technical preparations are most often produced from cultures of *A. niger* (Figs. 3.55 and 3.56). Activity determinations for these enzymes are standardized using CMC as substrate. The activity of the dry products ranges from 1000 to 20,000 CU/mg.

Those preparations are accompanied by enzymatic activity from pectinases, acidic fungal proteases, α- and β-glycosidases, β-glucanases, and pentosanases. Thus, these enzymes can be designated as multifunctional hydrolases. The cellulase preparation, Rohament CW [RM] from *A. wentii,* for example, has yielded several interesting enzymes, including a β-glycosidase (Legler and Gillis, 1970) (Table 3.18).

The pH dependence of CMCase stability is illustrated in Figure 3.57.

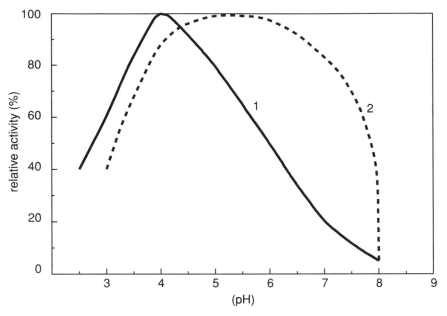

Figure 3.55   pH dependence of a technical-grade cellulase preparation from *A. niger* ([RM], 1987). The activity was determined viscosimetrically (CU/mg, curve 1); and by increase in reducing sugars (curve 2).

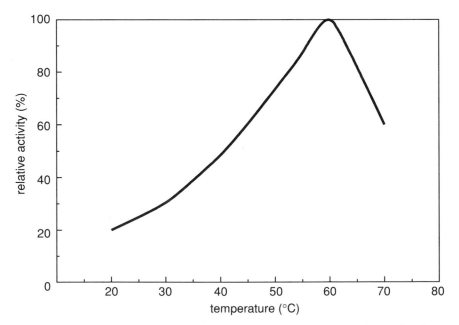

Figure 3.56   Temperature dependence of CMCase from *A. niger* (pH 4.0; reaction time 15 min [RM]).

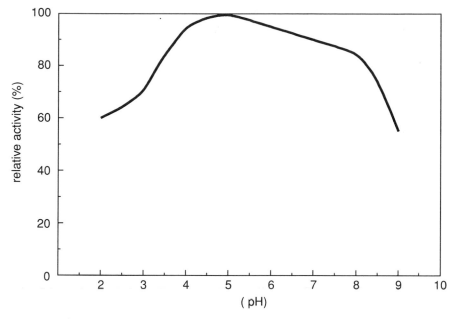

Figure 3.57 pH dependence of CMCase stability in aqueous solution (40°C, reaction time: 30 min [RM]).

Products and Manufacturers: Rohament CA and CW [RM], Cellulase AP [AM], Cellulase P 4000 [SO], Cellusin [UA], Pancellase [YT]

Cellobiase Preparations Cellobiase preparations are used in combination with cellulase preparations from *Trichoderma* to hydrolyze cellulose in order to obtain optimum conversion of cellulose into glucose. In silage, for example, the application of *Trichoderma* cellulases leads to the formation of cellobiose, which cannot be utilized by most microorganisms. (See Figs. 3.58 and 3.59.)

Manufacturer: [NO].

TABLE 3.18 Enzymes in ROHAMENT CW [RM]

Cellulase (C_x)	900 CU/mg
Pectin glycosidase (PG)	560 PGU/mg
Pectin esterase (PE)	95 PEmU/mg
α-Glucosidase	10 mU/mg
β-Glucosidase	200 mU/mg
β-Galactosidase	170 mU/mg
Galactomannanase	3 mU/mg
Xylanase	160 mU/mg
Xylobiase	2 mU/mg
Protease	30 HbmU/mg

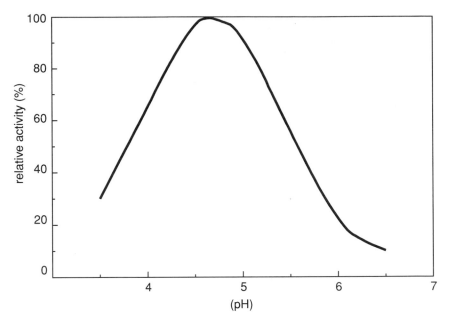

Figure 3.58   ph dependence of cellobiase [NO].

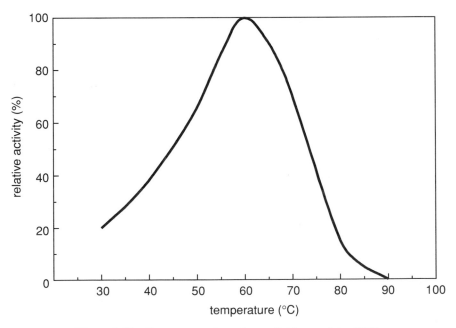

Figure 3.59   Temperature-dependent cellobiase activity [NO].

Cellulase from Humicola Insolens   The enzyme from this thermophilic fungal strain differs from other cellulases in its pH-activity profile, which extends into the alkaline range. Stability is surprisingly good at pH 9, even at a temperature of 60–65°C. The enzyme is used as an additive in detergents and is commercially available in granulated form. (See Figs. 3.60 and 3.61.)

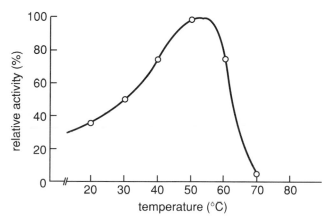

Figure 3.60   Temperature-dependent cellulase activity (substrate: 7.5g CMC/liter; phosphate buffer; 40°C; reaction time 20 min [NO]).

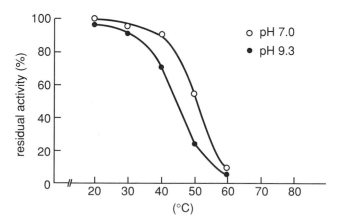

Figure 3.61   Temperature-dependent stability of *Humicola* cellulase at pH 7 and 9.3 [NO].

REFERENCES

Albersheim, P., "The walls of growing plant cells," *Sci. Am.* 81 (1975).

Alberts, B., et al., in *Molecular Biology of the Cell,* Garland, New York, 1983, p. 1108.

Almin, K. E., and Eriksson, K. F., *Biochem. Biophys. Acta.* **139,** 238 (1967).

Aspinal, G. O., in Prets, K. (ed.), *The Biochemistry of Plant Carbohydrates, Structure and Function,* Vol. 3, Academic Press, New York, 1980, p. 473.

Beldman, G., Voragen, A. G. J., Rambouts, F. M., Seaarle-Van Leeuwen, M. F., and Pilnik, W., *Biotech. Bioeng.* **31,** 160 (1988).

Berghem, E. R., Pettersson, L. G., and Axiö-Fredrikson, U. B., *Eur. J. Biochem.* **61,** 621 (1976).

Berghem, L. E., and Pettersson, L. G., *Eur. J. Biochem.* **46,** 295 (1974).

Bertin, G., and Hellings, P., in *Silage, New Biological Aspects,* Sanofi, Paris, 1986, pp. 59–86.

Bruchman, E. E., Schach, H., and Graf, H., *Biotechnol. Appl. Biochem.* **9**(2), 146 (1987).

De Bary, B., *Bot. Zeitschr.* 377 (1886).

Enari, T. M., Niku-Paavola, M. L., and Numimi, M., in Ghose, T. K. (ed.), *Proc. 2nd Internatl. Symp. Bioconversion and Biochem. Eng.,* Vol. 1, 1980, p. 87.

Enari, T.-M., in Fogarty, W. M. (ed.), *Microbial Enzymes and Biotechnology,* Applied Science Publishers, London, 1983, p. 183.

Eriksen, K. E., in *The Ekman-Days, Internatl. Symp. Wood and Pulp Chemistry,* Vol. III, *SPCI,* Stockholm, 1981, p. 60.

Eriksen, J., and Gosköyr, J., *Archiv. Microbiol.* **110,** 233 (1976).

Fagerstam, L. G., Pettersson, L. G., and Engstrom, J. A., *FEBS Lett.,* **167,** 305 (1984).

Feldmann, G., *Confructa* **31**(III/IV), 62 (1987).

Franz, G., in Buschard, W. (ed.), *Polysaccarides,* Springer, Heidelberg, 1985, pp. 1–14.

Gierschner, K., in *Pectin and Pectic Enzymes in Fruit and Vegetable Technology,* Gordian, 1981, p. 171.

Gosköyr, J., and Eriksen, J., in Rose, A. H. (ed.), *Microbiol. Biotechnol.* **8,** 1983 (1979).

Gum, E. K., and Brown, R. D., *Biochim. Biophys. Acta* **446,** 371 (1976).

Hagget, K. D., Gray, P. P., and Dunn, N. V., *Eur. J. Appl. Microbiol. Biotechnol.* **8,** 183 (1979).

Halliwell, G., and Riaz, M., *Biochem.* **7** 35, 116 (1970).

Horikoshi, K., and Akiba, T., *Alcalophilic Microorganisms,* Springer, Heidelberg, 1982.

Jermyn, M. A., and Sherwood, F. A. *Biochem. J.* **64,** 123–132, (1956).

Kawai, S., Okoshi, H., Ozaki, K., Shikata, S., Ara, K., and Ito, S., *Agric. Biol. Chem.* **52**(6), 1425 (1988).

Kawaji, S., Ishikawa, T., Satto, K., Inohara, T. Kubo, A., and Tejima, E., Jpn. Patent 2986 (1964).

Kikkoman, Jpn. Patent, 59.118.084 (1986).

Legler, G. H., and Gilles, H., *Hoppe-Seyler's Z. Physiol. Chem.* **351,** 741 (1970).

Leisola, M., and Linko, M., *Analyt. Biochem.* **70,** 592 (1976).

Mandels, M., *Biotech. Bioeng. Symp.* **5,** 81 (1975).

Mandels, M., and Weber, J., *Advances in Chemistry Series* **95,** 391 (1969).

Mega, T., and Matsushima, Y., *J. Biochem.* **85,** 335 (1979).

Meiji Co. Information, 1907, 1987.

Melchers, G., Sacristan. M., and Holder, A., *Carlsberg Res. Commun.* **43,** 205 (1978).

Montenecourt, B. S., and Eveleigh, D. E., *TAPPI, Ann. Proc. Techn. Assoc. Pulp and Paper Industry,* Atlanta, 1979, p. 101.

Nummi, M., Niku-Paavola, M.-L., Lappalainen, A., Enari, T. M., and Ronnio, V., "Cellobiohydrase from *Trichoderma reesei,*" *Biochem. J.* **215,** 677 (1983).

Olutiola, P. O., *Experientia* **38**(11), 1332 (1982).

Oshino, K., Kawai, S., Ogoshi, H., Ito, S., and Okamoto, K., Jpn. Patent 63.146.786 (1986).

Pettersson, L. G., Fägerstam, L., Bhikhabhai, R., and Leandoer, K., in *The Ekman-Days, Internatl. Symp. Wood and Pulping Chemistry,* Vol. III, SPCI, Stockholm, 1981, p. 39.

Pomeranz, Y., *Functional Properties of Food Components,* Academic Press, New York, 1985, pp. 91–118.

Poutanen, K., Ratto, M., Puls, J., and Viikari, L., *J. Biotechnol.* **6**(19), 41 (1987).

Reese, E. T., Siu, R. G. H., and Levinson, H. S., *J. Bacteriol.* **59,** 485 (1950).

Rehm, H.-J., in Reed, G. (ed.), *Biotechnology,* Vol. 3, VCH, Weinheim, 1983, p. 309.

Robinson, D. G., *Adv. Botanical Research* **5,** 89 (1977).

Röhm GmbH information, 1987.

Röhm GmbH information, 1988.

Ruyssen, R., and Lauers, A., *Pharmaceutical Enzymes,* E. Story, Scientia, Sci. Publ. Comp., Gent, Belgium 1978, p. 155.

Sarko, A., Muggli, R., in Albersheim, P., "The walls of growing plant cells," *Sci. Am.* 81 (1975).

Selby, K., and Maitland, C. C., *Biochem. J.* **94,** 578 (1965).

Van Belle, M., *Silage, New Biological Aspects,* Sanofi, Paris, 1986.

Whitaker, J. R., *Enzyme Microb. Technol.* **6,** 342 (1984).

Wieg, A., *Die Stärke* **36,** 135 (1984).

Williams, K. T., and Bevenue, A. J., *Agric. Food Chem.* **2,** 472 (1954).

Wood, T. M., and McCrae, S. I., *Biochem. J.* **128,** 1183 (1972).

Wood, T. M., and McCrae, S. I., *Advances in Chemistry Series* **181,** 181 (1979).

Wucherpfennig, K., and Dietrich, H., *Confructa* **31**(3/4), 80 (1987).

Yamada, K., *Biotechnol. Bioeng.* **19,** 1563 (1977).

3.1.5 Hemicellulases

3.1.5.1 Substrates The heteropolymer carbohydrates that are associated with cellulose in plants are, with the exception of pectic substances, hemicelluloses. Included are polymers of hexoses, such as glucose, rhamnose, or mannose, as well as the polymers of pentoses: xylose and arabinose. Furthermore, plant mucins and other polymers obtained from plants are included in this group and are utilized in the food industry as thickening agents. From this large area only those substances, the enzymatic degradation of which is of commercial importance, will be discussed. There are the β-glucans and pentosans from cereals, hemicelluloses from fruits, seeds, and legumes, and also the β-glucans produced by yeasts and fungi.

Wheat endosperm contains arabinoxylan and a 1,3- and 1,4-β-D-glucan, which constitute 2–3% of the weight of wheat grain. About one-third of this fraction is water-soluble. The other two-thirds can be extracted with alkali. Rye flour contains 8% pentosan, which contributes to the high water-binding capacity and viscosity of rye dough. The β-1,3;1,4-glucans are found in barley and are responsible for filtration problems encountered during brewing beer. In corn, there are mucins that reduce starch yield in the separation of starch from gluten. Undesirable turbidity, attributable to insoluble arabinose polymers, is encountered in the production of modern pear and apple juice concentrates. For processing of berry fruits it is necessary to degrade not only pectin but also the various polymers of unknown structure. During extraction of coffee, the degradation of mannan polymers by enzymes can improve soluble yield. Turbidity that occurs in coffee concentrates can also be prevented by enzyme treatment. Similar problems are encountered during the extraction of tea for instant-tea or liquid-tea production.

This area of application is generally an interesting one for enzymes; however, their targeted application is made difficult by the fact that very complex heteropolymers with various monomeric compositions and linkages are present. During the past few decades, basic research has made such progress in this area mainly by determining the structures of the natural polymers, as well as through the isolation, purification, and characterization of the necessary enzymes. Numerous problems still exist, and these can be solved by the cooperative efforts of natural product chemists, enzymologists, and last, but not least, the technical people working in the field.

In the following section, the enzymes will be emphasized and the natural products and their associated problems will be dealt with only briefly. More detailed information on those substances will be found in the chapters on industrial applications.

3.1.5.2   β-Glucanases
References: Steiner (1968), McCleary et al. (1987), Matheson and McCleary (1987)

β-Glucans constitute, on a weight basis, 2.8–5.5% of the total weight of the cell walls of barley endosperm (Fincher, 1974). They are heterogeneous, linear polymers of glucose that are linked by β-1,4 (75%) and β-1,3 glycosidic bonds. Depending on the cereal source, molecular weights are as follows: barley glucan ~55,000, rice glucan ~64,000, and wheat glucan ~34,000 daltons. Pentosan preparations from barley have molecular weights from 50,000 to 200,000 daltons. β-Glucans are of particular interest in the brewing industry because they can cause difficulties during clarification of wort and the filtration of beer. During the malting of barley, the β-glucans and pentosans are degraded by the enzymes endogenous to barley. If the activity of the endogenous β-1,4;1,3-glucanase is not high enough, then high-molecular-weight glucans remain in the mash and cause the difficulties described above. In addition, the degradation of the β-glucans is dependent on the temperature of the various mash processes.

The endogenous barley β-glucanases are inactivated above 45°C so that in the typical infusion processes in brewing at ~65°C, only a part of the β-glucans is hydrolyzed. An improvement in solubility can be achieved by addition of temperature-stable microbial β-glucanase preparations. Problems with β-glucans are also encountered in wine production. These β-1,3;1,6-glucans have structures different from those of the cereal β-glucans. They are produced by fungi and will be considered in more detail in Section 5.5 (on wine production). Commercial preparations involve a number of different types of β-glucanases, which are described in the following paragraphs.

Cellulases (EC 3.2.1.4)   The cellulase enzymes that attack amorphous cellulose and hydrocellulose can also attack barley β-1,3;1,4-glucans. They are capable of hydrolyzing highly branched β-1,4-glucans, such as the tamarind amyloid, a highly branched xyloglucan. Commercial bacterial β-glucanases do not have this enzyme, which appears in *Trichoderma* and *Penicillium* cellulases.

Endo-1,4-β-glucanase (EC 3.2.1.73; CAS 37288-51-0), Also Called Lichenase   This enzyme acts on 1,3;1,4-β-glucan but shows no activity with 1,4-β-glucan. It cleaves the β-1,4 linkage adjacent to the reducing end of a β-1,3 linkage, which is absolutely necessary for cleavage to occur. This cleavage pattern is characteristic of bacterial β-glucanases (lichenase) and the malt β-glucanases. These enzymes also show no action on linear β-1,3-glucans, as can be seen in Figure 3.62.

Endo-1,3-β-glucanase (EC 3.2.1.39; CAS 9025-37-0), Also Called Laminarinase and Endo-1,3(4)-β-glucanase (EC 3.2.1.6; CAS 62213-14-3)   The first enzyme cleaves the β-1,3 linkage at the reducing end of β-1,3-bound glucose chains. The 1,3(4)-β-glucanase, in contrast, is also capable of cleaving a 1-4-β linkage at the reducing end of a β-1,3-bound glucose chains so that it can also act on those β-glucans with both linkages similar to the action of lichenase.

Endo-1,6-β-D-glucanases   These enzymes attack the β-1,6-bound glucose residues that appear, for example, in yeast glucan. Lutean, a β-1,6-glucan, is degraded by this enzyme and can be used to test for it.

3.1.5.3   Technical Preparations

Technical β-glucanase preparations are prepared from different *Bacillus* and fungal species, such as *Penicillium, Aspergillus, Rhizopus,* and *Trichoderma*.

Sauter and Sprössler (1989) isolated pure β-glucanases from industrial preparations and characterized them according to their action on various substrates. The enzyme isolated form *B. subtilis* had a molecular weight of 29,000 daltons, a pH optimum at pH 7.5, and an optimum temperature of 40–50°C. The enzyme from *A. niger* also has a molecular weight of 29,000 daltons and optimum activity at pH 5.0 and 50–60°C. (See Table 3.19.)

β-Glucanases from B. subtilis   These enzymes have specificities similar to those of the glucanases from malt. They are active against lichenin, have an affinity for

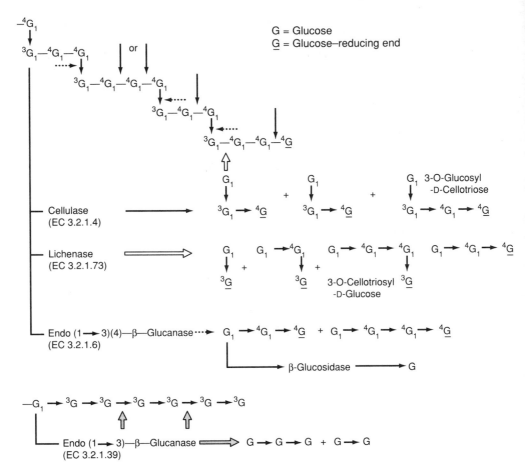

Figure 3.62   Action of various enzymes on β-glucan (McCleary, 1987).

TABLE 3.19   Comparison of Various β-Glucanases by Sugar Oligomers Generated with Different Substrates

Enzyme Source	CMC[a]	Laminarin	Barley β-Glucan
		Substrate	
B. subtilis	—	—	G_3, G_4
A. niger	G_2, G_3	G_3, G_4, G_5	G_3, G_4
T. reesei	G_2, G_3	G_2, G_3, G_4	G_2, G_4, G_5

[a]Carboxymethylcellulose

the trimeric glucose unit of a β-glucan chain, and cleave the linkage adjacent to the next β-1,4 linkage. Neither acts on laminarin or carboxymethylcellulose.

Molecular Weight: 26,000 daltons (Borriss, 1981); 29,000 daltons (Sauter and Sprössler, 1989).

Activity: pH optimum is 6.0–7.5. Other types show an optimum between 7.0 and 9. A third variation has a narrow optimum at pH 7.5.

K_m Values: Lichenin, 1.43 mg/ml; barley glucan, 1.15 mg/ml (Borriss, 1981).

Temperature Optimum: 58–60°C. Temperature-stable types are active up to 65°C. Inactivation at temperatures above 75°C. The temperature stability is improved by Ca^{2+} ions.

Another *B. subtilis* β-glucanase was described by Rafalovskaya et al. (1986).

Molecular Weight: 35,000 daltons, pI at pH 5.2.

Activity: Optimum pH 6 and 55–60°C.

Stability: pH range 6–9.

Inhibition: SH reagents inhibit, as well as Ca^{2+}, Cu^{2+}, Al^{3+}, Pb^{2+}.

β-Glucanase from B. licheniformis Lloberas et al. (1988) isolated and characterized an endo-β-1,3;1,4-D-glucanase from the organism described above. The molecular weight was between 27,000 and 28,000 daltons; the isoelectric point was at pH 4.7. The activity optimum was very broad, ranging from pH 4.0 to 10.5. The optimum temperature was 55°C.

β-Glucanase from Penicillium emersonii There are also at least two types, which differ from the bacterial enzymes with respect to both specificity and activity–stability range.

The enzyme complex is particularly heat-stable; at least 75% of the activity is retained after heating the enzyme solution to 95°C for 10 min.

See Figures 3.63 and 3.64.

β-Glucanase from A. niger This enzyme has an optimum activity at pH 5, with 80% of the maximum activity between pH 4 and 6.2. Optimum temperature is 65°C; maximum temperature is 75°C (Glucanase GV [GT]).

β-Glucanase from Trichoderma reesei A β-1,3-glucanase was isolated by Ogawa et al. (1988) from a fixed-bed culture.

Molecular Weight: ~40,000 daltons.

Activity: The maximum activity is at 55°C and pH ~5.

Stability: The enzyme is stable between pH 4 and 6 at ~50°C.

Specificity: Laminarintetraose is cleaved to glucose, laminarinbiose, and laminarintriose.

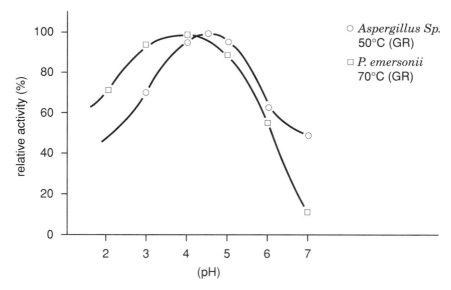

Figure 3.63   pH dependence of fungal β-glucanase from *P. emersonii* and *Aspergillus* sp.

β-Glucanases from Trichoderma harzianum   Enzymes from this organism contain a series of very active β-glucanases. After fractionation by Dubourdieu et al. (1985), two exo-1,3-glucanases and one endo-1,6-D-glucanase were found. This enzyme can degrade β-1,3;1,6-D-glucans, such as cinerean.

Molecular Weight: ~40,000 daltons; pI at pH 7.8.

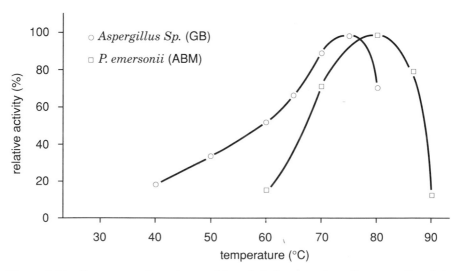

Figure 3.64   Temperature dependence of fungal β-glucanase from *P. emersonii* and *Aspergillus* sp.

β-Glucanase from Arthrobacter sp.   So-called yeast cell lytic enzymes are produced by some enzyme manufacturers from various organisms such as *Trichoderma* sp. These were not particularly effective when used for the lysis of yeast cell walls; thus enzymes that were more specific were sought. Various *Arthrobacter* sp. were found that formed a yeast-lysing enzyme system for yeast cells or yeast cell walls. β-1,3-Glucanase, protease, and mannanase activities were identified in the enzyme complex (Kitamura, 1982). "Tunicase" [DA] is a commercial enzyme preparation prepared from *Arthrobacter* cultures. It is a cellulase that acts specifically on the β-1,3-glucans of yeast cell walls. Cell walls form *Candida, Saccharomyces cerevisiae, Torula,* and *Kluyveromyces fragilis* are lysed. The enzyme has an optimum activity between pH 7 and 8 and is stable only below 35°C. The enzyme is stable in the pH range 4.0–9.0. A similar enzyme from *Archromobacter* is active at pH 7 up to 55°C.

PRODUCTS AND MANUFACTURERS   Technical preparations are standardized by using β-glucan as substrate. The amount of reducing sugar produced or the reduction in viscosity is determined. Another method is described by McCleary (1986) in which the activity of lichenase can be determined by using a soluble chromogenic substrate.

Most commercial preparations are in liquid form. They contain amylases and, depending on source of origin, protease side activities:

- Bacterial β-glucanases from *B. subtilis* [AM], [NO], [RM], [SOE], [GB]
- Fungal β-glucanase from *P. emersonii* [GX], [BN], [AM]
- *Trichoderma reesei* β-glucanase [GR]
- Fungal β-glucanase from *A. niger* [RM], [GT], [NO]
- Glucanase from *Arthrobacter* sp. [DA]
- Glucanase from *Achromobacter* sp. [AM]

3.1.5.4   *Mannanases and Galactomannases*

Mannans, linear α-1,2 or α-1,6 polymers of mannose, occur in yeasts and other microorganisms and also in higher plants. Mannose is, however, more frequently found as a component of heteropolymers such as glucomannan and galactomannan. Galactomannans occur as reserve carbohydrates in guar beans and in locust bean seeds. They consist of a main chain of β-1,4-linked mannose residues that carry side chains of α-1,6-bound-D-galactose residues. In addition to monomeric galactose residues, digalactosyl residues or gal-(3,1)-ara-(3,1)-ara side chains can also occur.

Depending on their origin, the galactomannans differ in their galactose:mannose ratios. The galactose content varies between 10 and 50%. Galactomannan containing more than 25% galactose (e.g., guar bean galactomannan contains 35% galactose) are readily soluble in cold water and give highly viscous solutions. Those galactomannans with a galactose content of 18–24% are soluble in hot water (e.g., locust bean meal, 23% galactose). When these polymers are mixed with other polysaccharides such as agarose or xanthan, gels with a very low polymer concentration (solid content) are formed.

Mannanase (EC 3.2.1.78; CAS 37288-54-3) Endo-β-1,4-D-mannanase   This enzyme cleaves the linear mannan chain of locust bean galactomannan to oligomers with galactose branches, which serve as a basis for the production of food additives. Guar galactomannan is only slightly attacked by this enzyme. The enzyme is produced by *B. subtilis, A. awamori, A. niger,* and *T. reesei* and can be isolated from commercial galactomannanase preparations.

Galactomannanase   Technical galactomannanase preparations that can be obtained from *A. niger* contain, in addition to the mannanase, a series of additional enzymes such as α-galactosidase (EC 3.2.1.22) and β-D-mannosidase (EC 3.2.1.25). These preparations are used to reduce the high viscosities of galactomannan solutions such as in the process used to extract secondary yields from crude oil. Galactomannan modification is also desirable for some applications in the food industry. For the extraction of coffee, the mannan-containing matrix of the coffee beans can be ruptured. This not only leads to an increase in extraction yield but also prevents the accumulation of mannan-containing sediments in the extracts that are difficult to filter.

Standardization is accomplished by determining the reducing sugar produced, calculated as mannose, from locust bean gum galactomannan (Fig. 3.65).

The hemicellulose preparation Rohament AC [RM] is stable in the absence of substrate over a wide pH range; 95–98% of the enzyme remains active after incubation at 20–25°C for 24 h between pH 3 and 8.5. The enzyme is stable up to 60°C at pH 4–5; in the presence of 2% substrate, up to 70°C.

See Figures 3.66 and 3.67.

MANUFACTURERS   [RM], [NO], [SN], and others.

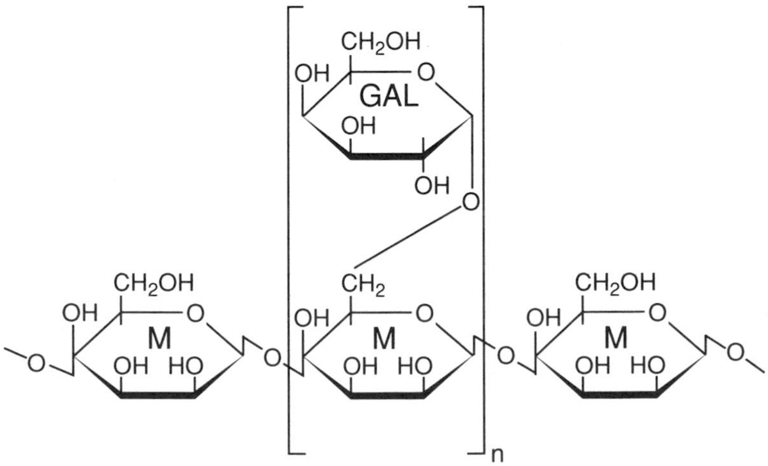

Figure 3.65   Structure of locust bean gum galactomannan.

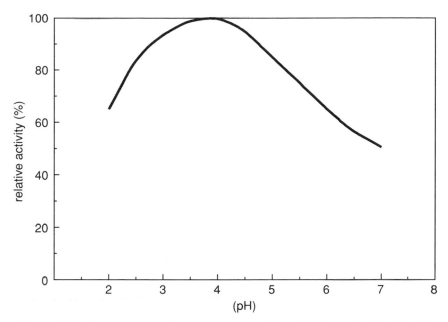

Figure 3.66   ph dependence of galactomannase activity (40°C; reaction time 10 min).

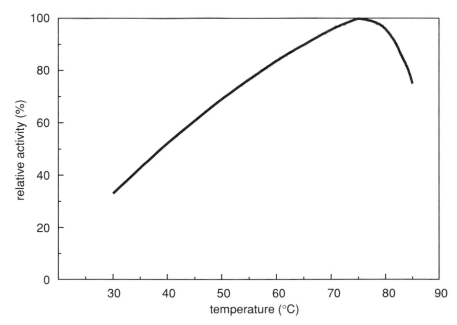

Figure 3.67   Temperature dependence of galactomannase activity (pH 4.0; reaction time 10 min).

3.1.5.5 Chitosanase
References: Methods in Enzymology, Vol. 161 (1988); Zikakis (1989)

Chitin and chitosan occur in crustaceans, where they are bound to calcium and co-valently linked to proteins. The proteins serve as binding links to other carbohydrates. Chitin and chitosan are copolymers of *N*-acetylglucosamine and glucosamine, which are linked in the β-1,4 position. The copolymer is referred to as *chitin* if it contains less than 7% nitrogen. The chitin deacylase, for example, from *Mucor rouxii,* catalyzes the hydrolysis of chitin to chitosan and acetic acid. The soluble chitosan has a molecular weight of several hundred thousand to two million daltons.

Chitinases are formed by higher plants and microorganisms. The enzymes have been isolated from plants such as tomatoes, soybeans, and wheat bran. *Streptomyces,* especially *S. griseus, Serratia,* and *Aeromonas* species are potential producers.

The enzyme from *S. griseus* has a molecular weight of 35,000 daltons, and optimum pH is 8.0. The enzyme is stable at 37°C between pH 6.0 and 8.0. Inactivation occurs at higher temperatures.

Chitinase, similar to other carbohydrases, consists of a complex of three enzymes: endochitinase, exochitinase, and chitobiase.

A commercial chitosanase is obtained by fermentation of *B. pumilus* (Chitosanase BP [MJ]). The activity of this chitosanase is determined by lowering the viscosity of a chitosan solution or determining of the reducing sugar groups according to the Somogyi–Nelson method. Chitooligosaccharides (DP = 2–10) and, eventually, after longer reaction times, small amounts of the glucosamine monomer are produced. The enzyme is also capable of degrading the cell walls of microorganisms that contain glucosamine polymers. The activation optimum ranges within pH 5–6.5 and 40–60°C. Another very active preparation is produced by *Aeromonas hydrophilia* subspecies *anaerogenes* cultures [GO].

MANUFACTURERS   [MJ], [GO].

REFERENCES

Borriess, R., *Zeitschr. Allg. Mikrobiol.* **21**(1), 7 (1981).

Dubourdieu, D., Desplanges, C., Villettaz, J.-C., and Ribereau-Gayon, P., *Carbohydr. Res.* **144**(2), 277 (1985).

Finscher, G. B., Sawyer, W. H., and Stone, B. A., *Biochem. J.* **139,** 535 (1974).

Kitamura, K., *Ferment. Technol.* **60,** 253 (1982).

Lloberas, J., Querol, E., Bernues, J., *Appl. Microbiol. Biotechnol.* **29**(1), 32 (1988).

Matheson, N. K., and McClery, B. V., in Aspinall, G. O. (ed.), *The Polysaccarides,* Vol. III., Academic Press, New York, 1987, p. 1.

McClery, B. V., and Matheson, N. K., *Advances Carbohydr. Chem.* **44,** 147 (1987).

McClery, B. V., *Carbohydr. Res.* **6**(4), 307 (1986).

Ogawa, K., Zoyama, N., and Sugita, K. *Hakko Kogaku Kaishi* **66**(5), 385 (1988).

Rafalovskaya, T. Y., Shishkova, E. A., and Oreshenko, L. I., *Prikl. Biokhim. Mikrobiol.* **22**(5), 622 (1986).

Sauter, O., and Sprössler, B., *International Conference Biotechnology and Food,* University Hohenheim—Stuttgart, Germany, poster (1998).

Steiner, K., *Schweiz. Brauerei-Rundschau* **79,** 153 (1968).

Zikakis, J. *ACS Symposium Ser.* **389,** 116 (1998).

3.1.6   Pectin Degradation

3.1.6.1   The Substrate: Pectins
References: Kertesz (1951), Rombouts and Pilnik (1980), Gierschner (1981), Whitaker (1984) (see also Section 3.1.4.1)

The following pectin substances can be distinguished (Kertesz, 1951):

Pectic Substances: Pectic substances are those materials that contain principally polygalacturonic acids, partially or completely esterified. A completely 100% esterified polygalacturonic acid has a theoretical methoxyl content of 16.3%.

Protopectin: According to Joslyn (1962), protopectins are water-insoluble pectins that are chemically and physically bound to other cell wall components such as cellulose, xylan, or glucomannan (see Section 3.1.4.1).

Pectic Acids: Very lightly esterified or nonesterified polygalacturonic acids are called *pectic acids*. They form rigid gels with sugars and acids or divalent ions. The English literature differentiates between pectinic acid, for low degrees of esterification (up to 7%); and pectic acid, for totally deesterified pectin.

As already mentioned, the texture of fruit and vegetables is regulated by the content and the properties of the pectic substances (Table 3.20). In immature fruits they

TABLE 3.20   Pectin Content of Some Fruits and Juices (Fogarty and Kelly, 1983)

Source	Pectin (%)	Esterification (%)
Grapes	0.2–1.0	10–65
Grape juice	0.01–0.09	—
Apples	0.5–1.6	70–90
Apple juice (nonclarified)	0.2	—
Black currant mash	1.6	—
Grapefruit	1.6–4.5	—
Lemon (rind)	32.0	50–65 (albedo)
Lemon (pulp)	25.0	—
Orange (rind)	20.0	—
(membrane)	29.0	—
(juice sacs)	16.0	—
Turnips	10.0	—
Sugarbeet pulp	30.0	—

appear principally as insoluble protopectin. This converts to soluble pectin during ripening (Hulme, 1958) and is associated with the softening of the fruit. The loosening of the rigid cell wall can be counteracted by using calcium salts (Doesburg, 1965).

3.1.6.2 Pectinases   Pectinases are important in processing fruit and vegetables and in wine production. Specifically, these enzymes are used for clarification of fruit juice, enzymatic fruit mash treatment, improvement of juice yield and color in wine production, and the maceration and liquefaction of plant tissue. The pectinases are also used to improve the extraction yield of oils from citrus fruit and olives.

Pectinases are usually mixtures of several enzymes, which are capable of cleaving high-molecular-weight pectin within the polygalacturonic chain (depolymerases) and deesterifying pectins (esterases). Depolymerization can progress in one of two ways: either by a hydrolytic reaction, as occurs generally in the hydrolysis of glycosidic linkages, or by means of a very specific β-transelimination reaction. This class of depolymerases are called *lyases*. These different cleavage reactions can be used to classify the pectinases (Fig. 3.68 and Table 3.21).

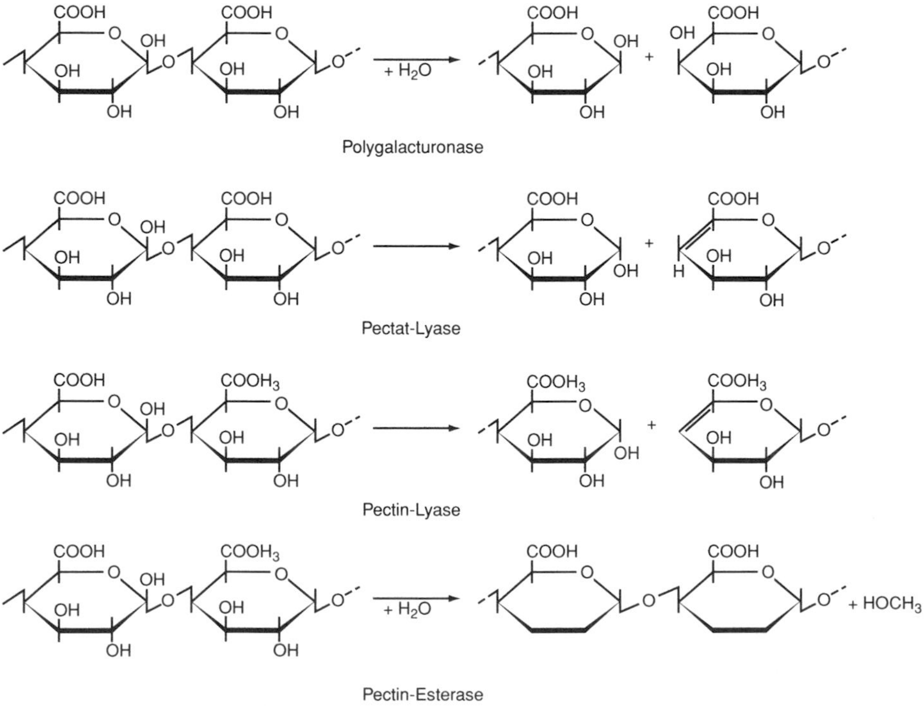

Figure 3.68   Ester hydrolysis and depolymerization of pectin.

TABLE 3.21 Pectinase Classification (Rombouts and Pilnik, 1980)

Name	EC Number	Preferred Substrate	Cleavage Site
Hydrolases			
Endopolygalacturonase	3.2.1.15	Pectate	Randomly within the chain
Exopolygalacturonase	3.2.1.67	Pectate	Terminal
Lyases			
Endopectate lyase	4.2.2.2	Pectate	Randomly within the chain
Exopectate lyase	4.2.2.9	Pectate	Terminal
Endopectin lyase	4.2.2.10	Pectin	Randomly within the chain
Esterase			
Pectinesterase	3.1.1.11	Pectin	Deesterification

Endopolygalacturonases (Endo-PG; EC 3.2.1.15; CAS 9032-75-1) Endopolygalacturonases (PG) (Table 3.22) are formed by a large number of microorganisms such as fungi, bacteria, and yeasts. They are found in many plants. Tomatoes are particularly rich in PG, which is responsible for the steady decrease in the viscosity of cold-pressed tomato juice.

The best substrate for PG is pectate from which random cleavage leads to oligogalacturonides, such as mono-, di-, and trigalacturonides. Rapid reduction in the viscosity of pectate solutions is an indication of PG activity. The cleavage of only a few linkages causes a viscosity reduction of 50% (Kelly and Fogarty, 1978). PG activity drops with increasing degree of esterification of the substrate, as free carboxyl groups are necessary for catalytic action. In addition, PG activity also decreases with reduction of the degree of polymerization. Digalacturonides are rarely cleaved, and trigalacturonides are cleaved only slowly (Rexová-Benkovà and Marcovic, 1976).

Aspergillus, Penicillium, and *Rhizopus* form different PG isozymes. In addition, there are two types of PG: a group of liquefying PGs and macerating PGs. The lat-

TABLE 3.22 Properties of Some Industrially Important Endopolygalacturonases

Enzyme Source	Molecular Weight (daltons)	pH Optimum	pI	Reference
A. niger	35,000	4.1	—	Cooke et al. (1976)
	85,000	3.8	—	
A. niger	46,000	5.0	—	Henrichovà and Rexová-Benková (1977)
A. japonicus	35,500	4.5	—	Ishii and Yokotsuka (1976)
Trichoderma koningii	32,000	5.0	6.41	Fanelli et al. (1978)

ter reduce the viscosity of pectate solutions and are capable of disintegrating plant tissues. An industrial process for the separation of macerating PG from *A. niger* from a liquefying PG on an ion exchanger has been described by Löffler et al. (1989). The macerating PG from *A. saitoi* (Yamasaki et al., 1966) showed optimum activity at pH 4.8–5.0 and 45°C and cleaved 60% of the linkages in polygalacturonic acid and 17% of the linkages in pectin. The enzyme was inactivated in 10 min at 50°C.

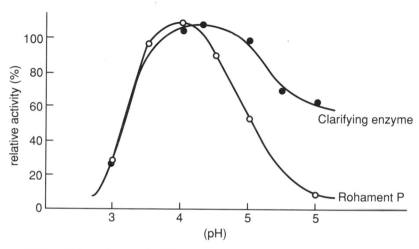

Figure 3.69   pH dependence of a PG enzyme activity (Rohament P [RM]) and a commercial clarifying enzyme; reaction temperature 30°C.

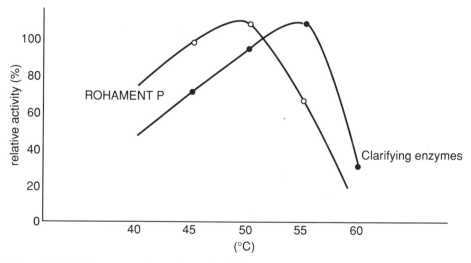

Figure 3.70   Temperature-dependence of a PG activity (Rohament P [RM]) and a commercial clarifying enzyme; reaction at pH 4.2.

The macerating activity of a PG from *A. alleacus* (Rohament P [RM]) is, according to Bock et al. (1972), distinguishable from other PGs (see Figs. 3.69 and 3.70) by its inhibition by various plant extracts. It is, in contrast to PG from tomatoes, almost totally inhibited by compounds found in cucumbers and pears. The macerating PG from *A. alleacus* (Rohament P) cleaves protopectin into pectin fragments with molecular weights of about 100,000 daltons (Radola, 1986). These are then only slowly further degraded. A similar observation was made by Kaiji (1956) for the PG isolated from *C. felsineum*. Under strong macerating conditions, the reduction in viscosity of pectin at pH 4.7, as well as the production of reducing sugars, was small. Various macerating PGs were also found in *Rhizopus* sp.

ANALYSIS OF PG   Polygalacturonases can be analytically determined by the reduction of the viscosity of pectate solutions (Wimborn and Richards, 1978). The pH used in the measurements should lie near the PG pH optimum. This optimum is dependent on temperature (see also section on technical pectinase preparations).

The activity of PG is determined by the amount of reducing sugars formed. This amount can be measured with the dinitrosalicylic acid method of Miller (1959).

Exopolygalacturonases (Exo-PG, EC 3.2.1.82)   Monomeric galacturonic acids from pectate, oligogalacturonates, and digalacturonates are cleaved in stepwise fashion from the nonreducing end of the polymer chains by exopolygalacturonases (exo-PG). These enzymes occur in higher plants, some bacteria, and fungi. An extracellular exopolygalacturonase produced by *A. niger* was described by Heinrichovà and Rexovà-Benkovà (1976). The formation of monomeric galacturonic acids by the action of commercial enzymes from *A. niger* or *A. wentii* on pectate can be attributed to the action of exoenzymes. Omran et al. (1986) were able to obtain and subsequently characterize an exo-PG from the commercial preparation Pectinol C [RM] by inactivating the endopolygalacturonase at pH 8.5 and purifying by chromatography. After cleaving 75% of the glycosidic linkages in polygalacturonic acid (PGA) by the isolated enzyme, Omran et al. observed that the viscosity of the PGA solution fell to only two-thirds that of the starting material.

Nevertheless, the molecules remaining after the degradation still had a high degree of polymerization. During hydrolysis of oligogalacturonides, the trimer is cleaved at the highest maximum reaction rate. The optimum pH of this exo-PG is about 4.32. The stability (80% residual activity) extends from pH 3.6 to neutral. The maximum temperature for activity and stability is 45°C.

Endopectate lyases (PATE; EC 4.2.2.2; CAS 9015-75-2)
Reference: Rombouts and Pilnik (1980)

These lyases, which are produced by some bacteria and phytopathologic fungi, show a pH optimum above pH 8.0 and require CA^{2+} ions to be active.

The enzyme from *B. subtilis* isolated by Cesson and Codner (1978) has a molecular weight of 33,000 daltons, an isoelectric point at pH 9.2, and an optimum pH of 8.5.

The endopectate lyase from an alkalophilic *B. subtilis* has an optimum pH of 10.0 and is completely inhibited by EDTA (Kelly and Fogarty, 1978).

ANALYSIS OF PATE   Endopectate lyase is best determined by the method of Macmillan and Pfaff (1966) using 1% polygalacturonate solution in buffer in the presence of Ca^{2+} ions. The increase in extinction at 235 nm at 40°C is measured using a spectrophotometer.

Endopectin Lyases (PTE; EC 4.2.2.10; CAS 9033-35-6), Endopolymethylgalacturonate Lyases   PTE was first found in the technical preparation Pectinol R 10 [GR] by Albersheim et al. (1960) and was later isolated and characterized. These types of depolymerases are important components of commercial enzymes and are especially important in practical applications. They are formed mainly by fungi and depolymerize highly esterified pectin in a random manner. They cause rapid decrease in the viscosity of pectin solutions. They specifically cleave at the glycosidic linkages adjacent to a methylated galacturonic acid residue. Highly esterified pectins are the best substrate, while pectate, pectin acid amide, and the glycyl ester of polygalacturonic acid are not degraded (Pilnik et al., 1974). Activity decreases at lower chain lengths and lower degrees of esterification. The smallest substrates still being degraded are the completely esterified tetra- and trigalacturonates (Van Houdenhoven, 1975). Most pectin lyases are activated by calcium. This activation is dependent on pH and the degree of esterification of the pectin (Fig. 3.71).

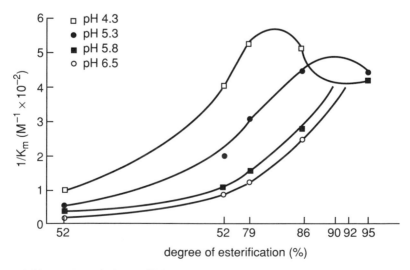

Figure 3.71   Endopectin lyase affinity ($1/K_m$ as a function of the amount of esterified pectin substrate at various pH (Voragen et al., 1971).

TABLE 3.23   Properties of Some Pectinlyases

Enzyme Source	Molecular Weight (daltons)	pI	pH Optimum	Reference
A. niger	35,000	3.65	6.0	Van Houdenhoven (1975)
A. niger	33,100	3.75	6.0	—
A. japonicus	32,000	7.7	6.0	Ishii and Yokotsuka (1975)
A. sojae	32,000	—	5.5	Ishii and Yokotsuka (1972)
A. niger	—	3.9	6.0[a]	Zhou et al. (1989)
			8.5[b]	

[a]Containing 71% esterified pectin.
[b]Containing 94% esterified pectin.

ANALYSIS OF PTE   Pectin lyase (Table 3.23) can be determined with a spectrophatometric test by using pectin that is at least 70% esterified as substrate and measuring the increase in absorption at 235 nm. The presence of pectinesterase strongly influences the analytic results. A less specific but more commonly used method is a viscometric determination with solutions of highly esterified pectin as described by Bush and Codner (1970).

In PTE analysis, the dependence of the pH optimum on temperature and the degree of esterification of the substrate must be considered.

The optimum activity (viscometric) is at pH 6.5 for 90% esterified pectin and a reaction temperature of 30°C. The optimum is at pH 4.3 for 80% esterified pectin. The pH optimum for citrus pectin is at pH 5.1–5.3 (30°C).

The PTE isolated by Zhou et al. (1989) from the commercial preparation Pectinex Ultra [NO] showed an optimum activity with apple pectin (71% esterified) at pH 6.0. Optimum activity with a synthetic highly esterified (94%) pectin is at pH 8.5. The optimum temperature is 60°C; higher temperatures cause rapid inactivation.

Pectinesterase (PE; EC 3.1.1.11; CAS 9025-98-3)   Pectins are enzymatically deesterified by PE, and converted to low esterified pectin or pectic acid with release of methanol. PE occurs in many higher plants, in various fungi, and in some yeasts and bacteria. PE preparations contain a large number of isozymes. These have similar action, but their isoelectric points and molecular weights differ. The preferred point of attack of PE in the pectin chain is at an esterified galacturonic acid residue adjacent to a free carboxyl group.

Plants, especially tomatoes and citrus fruits, show very high PE activity (see Table 3.24). This can have a major effect during the production of canned foods and juices. For example, citrus pectin is rapidly deesterified in the manufacture of orange juice by endogeneous PE. This causes coagulation of particulates and subsequent undesirable clarification of the juice. This can be prevented only by inactivation of the enzyme.

Pilnik and Rombouts (1981) have reported on the different pectinesterases that occur in oranges and their mutual synergistic interaction with depolymerases. Thus,

TABLE 3.24   **Properties of Some Pectinesterases**

Enzyme Source	Molecular Weight (daltons)	pI	pH Optimum	K_m Pectin (mg/ml)	Reference
Tomatoes	27,500	—	6–9	0.74	Lee and Macmillan (1970)
Oranges	36,200	10.0	7.6	0.083	Versteeg et al. (1978)
Fusarium sp.	35,000	—	7.0	—	Miller et al. (1971)
A. niger	—	—	4.32	0.215 with pectin 68.4% VE	Dörreich (1982)

orange PE shows optimum activity at pH 7, but is still effective at pH 2–2.5 and is very heat-stable. The affinity of this PE for pectin increases with reduction in degree of esterification of the pectic substrate.

Ishii et al. (1979) have described a PE that they isolated from *A. japonicus.* Commercial PE preparations have been prepared from cultures of mold fungi, particularly *A. niger,* by inactivation of the accompanying depolymerase and chromatographic purification. As many of the previous characterization and kinetic studies had been undertaken on crude or only partly purified preparations. Döerreich et al. (1986) prepared a highly purified PE from *Aspergillus* using ion exchangers in an attempt to characterize the enzyme (Fig. 3.72).

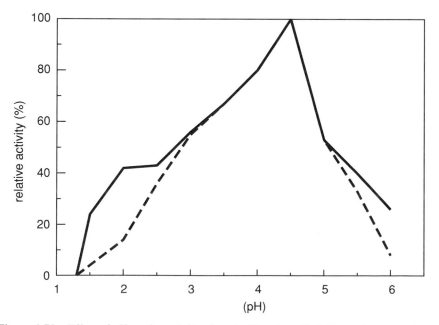

Figure 3.72   Effect of pH on the activity of crude (1) and purified (2) pectin esterase in Rohapect C [RM] [after Dörreich et al. (1986)].

The crude enzyme shows optimum activity at pH 4.4, while that of the pure enzyme is pH 4.3. Under acidic conditions, the enzyme is active to pH 1.5–2 and shows relatively good stability.

The crude enzyme is fully inactivated in 5 min at 65°C; the purified enzyme, at 70°C. The phenolic compounds tannin and gallic acid inhibit activity only slightly. The affinity of this PE for pectin with a degree of esterification between 50 and 80% is considerably less than that for pectin with 33% esterification.

ANALYSIS OF PE   A rapid method for the determination of PE involves the monitoring of the color change of a pH indicator such as bromocresol green. Agar plates can be used for the test. For quantitative assays, the free carboxyl groups released can be measured titrimetrically or, alternatively, the methanol produced can be estimated by gas chromatography.

3.1.6.3 Technical Preparations   Technical preparations are obtained principally from fungi, including *A. niger, A. alleacus, A. wentii, A. aculeatus, A. usamii,* and *Rhizopus* sp. The preparations have been obtained since the 1930s by the cultivation of fungi in fixed-bed cultures. The major portion of the culture medium consists of wheat bran with a moisture content between 45 and 60% (Beckhorn et al., 1965). Following extraction, ultrafiltration is used to remove metabolites and culture medium fragments and at the same time to concentrate the extract. After sterile filtration, liquid preparations are standardized with water and stabilizers such as glycerol or sorbitol. The addition of approved food preservatives is allowed in most countries.

Solid, powdered, or granulated preparations are produced by spray or agglomeration drying with the addition of glucose, sucrose, or maltodextrin.

Fixed-bed or surface processes are currently used by various international enzyme manufacturers for the preparation of pectinases. At one time it was assumed that a multienzyme complex of pectinases could be prepared only by the fixed-bed process. The work of Zetelaki (1976) and Nyiri (1968) showed that the enzyme complex necessary for industrial applications could also be prepared using submerged fermentation.

The composition of the enzyme complexes with respect to the activities of the individual enzymes—PG, PTE, and PE—is dependent on the fungal strain as well as the culture conditions. Specific conditions are required in preparations for various applications of pectinases in the production of fruit and vegetable juices. This is relevant for the characteristics of the pectinase complex, as well as for the side activities. Because of the different amounts of PG, PTE, and PE, the characteristics of the preparations vary with respect to stability at various pH values and to dependence of activity on pH and temperature.

In addition, enzymatic activities other than those of the pectinases are necessary. For maceration, the presence of special PG enzymes and cellulases is required. In order to remove araban precipitation in apple juice concentrates, it is important to have standardized arabanase activity in commercial preparations. Special proteases work synergistically with the pectinases and hemicellulases in clarification of red wine. Because of these requirements, the first generation of commercial preparations that were initially used in fruit processing have been developed for more specific applications. Standardization of the side activities is difficult because various manufacturers use their own analytic methods; therefore, it is not easy to compare commercial preparations. In practice, preparations are selected on the basis of one or more analytically controlled production tests.

In addition to the specific practical applications, an analytic comparison of technical products is of interest to establish a correlation between analysis and practice.

Gierschner's group (Tabeleros et al., 1987) investigated a series of technical preparations and attempted to define the relationship between the analytically determined product composition and the results of practical trials. Some interesting qualitative data but no quantitative relationships resulted. Some of the analytic results are reproduced in Figure 3.73.

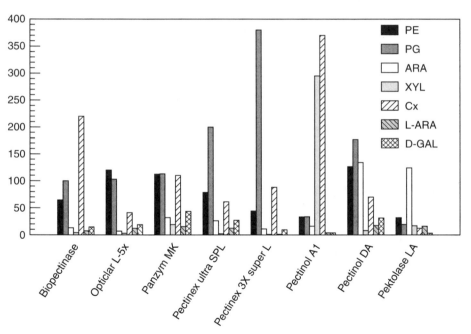

Figure 3.73   Glycosidic activity of commercial preparations (from Tabeleros et al., 1987).

Practical information on the properties of technical products include the relationships of activity and enzyme stability to various reaction parameters such as pH, temperature, and reaction time. Descriptions of these properties can differ according to the particular enzyme manufacturer. For example, some of the factors on which reaction rates depend are given in Figures 3.74–3.76.

Preparations with Specific Activity

MACERASE (ROMAMENT P [RM])   This preparation (see also Figs. 3.77 and 3.78), in which the primary activity is a polygalacturonase with special macerating

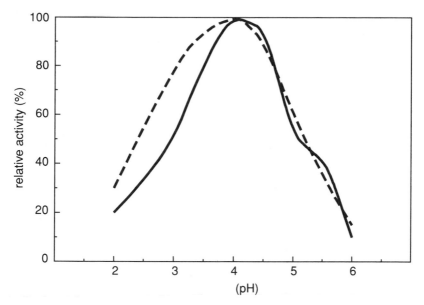

Figure 3.74   Effect of pH on the activity of two technical preparations.

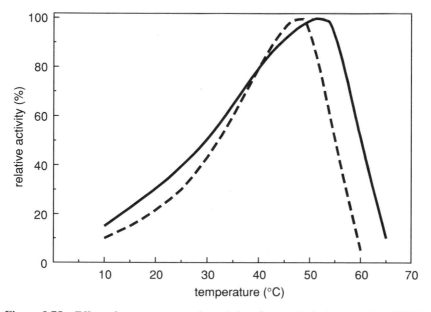

Figure 3.75   Effect of temperature on the activity of two technical preparations [RM].

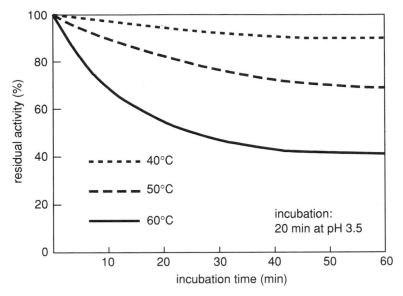

Figure 3.76   Effect of temperature on enzyme stability [RM].

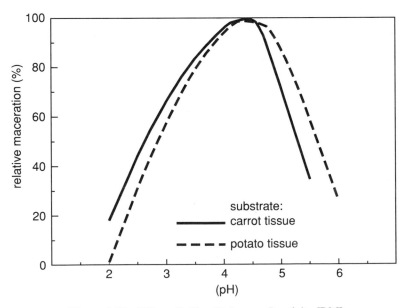

Figure 3.77   Effect of pH on Rohament P activity [RM].

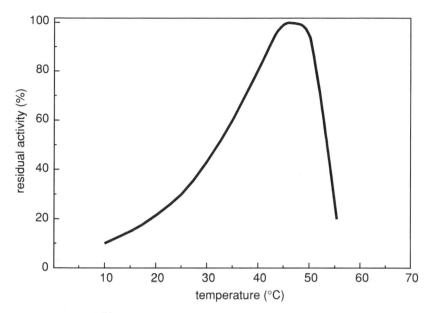

Figure 3.78   Effect of temperature on Rohament P stability [RM].

properties, was prepared from *A. alleacus* prior to 1988. Today it is prepared from *A. niger.* Various authors, including Grammp (1969) and Radola (1986), have described the properties of this preparation, which is used for the production of macerates from vegetables, fruit concentrates, and pulpy fruit juices. The macerating properties depend principally on the activity of a particular PG and less on the accompanying enzymes such as cellulases and hemicellulases. Because of the lack of pectinesterase, the preparation is not capable of clarifying apple juice or depectinizing fruit juice. The temperature stability maximum is between 45 and 50°C and can be increased to 65°C in the presence of substrate.

PECTINESTERASE PREPARATIONS   It is difficult to compare information on the activity of technical preparations as given by the manufacturers because their standardization is specific for the company's own methodologies. Their specifications depend on the particular reaction parameters used and the nature of the substrates. Thus, for a particular application, only on use can it be determined whether the material is an enzyme concentrate or a dilute preparation. (See Fig. 3.79.)

Products and Manufacturers: Rohapect [RM], Rohament [RM], Pectinex [NO], Ultrazym [NO], Biopectinase [BN], Panzym [BÖI], Pectolase [GT], Klerzyme [GB], Spark [SOE], Sumizyme [SN], Pectinol [GR], [AM].

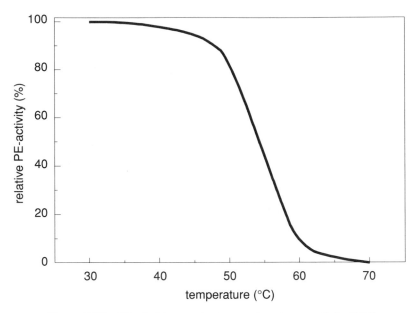

Figure 3.79 Effect of temperature on pectin esterase activity [RM].

REFERENCES

Albersheim, P., and Neukom, H., *Deul. Arch. Biochem. Biophys.* **90,** 46 (1960).

Beckhorn, E. J., Labee, M. D., and Underkofler, L. A., *J. Agr. Food Chem.* **13,** 30 (1965).

Bush, D. A., and Codner, R. C., *Phytochemistry* **9,** 87 (1970).

Cesson, A., and Codner, R. C., *J. Appl. Bacteriol.* **44,** 347 (1978).

Cooke, R. D., Ferber, E. M., and Kanagasabapathy, L., *Biochimica et Biophysica Aeta* **452,** 440 (1976).

Doesburg, J. J., in *The Pecti Substances in Fresh and Preserved Fruits and Vegetables,* I.B.V.T. Communication No. 25, Wageningen, the Netherlands (1965).

Dörreich, K., Omram, H., and Gierschner, K.-H., *Lebensmittel Wissensch. und Technik* **19,** 464 (1986).

Fanelli, C., Cacace, M. G., and Cervone, F., *J. Gen. Microbiol.* **104,** 305 (1978).

Fogarty, W. M., and Kelly, C. T., in *Microbial Enzymes and Biotechnology,* Appl. Science Publ., London, 1983, pp. 131–173.

Gierschner, K.-H., *Gordian* **81,** 171 (1981).

Grampp, E., *Dtsch. Lebensmittel-Rundsch.* **65,** 343 (1996).

Heinrichova, K., and Rexova-Benkova, L., *Biochim. Biophys. Acta* **422,** 349 (1976).

Hulme, A. C., *Adv. Food Res.,* 197 (1958).

Ishii, S., and Yokotsuka, T., *Agric. Biolog. Chem.* **36,** 146 (1972).

Ishii, S., and Yokotsuka, T., *Agric. Biolog. Chem.* **39,** 313 (1975).

Ishii, S., Kiho, K., Sugiyama, S., and Sugimoto, H., *J. Food Sci.* **44,** 611 (1979).

Joslyn, M. A., *Adv. Food Res.* **11,** 1 (1962).

Kaiji, A., *Bull. Agr. Chem. Soc Japan* **20,** 8 (1956).

Kelly, C. D., and Fogarty, W. M., *Canad. J. Microbiol.* **24,** 1164 (1978).

Kertesz, Z. I., *The Pectic Substances,* Interscience Publishers, New York, 1951.

Lee, M., and Macmillan, J. D., *Biochemistry* **9,** 1930 (1970).

Löffler, F., Sprössler, B., and Reiner, R., Europ. Pat. 324,399 (1988).

Macmillan, J. D., and Pfaff, H. J., in *Methods in Enzymology,* Vol 8, 1966, p. 602.

Miller, L. G., *Anal. Chem.* **31,** 426 (1959).

Nyiri, L., *Process Biochem.* **3,** 27 (1968).

Omram, H., Dörreich, K., and Gierschner, K.-H., *Lebensmittel Wiss. und Technol.* **19,** 457 (1986).

Pilnik, W., and Rombouts, F. M., in Birch, G., Blakebrough, N., and Parker, K. J. (eds.) *Enzymes and Food Processing,* Appl. Science Publ., London, 1981, p. 105.

Pilnik, W., Voragen, A. G., and Rombouts, F. M., *Lebensmittel Wiss. und Technol.,* 353 (1974).

Radola, B. J., *J. Ferment. Techn.* **64**(1), 37 (1986).

Rexova-Benkova, L., and Marcovic, O., in Tipson, R. S., and Horton (eds.), *Advances in Carbohydrate Chemistry and Biochemistry,* Vol 33, Academic Press, New York, 1976, p. 323.

Rombouts, F. M., and Pilnik, W., *Pectic Enzymes,* in Rose, A. H. (ed.) *Microbial Enzymes and Bioconversion,* Academic Press, New York, 1980, p. 227.

Tabeleros, M. A., Bannert, E., and Shen Z., Symp. Wissensch.-Techn. Kommission der Internat. Fruchtsaft-Union, Report Vol. 19, Den Haag, Juris Publ., Zürich, 1987, p. 463.

Van Houdenhoven, F. E. A., Thesis, Biochemistry Department, Agriculture University Wagenien, the Netherlands (1975).

Versteeg, C., Rombouts, F. M., and Pilnik, W., *Lebensm. Wissensch. und Techn.* **11,** 267 (1978).

Wthaker, J. R., *Enzyme Microb. Technol.* **6,** 341 (1984).

Wimborn, M. P., and Richards, P. A., *Biotechn. Bioengin.* **20,** 231 (1978).

Yamnasaki, M., Yasui, T., and Arima, T., *Agric. Biol. Chem.* **30,** 1119 (1966).

Zetelaki, K., *Process Biochem.* **11**(7), 11 (1976).

Zhou, R., Omram, H., Bai, Y., Tableros, M., and Gierschner, K.-H., unpublished data, Institut for Food Technology, University Hohenheim, Stuttgart, 1989.

3.1.7 Pentosanases

3.1.7.1 Arabanases   In addition to cellulose and pectin, the arabinogalactans and the arabans belong to the major components of the plant cell wall (see Section 3.1.4). Here, similar to the other carbohydrases, are endoenzymes, which cleave the linkages within the polymer chain, and exoenzymes, which release monomeric units from the end of the chains. The endoarabinases cleave α-1,5 linkages within the araban chain to give di- and trisaccharide products. In addition, there are numerable exoenzymes, α-L-arabinofuranosidases, some of which attack only the α-1,3 linkages in the side chains and others, that cleave α-1,3 as well as α-1,5 linkages.

Following developments in fruit juice extraction technology, the arabanases have become important in recent years, especially in the production of pear and apple juice concentrates. By treating the apple mash with complex pectinase preparations, low-solubility arabanases are released. These cause turbidity in the juice concentrates (see Section 5.5). According to Pilnik and Voragen (1984), the native araban in apples is a branched water-soluble α-araban. Exoarabanase, which occurs in most commercial pectinase preparations, cleaves off the α-1,3 side chains. The residual α-1,5-araban then aggregates into an insoluble precipitate, which is analogous to the retrogradation of starch (Fig. 3.80).

The araban in apple juice was isolated by Churms et al. (1983). The degradation of araban from pear concentrate was described by Babsky and Scobinger (1986). Rombouts et al. (1987) investigated the sugar composition as well as the distribution of linkages in several arabanases. They showed that 90% of linear α-1,5-L-araban consisted of arabinose. Eighty-eight percent of the arabinose residues are α-1,5-glycosidically linked. The araban in beets (beet araban) has about the same arabinose content, but only 28% of the arabinose residues are α-1,5-glycosidically linked.

According to Ducroo (1987), the molecular weight of the insoluble araban in apple juice concentrates is about 10,000 daltons, but molecular weights of 3000 daltons were also found.

Endoarabanases (EC 3.2.1.99)   Some endoarabanases from *B. subtilis* with preferred specificity for α-1,5 linkages were described by Weinstein and Albersheim (1978) and Kaji and Saheki (1975). The molecular weights of these enzymes lie between 32,000 and 35,000 daltons; optimum activity is between pH 5 and 6. Because the stability of these enzymes is low at acidic pH, they are not suitable for the manufacture of fruit juice.

The endoarabanase isolated from *A. niger* by Rombouts et al. (1988) has the following characteristics:

Molecular Weight: (SDS electrophoresis) 35,000 daltons.
Isoelectric Point: pH 4.5–5.5.
pH Optimum: 5.0.
Temperature Optimum: 50°C.

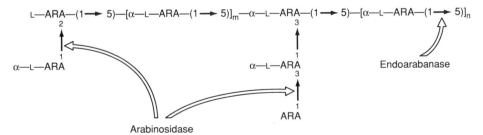

Figure 3.80   Sites of enzymatic attack in arabinan (Pilnik and Voragen, 1984).

TABLE 3.25   Properties of Arabinofuranosidases A and B

Property	Enzyme	
	A	B
Molecular weight (daltons)	128,000	60,000
pI	6–6.5	5.5–6
pH optimum	4.1	3.7
Temperature optimum	50°C	60°C
Cleaves		
p-Nitrophenyl arabinoside	+++	+++
Arabinose disaccharide	++	++
a-1,5-Araban	—	+
Araban from beets	+	+++

The linear α-1,5-arabans, beet araban, potato araban, and the arabinose–pentasaccharides are also hydrolyzed. Disaccharides are less readily attacked; p-nitrophenylarabinoside remains intact.

Exoarabanases, α-L-Arabinofuranosidase (EC 3.2.1.55)   This enzyme is produced by fungi and bacteria. Currently, the enzymes from *A. niger* are of commercial importance. One of these was isolated by Kaji and Tagawa (1970). It resembles one of the enzymes isolated by Rombouts et al. (1987) from the commercial preparation Pectinase 29 [GB] (enzymes A and B, Table 3.25).

ANALYSIS OF ARABANASES

Technical Preparations: Although a pure arabanase is not available at present, different industrial enzyme preparations contain various amounts of arabinase activities such as, for example, Rohapect DA 3L [RM] and Pectinex AR [NO], [GB].

3.1.7.2   Xylans and Xylanases
Reference: Puls et al. (1988)

Xylan is one of the most widely occurring polysaccharides. In terms of abundance, xylan occupies third place after cellulose and starch. It is present in wood and grasses and is, as was described in Section 3.1.4.1, a component of the plant cell wall. Xylans from different plant sources are commercially available, although some of these products contain significant quantities of starch and other carbohydrates.

Xylans have chains of 1,4-linked β-D-xylopyranose residues. Xylans from hardwoods, softwoods, and grasses are substituted differently. In grasses, especially in straw, single arabinofuranosyl residues are linked at the C-3 positions in

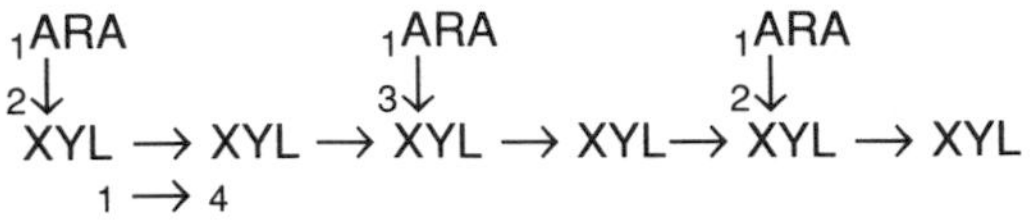

1ARA
2↓
1ARA
3↓
1ARA
2↓
XYL → XYL → XYL → XYL→ XYL → XYL
1 → 4

Figure 3.81 Structure of the branched arabinoxylan (D'Appolonia and Kim, 1976).

the xylan chain. Xylans are also substituted in the C-2 position with D-glucopyranosyl–uronic acid residues or their 4-methyl esters. There are also a few arabinofuranosyl residues bound to the chains of softwood xylan. Hardwood xylan is acetylated.

As shown in Section 3.1.4.1, substituted xylans are components of the plant cell wall. D'Appolonia and Kim (1976) have suggested a structural scheme for the water-soluble pentosans in wheat (Fig. 3.81).

Anhydro-L-arabinofuranose residues are bound at the 2- or 3-position of the β-1,4-xylopyranose chain. The coupling of such arabinoxylans with the protein in the cell wall leads to glycoprotein complexes that are dispersible in water to give highly viscous solutions (Pomeranz, 1985). (See Fig. 3.82.)

The phenolic groups of ferulic acid can be oxidatively cross-linked to yield diferulic acid residues. Gels are formed (Fig. 3.83; Geissmann and Neukom, 1973). In this manner, horseradish peroxidase and H_2O_2 can effect the cross-linking of proteins in wheat doughs, thereby raising dough viscosity. According to Kiefer et al. (1981), this improves the baking properties of flour (see Section 5.4).

OH
OCH3
CH
CH
C=O
Ferulic acid
Protein
ARA
XYL—XYL—XYL—XYL—XYL—XYL—XYL—XYL
ARA
[Galactose/Arabinose]n

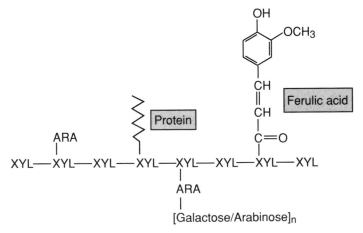

Figure 3.82 Hypothetical structure of wheat flour glucoprotein (after Neukom et al., 1967; Finscher et al., 1974).

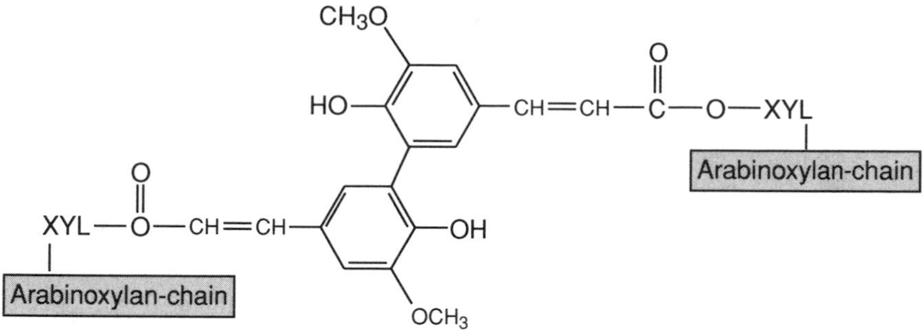

Figure 3.83 Oxidative phenolic cross-linking of arabinoxylan chains (Neukom and Markwalder, 1978).

There are various types of xylanases with different specificities:

I.   *Endo-1,4,-β-xylanase* (EC 3.2.1.8; CAS 9025-57-4); (1,4-β-D-xylan-xylanohydrolase)

II. *Endo-1,3-β-xylanase* (EC 3.2.1.32; CAS 9025-55-2); (1,3-β-D-xylan-xylanohydrolase)

III. *Exo-1,4-β-xylosidase* (EC 3.1.2.37; CAS 9025-53-0); (1,4-β-D-xylan-xylo-hydrolase)

Xylanases are formed by many *Aspergillus, Bacillus,* and *Trichoderma* species. In technical enzyme products, they are accompanied by many different cellulases and hemicellulases. Only a few of these organisms are used for the technical manufacture of such complex preparations.

A complete survey of the chemical and physical characteristics of xylanases from different microorganisms was published by Kornelink (1992). See also Tables 3.26 and 3.27 and Figures 3.84–3.86.

Xylanases from Bacillus sp. Numerous bacteria produce xylanases; these have been studied in connection with the microbial degradation of wood. Technically, these enzymes are currently of little importance. A few characteristics of bacterial xylanases are given in Table 3.28.

In addition, wood-destroying fungi belonging to the *Basidiomycetes* produce xylanases that are manufactured technically in small amounts and used as components in digestive aid preparations. The xylanases from alkalophilic *Bacillus* sp. (Horikoshi and Atsukawa 1974) are active in the pH range 5.5–9 and stable for a duration of 10 min at 60°C. More xylanases of this type are described by Akiba and Horikoshi (1988). A new thermostable xylanase (type I) from *B. stearothermophilus* was produced by Grueninger and Fichter (1986). This enzyme has a broad activation optimum from pH 5.5 to 8.5 and a half-life of 15 h at 75°C and is stable at 68°C for 5 days.

TABLE 3.26   Xylanases from *Trichoderma* sp.

Enzyme No./ Source	Molecular Weight (daltons)	Carbohydrate (%)	pI	pH Optimum	K_m (%)	Reference
I/*T. reesei*	32,000	14	4.1–4.2	4–5	—	Lappalainen (1986)
II[a]	23,000	8	6.4–6.5	4–5	—	Lappalainen (1986)
I/*T. harzianum*	20,000	—	—	—	0.16	Tan et al. (1985)
I/*T. harzianum*	29,000	—	—	—	0.066	Tan et al. (1985)
I/*T. harzianum*[b]	22,000	—	8.5	4.5–5 (50°C)	—	Wong et al. (1986)
III/*T. viride*	102,000	4.5	4.45	4.5 (55°C)	—	Matsuo and Yasui (1984)

[a]This enzyme produces xylobiose and large xylose oligomers from xylan. Laminarin is also hydrolyzed.
[b]Three isoenzymes with molecular weights of 20,000–29,000 daltons were detected. It has been reported that all three enzymes are exclusively xylan-specific. They do not digest xylobiose and do not convert arabinoxylan to arabinose.

A new xylanase produced by a recombinant strain of *B. subtilis* (molecular weight 20,000 daltons) was reported in a patent of Röhm (1994). The enzyme is produced for a baking application, especially to obtain a high volume in bread.

Analysis of Xylanases
Reference: Ghose and Bisaria (1987)

1. Determination of reducing sugars. Xylan from wood or straw is swelled by incubation for approximately 12 h in 0.5 N NaOH and adjusted with buffer to the desired pH. About 2 ml of a solution containing approximately 1 g of xylan in 100 ml of buffer is required for hydrolysis. Following hydrolysis (30°C, 15 min), the reducing sugars produced are analyzed according to the method of Somogyi–Nelson.

TABLE 3.27   Xylanases from *Aspergillus* sp.

Enzyme No./ Source	Molecular Weight (daltons)	Carbohydrate (%)	pI	pH Optimum	K_m	Reference
I/*A. niger*	14,000	—	—	4–4.5	2.4×10^{-2}	—
II/*A. terreus*[a]	11,000	—	—	4–5.5 (40–55°C)	—	Chen et al. (1986)
	14,500	—	—	—	—	
	20,000	—	—	—	—	
III/*A. niger*[b]	122,000	0	4.92	3.8–4 (70°C)	—	Radionovo et al. (1983)

[a]Six enzymes of varying molecular weight but with strict β-1,3 specificity were isolated. Pure β-1,3-xylan from seaweed is hydrolyzed. β-1,4-Xylan, laminarin, cellulose powder, and β-1,3-xylobiose are not digested.
[b]The enzyme is stable between pH 3–8 and shows no loss in activity during heating at 50°C for 1 h. D-Xylose is a competitive inhibitor.

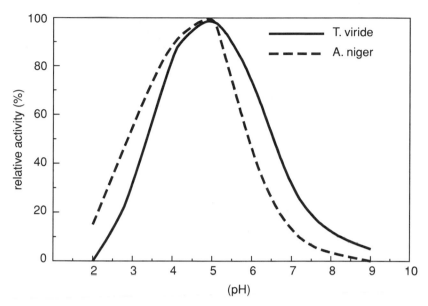

Figure 3.84 Effect of pH on activity of xylanase from *A. niger* [RM] and *Trichoderma viride* [UA].

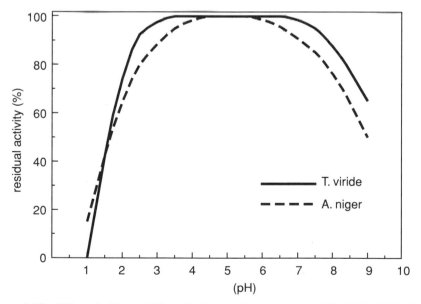

Figure 3.85 Effect of pH on stability of xylanases from *A. niger* and *T. viride* (1% buffered enzyme solution, 4 h, 30°C).

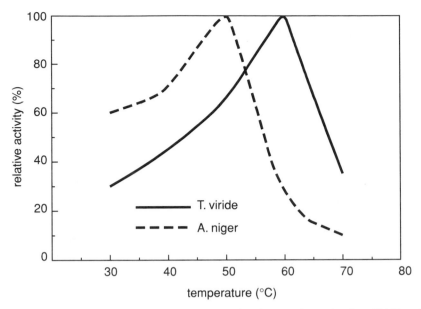

Figure 3.86   Effect of temperature on activity of xylanases from *A. niger* [RM] and *T. viride* [RM].

Enzyme unit: A xylanase unit is determined as the amount of enzyme needed to liberate from xylan the amount of reducing sugars equivalent to 1 μM xylose per minute at 30°C under standard conditions.

2. Xylanase determination with a nephelometric method. Nummi et al. (1985) established a method that allows the determination of endoxylanase as well as exoxylanase using freshly precipitated insoluble birch wood xylan.

3. A valuable analytical method for the xylanase has been commercialized by Megazyme Ltd (2711 Ponderosa Parade, Warriewood (Sydney) NSW 2102, Australia).

Manufacturers: [RM], [SN], [GT], [NO] and others.

TABLE 3.28   Xylanases from *Bacillus* sp.

Enzyme No./Source	Molecular Weight (daltons)	pI	pH Optimum (Temperature)
I/*B. pumilus*	20.000	—	6.5 (40°C)
I/*B. coagulans*	22,000	10	6.0 (37°C)
III/*Bacillus* sp.	190,000	4.9	6.5 (30°C)
I/*B. subtilis*	20,000	—	6–7

REFERENCES

Akiba, R., and Horikoshi, K., *Methods in Enzymology* **160,** 665 (1988).

Barsky, N. E., and Scobiner, U., "An arabinan in pear juice concentrate," *Alimenta* **25**(5) 125 (1986).

Chen, J., Matsuo, M., and Yasui, T., "Xylanase from *Aspergillus* in *Methods in Enzymology,* G. Perlmann and L. Lorand (eds.), Vol. 160 (1986).

Churms, S. C., Merrifield, H. E., Stephen A. M., Walwyn, D. R., Polson, A., van der Merwe, K. J., Spiess, H. S. C., and Costa, N. "An L-Arabinan from apple juice concentrates," *Carb. Res.* **113,** 339 (1983).

D'Appolonia, B. L., and Kim, S. K., "Recent developments in wheat flour pentosans," *Baker's Dig.* **50**(3), 45 (1976).

Ducroo, P., "Erprobung einer Vorherbestimmungsmethode für Arabantrübungen in Apfelsaft," *Flüssiges Obst* **5** 265 (1987).

Geissmann, T., and Neukom, H., "Composition of the wheat flour pentosans and their oxidative gelation." *Lebensm.-Wiss. u. Technol,* (2), 59 (1973).

Ghose, T. K., and Bisaria, V. S., "IUPAC Comm. on Biotech. Measurement of Hemicellulase Activities. Part 1 Xylanases," *Pure Appl. Chem* **59**(12) 1739 (1987).

Grueninger, H., and Fichter, A., *Enzyme Microb. Technol.* **8**(5), 309 (1986).

Horikoshi, K., and Atsukawa, Y., *Agr. Biol. Chem.* **37**(9), 2097 (1973); Jpn. Patent 7,480,287 (1974).

Kaji, A., and Saheki, T., *Biochem. Biophys. Acta* **410,** 354 (1975).

Kaji, A., and Tagawa, K., "Purification, cristallisation, amino acid composition of α-arabinofuranosidase from *Aspergillus niger,*" *Biochem. Biophys. Acta* **207,** 456 (1970).

Kiefer, R., Matheis, G., Hofmann, H. W., and Belitz, H.-D., "Improvement of baking properties of wheat flours by addition of horse radish peroxydase, H_2O_2 and catechol," *Z. Lebensm. Untersuch. Forsch.* **113,** 376 (1981).

Kormelink, F. J. M., *Characterisation and Mode of Action of Xylanases and Some Accessory Enzymes,* doctoral thesis, University of Wageningen, CIP-DATA Koninkl. Bibliotheek, The Hague, 1992, pp. 7–34.

Lappalainen, A., *Biotechn. Appl. Biochem.* **8**(5), 437 (1986).

Matsuo, M., and Yasui, T., *Agric. Biol. Chem.* **48**(7), 1845 (1984).

Neukom, H., and Markwalder, H. U., "Oxidative gelatin of wheat flour pentosans," *Cereal Foods World* **23,** 374 (1978).

Nummi, M., Perrin, J., Niku-Paavola, M.-L., and Enari, T. M., "Measurement of xylanase activity with insoluble xylan substrate," *Biochem. J.* **226**(2), 617 (1985).

Pilnik, W., and Voragen, A. G. J., "Polysaccharides and food," *Gordian* **84**(9), 166 (1984).

Pomeranz, Y., Function Properties of Food Components, Academic Press, New York, 1985, p. 138.

Puls, J., Brockmann, A., Gottschalk, D., and Wiegel, J., *Method in Enzymology* **160,** 528 (1988).

Rodionova, N. A., Tavobilov, I. M., and Bezborodov, A. M., *J. Appl. Biochem.* **5**(4–5), 300 (1983).

Röhm GmbH, Darmstadt, Inventors: Gottschalk, M., Sprössler, B., Schuster, E., "Bacterial-xylanase, process for the production with a suitable bacterial strain, plasmid and its struc-

ture. Use of this enzyme in baking additives for the production of baking goods,"
Deutsche Offenlegungsschrift 1994, DE 4226528 A1 (1994).

Rombouts, F. M., Voragen, A. G. J., Searle van Leuven, M. F., Geraeds, D. S., Scholz, H. A., and Pilnik, W., *Carbo. Polymers* **9**, 25 (1988).

Tan, L. U., Wong, K. Y., and Saddler, J., *Enzyme Microb. Technol.* **7**(9), 431 (1985).

Weinstein, L., and Albersheim, P., "Structure of plant cell walls," *Plant Physiol.* **63**, 425 (1978).

Wong, K. Y., Tan, L. U., Saddler, J., and Yaguchi, M., *Can J. Microbiol.* **32**(7), 570 (1986).

3.2 PROTEASES

3.2.1 Brief Overview

The proteins from plants and animals are important for human and animal nutrition as well as for clothing (leather, wool, and silk). They are used as organic fertilizer in the cultivation of plants and are a necessary component of culture media for industrial fermentations. Some proteins serve as raw materials for industrial products as a paint base or for the manufacture of synthetics.

In nutrition, proteins are usually "naturally modified," that is, processed using physical methods such as heating or homogenization. The changes induced in the native proteins serve to improve digestibility or storage stability.

It is also desirable in different branches of food technology to change the physical and chemical properties of proteins. This can be done by the physical methods mentioned above, as well as by chemical and biological fermentation methods, such as in the production of milk and cheese. Many of these traditional processes are based on the activity of naturally occurring enzymes, including those produced by the organisms in the original product. The enzymatic activities in raw foods, such as in fruit, vegetable, or cereal products, depend on the variety and the climate and are thereby subject to natural variation. To achieve targeted and specific changes in proteins, commercial protease preparations with different specificities are used. Such preparations are currently employed to produce many foods in which enzymes can replace potentially carcinogenic or otherwise harmful chemicals. Table 3.29 shows some processes used in the manufacture of foods in which enzymes have specific applications.

TABLE 3.29   Application of Proteases in Food and Industrial Technology

Product or Process	Use
Beer	To solubilize grain proteins; to stabilize beer
Cheese	To coagulate milk proteins and to ripen cheese
Meat tenderizing	To partially separate connective tissues
Bread	To increase gluten elasticity
Cookies and crackers	To improve crispness
Leather	To remove wool, hair, and pigments; to soften skins
Laundry detergents	To remove protein stains

In addition, there are numerous processes used in Asia for the modification of fish or soy protein with living microorganisms or with microbially produced proteases. For example, there is the production of soy sauce using *Aspergillus sojae* or *Aspergillus oryzae,* and the production of tofu or tempeh with *Mucor* or *Rhizopus,* which form proteases during their growth.

Protease preparations are, on a commercial basis, the most important of the currently produced enzymes. They are obtained from plants and animal organs and microorganisms, with the majority obtained from bacteria.

3.2.2 Plant Proteases

References: Caygill (1979), Schwimmer (1981), Hwang and Ivy (1951)

Plant proteases occur in numerous, mainly tropical, plants. Proteases are found in papaya *(Carica papaya),* pineapple *(Anana sativa),* figs *(Ficus carica, Ficus glabrata),* artichokes *(Cynera cardunculus),* and soybeans *(Soya hispidus).* All of these proteins belong to the same group of enzymes. The juice of some agave species is also strongly proteolytic.

Plant proteases are characterized by sulfhydryl groups in the active site. These groups are responsible for the catalytic activity.

Proteolytic activity can be impressively demonstrated by the action of fresh pineapple on meat. During the preparation of Indonesian dishes of meat with fresh pineapple, pieces of chicken are amazingly liquefied after warming for only 1–2 h. The main source of industrial enzymes is green papaya, from which papain is obtained. Bromelain is produced by alcohol precipitation from extracts of pineapple or pineapple stalk. The ficin preparations, in the form of ficus latex, are of limited commercial importance. The plant proteases are used as meat tenderizers and in other areas of food production. Examples including brewing, cookie baking, and the production of protein hydrolysates. Other applications are in tanning and pharmaceuticals (digestive-aid preparations).

3.2.2.1 *Papaya proteases*

A Brief Note on Papain Manufacture Papain is obtained from papaya tree plantations in India, Sri Lanka, Zaire, and Uganda. The still-green young fruit is scratched with a sharp knife along the surface. A colorless latex immediately appears and turns milky white after a few minutes before coagulating. The dripping latex is caught in pieces of cloth or tubs and is air-dried. One kilogram of this latex yields approximately 200 g of crude papain. On average, 0.45 kg of latex per tree is obtained (Schwimmer 1981). Newer production methods utilize purification of latex at the point of origin. This is accomplished by adding sodium hydrogen sulfite to raw juice to prevent oxidation, followed by centrifugation or filtration. Hemicellulose addition can facilitate filtration. After concentrating the juice in vacuo and then spray drying, a powderlike product is obtained. Production of crude papain amounts to about 500 tons per year, about $15 million in value.

Crude papain contains, in addition to papain (10% of the protein in the latex), a number of other enzymes—namely, 45% chymopapain A and B (Kunimastu and Yasunobbu, 1967), which are similar to papain but differ in their specificity. Besides endo-β-1,4-glucanase, a chitin-splitting hydrolase and a carboxypeptidase, a very active lysozyme ($\sim$10%) (Dahlquist et al., 1969) is also present. The milk-curdling activity of crude papain seems to be a characteristic of the protease. This activity is not destroyed by heating to 70°C for 90 min. Hydrogen peroxide inactivates this activity; cystein reactivates it. Lipase is also found in papain. The activity optimum with olive oil is 40°C at pH 5.8–6.2. Amylase activity can be measured in fresh papaya latex.

The plant lysozyme has an activation optimum between pH 3.2 and 4.2, while the optimum for egg lysozyme lies at pH 5.3.

Papain (EC 3.4.22.2; CAS 9001-73-4)

Molecular Weight: 21,000 daltons.

Structure: A chain of 159 amino acids including three cystein bridges. The presence of cystein at the active site is required for activity. The SH group of cystein easily reacts with oxygen or heavy metals. Consequently, to obtain full activity, the presence of reducing agents or heavy-metal-complexing agents is necessary.

Isoelectric Point: pH 8.5–9 (different values are found; these are presumably a result of autolytic processes).

Activity: Ninety percent of maximum enzymatic activity is obtained in a pH range of 6.5–7.8 with hemoglobin denatured by urea. With native hemoglobin, activity is pH 3–4. If the substrate is casein, maximum activity lies between pH 7.5 and 8; and with gelatin, between pH 4.8 and 5.2. Optimum activity with ovalbumin is pH 7–7.5; with serum albumin, pH 5.7–5.9; with peptone, pH 5.0. At pH 6–7, the maximum temperature tolerated is 60–70°C (casein). The proteolytic activity of papain is given in Table 3.30.

TABLE 3.30   Determination of Proteolytic Activity of Papain with Various Methods

Method	Crude Papain	Purified Papain
BAPA[a] U/mg	—	420–470
BAPA U/mg	2000–3000	5000–6000
Anson–hemoglobin units	1–2	5–6
Tyrosine units	100–200	800–900
Hemoglobin units (HbmU/mg)	2000–3500	≤11,000
USP/mg	10,000	56,000

[a]Na-benzoyl-L-arginine-p-nitranilide.
[b]*U.S. Pharmacopoeia* units.

Inhibitors: Oxidizing substances, H_2O_2, O_2, and heavy metals; for example, Hg^{2+}, Fe^{3+}, and Cu^{2+}.

Reactivation: Cystein, $NaHSO_3$, glutathione, EDTA.

Specificity: Papain is a relatively nonspecific protease. Proteins, peptides, amides, and esters are cleaved by the enzyme. Benzoyl-L-arginine-*p*-nitroanilide (BAPA) is a good substrate that is used for analytic purposes.

Chymopapain (EC 3.4.27.6; CAS 9001-09-6) Chymopapain has characteristics similar to those of papain but differs in its activity toward proteins. It is less heat-sensitive and is more stable than papain at pH 2.

Molecular Weight: 36,000 daltons (three isozymes are known).

Isoelectric Point: pH 10.1.

Specificity: Very similar to papain. On the basis of equal BAPA activity, the milk-curdling activity of chymopapain is twice as high as that of papain.

Technical Papain Preparations Products in powder form are not desirable today because of the danger of allergic reactions. The production of liquid preparations that contain sorbitol or other sugars and glycerol as stabilizers is described in numerous publications. In order to stabilize enzymatic activity, reducing agents such as $NaHSO_3$, cystein, or thiosulfate are employed. The effect of pH on the enzymatic activity of papain is shown in Figure 3.87.

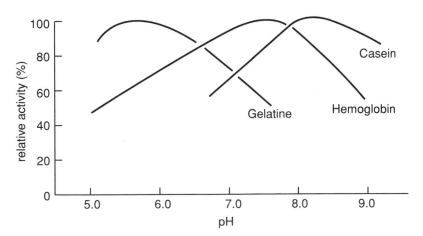

Figure 3.87 Effect of pH on the activity of papain with various substrates (Reisner, technical information [RM]).

Activity: 500–1000 BAPA units.

Analysis: Papain is standardized with casein or hemoglobin or more commonly, as in brewing, for example, with BAPA. In addition, a very sensitive immunologic determination that uses papain antibodies is available.

Stability: Liquid products are more stable than powder. At a storage temperature of 4–10°C, activity is reduced by approximately 10% per year. Binary combinations of papain with microbial proteases are not compatible in various proportions. When papain is combined with fungal proteases in aqueous solution, an inactivation of fungal protease of ≤40% was observed in 1.5 h (Bavisotto et al., 1958).

Products and Manufacturers: Colorase L 10 [RM], Auxillase [MK], Collupin [GB], and others.

3.2.2.2   Other Plant Proteases

Bromelain (EC 3.4.22.4; CAS 9001-00-1)   There are two enzymes in pineapples that show very similar specificities. Most industrial preparations are obtained from the pulverized stems of the whole plant (stem–bromelain). The fruits also have a high protease activity (juice–bromelain, EC 3.4.22.5). A commercial enzyme concentrate can contain up to 0.1% of this enzyme. Technical preparations are used for the production of digestive aids, for healing preparations, in diet foods, and for the stabilization of beer.

Molecular Weight: 33,000 daltons.

Structure: Glycoprotein, SH protease; five isoenzymes have been found.

Activity: See Table 3.31.

Optimal pH: The optimal pH for hemoglobin and casein is 6–8 with gelatin; with papain, pH 5. The optimal temperature is 50–55°C for casein, and 60°C for gelatin.

Products and Manufacturers: [TMKB]: Preparations with 1200 and 1400 CDU/mg; [MK], [GB].

Ficin (EC 3.4.22.3; CAS 9001-33-6)
Reference: Whitaker (1957)

Figs have a very high protease activity. A green fig weighing 10–15 g contains about 100–150 mg ficin, based on a commercial preparation (Yamamoto, 1975). On

TABLE 3.31   Comparison of Relative Activity of Various Sulfhydryl Proteases (Ota et al., 1964)

Substrate	Stem Bromelain	Fruit Bromelain	Papain	α-Chymotrypsin
BAA[a]	0.14	6.3	—	—
BAPA[b]	0.9	14.7	—	—
Casein	9.2	23.7	24.2	23.8

[a]Benzoyl-L-arginine amide.
[b]Defined in Table 3.30.

drying, most of the activity is lost. To the best of the author's knowledge, ficin is currently produced in only a few South American countries. At Röhm, calabashes were received from Peru that contained a very active ficin that stabilized with a high concentration of SO_2. The hides in the beamhouse of the tannery were strongly attacked. Glycoproteins were hydrolyzed and the collagen also degraded.

Applications are similar as those for papain and bromelain. These include the hydrolysis of fish waste and for plant proteins.

Molecular Weight: 26,000 daltons.

Structure: The active site is similar to that in papain.

Activity: The pH optimum against casein is pH 6.7. Stability is high in the pH range 3.5–9 (Whitaker, 1957).

Inactivation: At 80°C.

Inhibitors: Oxidants, heavy metals, 10^{-4} M sorbic acid.

Manufacturers: [MK], [GB].

REFERENCES

Bavisotto, V., Miller, C., and Dewane, R., Laboratory Communications 9, Paul-Lewis-Lab. Inc., Milwaukee, WI, (1958).

Caygiil, J. C., *Enz. Microbial. Technol.* **1**, 233 (1979).

Dahlquist, F. W., Broders, C. L., Jakobson, G., and Raftery, M., *Biochemistry* **8**(12), 694 (1969).

Hwang, K., and Ivy, A. C., in *Miner, R. W.* Papain (ed.), Annals New York Acad. Sciences, Vol 54, p. 143, (1951).

Kunimatsu, D. K., and Yasunobbu, K. T., *Biochim. Biophys. Acta* **139,** 405 (1967).

Ota, S., Moore, S., and Stein, W. H., *Biochemistry* **3,**180 (1964).

Schwimmer, S., *Source Book of Food Enzymology,* Avi Publ. Westport, CT, (1981).

Takahashi, N., Yasuda, Y., Goto, K., Mitake, T., and Murachi, T., J., *Biochemistry* **75,** 355 (1973).

Whitaaker, J. R., *Food Res.* **22,** 483 (1957).

Yamamoto, A., in G. Reed (ed.), *Enzymes in Food Processing* 2nd ed. Academic Press, New York, 1975, p. 144.

3.2.3 Animal Proteases

References: Ammon and Dirscherl (1959), Desnuelle et al. (1969)

The protein-hydrolyzing enzymes of the mammalian digestive tract have been known for a long time and have been researched intensively since the beginning of enzymology. Even before the chemical nature of proteins and specifically enzymes were determined, the protein-cleaving action of pepsin and pancreas proteins was known.

According to the current classification, which relates to the structure of the active site of proteases, the animal proteases belong to various classes. These en-

TABLE 3.32   Animal Proteases

Protease	Proenzyme	pH Range	Activator
Pepsin	Pepsinogen	1.5–4.0	H^+, proteases
Chymosin (gastric)	Prochymosin	5–6	Ca^{2+}
Trypsin	Trypsinogen	6.5–9	Enderokinase, Ca^{2+}
Chymotrypsin	Chymotrypsinogen	7–8.5	Ca^{2+}
Carboxypeptidases	Procarboxypeptidase	7.4 (6–8.5)	Trypsin
Aminopeptidases	—	7.5–9.0 (6.5–8)	—

zymes are considered together here because of historical developments. Typical for these enzymes is the fact that they occur as inactive precursors, the proenzymes, in the organs in which they are produced. The precursor is not activated to the active enzyme until the enzyme is secreted into the digestive tract. In addition to the endoenzymes, proteases that cleave the proteins inside the peptide chain, a series of exoenzymes that are important in certain industrial applications also exist.

Uses for these enzymes are found in food production, in the manufacture of protein hydrolysates, and for processing meat and fish residues. Today large amounts of pancreatic proteases are needed for the leather manufacturing industry. In medicine, these enzymes are a part of digestive medical aid preparations. The animal proteases and peptidases that are still used in the industry are summarized in Table 3.32.

The caboxypeptidases and aminopeptidases include a number of very specific single enzymes, which are active at different pH ranges. Some proteases and peptidases are called *cathepsins.* Today, this terminology is used only for the protein-cleaving enzymes that occur in tissue (liver, kidneys, and muscle).

3.2.3.1   Endoproteases

Pepsin (EC 3.4.23.1-3; CAS 9001-75-6)
Reference: Fruton (1971)

Pepsin, which was named by its discoverer, Ch. Schwann in 1836, is secreted from the gastric mucous membrane as the nonactive proenzyme, pepsinogen. This is converted into active protease by the action of stomach acid. This autocatalytic action occurs below pH 6.0, with the maximum rate of conversion at pH 2.0. Activity can be induced by unrelated pepsins or microbial acidic proteases. At alkaline pH levels, the pepsinogen is stable, whereas pepsin is rapidly inactivated. In a procine stomach, different proteins with similar action but different molecular weights are found. Pepsins of various animal species are immunologically different. In addition, an enzyme that resembles pepsin is found in the stomach mucous membrane. This is gastricin, which has a pH optimum at pH 3.2.

PROPERTIES OF PEPSIN

Molecular Weight: 35,000 daltons; pepsinogen, 41,000 daltons (hog).
Isoelectric Point: pH ~1.0. Values of pH 3 and 1.7 are also mentioned in the lit-

erature. A pI at pH 1.7 was determined following removal of a covalently bound phosphoryl group.

Structure: A single-chain polypeptide with 321 amino acids. The amino acid sequence and the tertiary structure are well known (Fruton, 1971).

Activity: Many proteins, including keratin, are cleaved at pH 1.0–2.5. Mucoids such as those of the stomach mucous membrane or ovomucoid are not attacked and are strong inhibitors. Silk fibroin and protamine are not hydrolyzed. The activity optimum of short peptides and synthetic peptide derivatives lies between pH 2 and 4.

Stability: Pepsin is relatively stable in aqueous solution from pH 5–5.5. It is more stable in glycerol, in which it dissolves. It is irreversibly inactivated at alkaline pH.

Specificity: Peptide linkages with aromatic amino acids as well as glutamic acid, cysteine, and cystine peptides are preferentially cleaved. Polyglutaminic acid is a good substrate, whereas oligomers such as Glu_2 to Glu_4 are not attacked. Relatively specific synthetic substrates include Ac–Phe–Tyr, with optimum at pH 2.0. Z–His–Phe–Phe, optimum at pH 3.0.

Technical Preparations: Pepsin 1:10,000 is equivalent to NF 11, or ~10,000 mHbU/mg (pH 3).

Manufacturers: Anex R. Werner Co., Germany; Orthana, Denmark.

Applications: In digestive-aid medicine, for the preparation of protein hydrolyzates, in mixtures with chymosin in rennet preparations, and occasionally for stabilization of beer. (See also Figure 3.88.)

Chymosin (EC 3.4.23.4; CAS 9001-98-3), Chymase, Renin, Rennet
References: Cheeseman (1981) (further references are listed in Section 5.6)

PROPERTIES AND MANUFACTURERS   Chymosin, like pepsin, is an acidic protease that occurs in the stomach and is required for the manufacture of cheese. Chymosin is produced in stomach mucous membranes as its inactive precursor, prorennin (molecular weight 36,000 daltons). Activation occurs at hydrogen ion concentration below pH 5.

Extraction: The enzyme is obtained by extraction from stomachs of young calves, goats, or lambs. The extraction from older animals contains significant quantities of pepsin.

Molecular Weight: 31,000 daltons.

Activity: One part of a highly purified chymosin can coagulate 72 million parts of milk in 10 min (Hankinson and Palmer, 1942). Optimum proteolytic activity is at pH 3.8. The optimum of the preparation from rennet is above pH 6.5, at an optimum temperature of about 40°C. Comparison of X-ray structural studies of homologous enzymes have shown that chymosin belongs to the

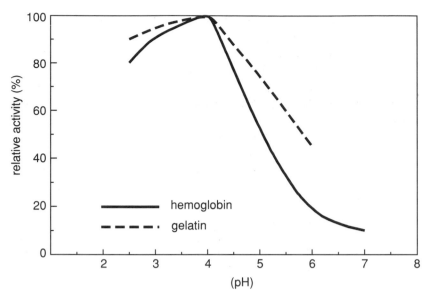

Figure 3.88   Effect of pH on the activity of pepsin with hemoglobin and gelatin substrates (25°C; 10 min [RM]).

carboxyl proteases. Fractionation of chymosin has revealed two to three isozymes with similar properties.

Specificity: The peptide linkage between phenylalanine and methionine is preferentially cleaved in soluble calcium caseinate in milk (see Section 5.6). This is the first stage of hydrolysis; in the second stage, calcium-induced coagulation occurs. Apart from the peptide linkage between phenylalanine and methionine, 22 other linkages in α-s_1-casein are cleaved, although the rate of hydrolysis is much slower.

Analysis: The rennet strength or the Soxhlet unit (Soxhlet, 1877) per milliliter is defined as the number of milliliters of milk that can be coagulated in 40 min at 35°C. The standard rennet in Great Britain has 15,000 rennet units per milliliter and contains ~75–80% pure rennet. The rennet preparations are mostly liquid and contain salt, glycol, and benzoate as stabilizers. Because milk is a difficult substrate to standardize, Berridge (1955) suggested that milk powder as substrate in 0.01 M calcium chloride solution should be used. One rennet unit is defined as follows:

$$\text{Rennet activity} = \frac{10 \times \text{volume of milk}}{\text{coagulation time} \times \text{rennet volume}}$$

Manufacturers: [CH], [GB], [SOE] (microbial rennett, see Section 3.2.4.5).

Trypsin (EC 3.4.21.4; CAS 9002-07-7)
Reference: Keil (1971)

Trypsin is a serine protease. The precursor of this enzyme is produced in the pancreas. This precursor, trypsinogen, is transformed into active trypsin by a protease in the intestinal mucous membrane known as *enterokinase*. This transformation of trypsinogen, with a molecular weight of 24,000 daltons, occurs on the hydrolysis of a single peptide linkage between lysine and isoleucine with the release of a hexapeptide (Val–Asp$_4$–Lys). Activation by trypsin also occurs autocatalytically. The reaction is strongly accelerated by calcium (Radhakrishnan et al., 1969).

PROPERTIES AND APPLICATION

Molecular Weight: 24,000 daltons.

Structure: Trypsin consists of a polypeptide chain with 233 amino acids.

Activity: The activation optimum for most proteins and synthetic substrates lies between pH 7 and 9. In terms of cleavage reactions, trypsin is very specific. Only peptide linkages in which the carboxyl group of lysine or arginine is linked with the amino group of other amino acids are attacked. This strict specificity was very helpful in determining the amino acid sequences of several proteins. In the case of the insulin B-chain, only two peptides are formed. These can be easily separated and further sequenced.

Stability: Optimal stability of trypsin is at pH 3. Incubation of trypsin preparations at pH 1.5 for several hours destroys other proteolytic activities; only about 20% of the trypsin activity is lost. The stability of aqueous solutions in alkali is low because of self-digestion.

Inactivation and Inhibitors: Alkylating reagents, such as DFP (diisopropyl fluorophosphonate) or PMSF (phenylmethylsulfonyl fluoride), react stoichiometrically with a seryl residue at the active site of trypsin. This leads to rapid inactivation. Natural inhibitors are found in soybeans (two soybean inhibitors), egg white (ovomucoid), potatoes, and cereals. The soybean inhibitor is destroyed during toasting. Kunitz (1946) crystallized a soybean inhibitor with a molecular weight of 20,100 daltons.

Application: Used medically in pure or crystalline form in wound treatment.

Chymotrypsin (EC 3.4.21.1; CAS 9004-07-3)
Reference: Blow (1971)

Several forms of chymotrypsin are known. Chymotrypsin is a serine protease and, like trypsin, is found in the pancreas as an inactive precursor. Trypsin converts chymotrypsin into the first active form: the unstable α-chymotrypsin. Further proteolytic processes subsequently lead to the α-, β-, and γ-chymotrypsins (Northrop, 1940). (See Fig. 3.89.)

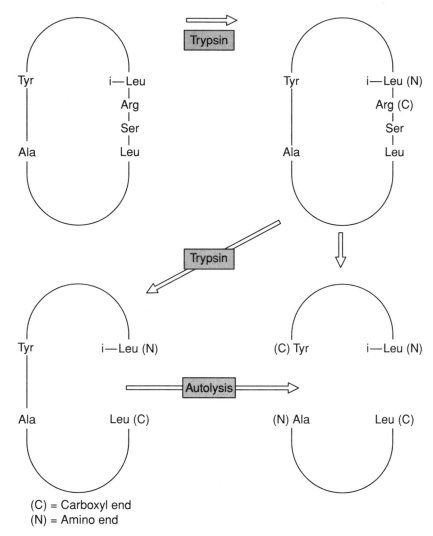

Figure 3.89   Conversion of α-chymotrypsinogen into α-chymotrypsin (after Rovary et al., 1957).

PROPERTIES

Molecular Weight: α-Chymotrypsin, 25,000 daltons.

Activity: Activation optimum is at pH 8–9, with, at pH 7, ~50% maximum activity at pH 7. Maximum temperature is 40–45°C; at high substrate concentrations, up to 55°C.

Stability: Chymotrypsin is less stable in acid than trypsin. Aqueous solutions are stable in the pH range 2.5–6 at 10°C for 20 h; at 40°C, for 1 h.

Inhibitors: Heavy metals, natural trypsin inhibitors, polyphosphates, strong oxidation agents, DFP and PMSF.

Application: Pure forms are used in medicine. In some countries, α-chymotrypsin, because of its antiinflammatory action, is prescribed to be taken orally against infections, broken bones, and so on. Technical applications include mixtures with other pancreas enzymes to be used, for example, in the leather industry.

Pancreatic elastase (EC 3.4.21.11; CAS 9004-06-2), Pancreatopeptidase E
Reference: Hartley and Shotton (1971)

Elastin is a component of animal connective tissue. It differs from other proteins in that it has a very high content of amino acids with nonpolar side chains, such as Leu, Val, and Ala. The cervical ligament of steers (ligamentum nuchae) has a high elastin content, from which preparations can be extracted, which may be of value in enzyme research.

The elastase, a serine protease, is formed in pancreas as an inactive proelastase. Pure elastase (Smilli et al., 1966) consists of a polypeptide chain with 240 amino acids and resembles the active center of a trypsin or chymotrypsin (Shotton and Hartley, 1970).

PROPERTIES

Activity: Optimum pH 6–8.

Specificity: Peptide linkages between aliphatic amino acids, such as Ala–Leu, Leu–Val, Val–Glu, and Val–Cys (SO$_3$H), are preferentially cleaved.

Synthetic Substrates: N-Benzoyl-L-Ala–methylester and N-acetyl-L-Ala–L-Ala–L-Ala–methylester.

Stability: Elastase is relatively stable below pH 6 and is degraded by autolysis above pH 8.

3.2.3.2 Peptidases Peptidase activity is found in technical pancreatic enzyme preparations, accompanied by proteases. The carboxypeptidases cleave amino acids from the carboxyl end and the amino peptidases amino acids from the amino end of the peptide chain. The reduction of the chain length of peptides by the action of endoproteases is of interest commercially. In other applications, protein hydrolysis is used to produce the maximum number of amino acids, or di- or tripeptides. A high degree of hydrolysis (DH) can be achieved only by the combined action of numerous proteases and peptidases. The enzymatic release of amino acids plays an important role in fermentation processes such as in cheese ripening. In this case, the peptidases are produced by *Lactobacillus* or fungi. These peptidases are described in Section 3.2.4.6.

Highly purified peptidases are used in research for peptide sequencing. Because of their importance in providing side activities in technical preparations, some information on the most important peptidases is given here.

Carboxypeptidase (EC 3.4.2.1; CAS 11075-17-5)
References: Quiocho and Lipscomb (1971)

Numerous carboxypeptidases occur in pancreatic glands. Carboxypeptidase A is formed by the action of trypsin on the inactive precursor of the procarboxypeptidase A. The enzyme was first crystallized by Anson (1937) and contained several other proteases.

PROPERTIES

Molecular Weight: 34,000 daltons.

Structure: The single chain with 307 amino acids contains one tightly bound zinc atom per mole of enzyme (Vallee and Neurath, 1959). Following elucidation of the primary structure, the tertiary structure was determined by Lipscomb (1970).

Activity: Optimum pH 7–8.

Specificity: Amino acids other than Pro, Arg, Hyp, and Lys are cleaved from the C-terminal of peptides. In the case of the amino acids mentioned, sequential cleavage is interrupted. Peptide linkages that include proline (Pro) or hydroxyproline (Hyp) are cleaved by prolidase and prolinase.

Aminopeptidases (EC 3.4.1.2; CAS 9054-63-1)
Reference: Delange and Smith (1971)

Aminopeptidases can be obtained from animal organs, especially pig kidneys. These animal aminopeptidases are of little commercial importance, although those produced microbiologically could become important for food production (see Section 5.8.4).

3.2.3.3 Technical Pancreatic Protease Preparations   The commercial product "pancreatin" contains all the enzymatic activities of the pancreatic glands. It is produced by grinding, defatting, and drying the glands of calves and pigs. The insoluble pancreatin, where the enzymes are still bound to tissue fibers, is standardized against different activities, such as protease or lipase. For pharmaceutical applications, standardization is done according to the analytic guidelines of the Federation International Pharmaceutique (FIP) or the *U.S. Pharmacopoeia* (USP).

Other products were prepared from the glands of pigs, or also from calves and

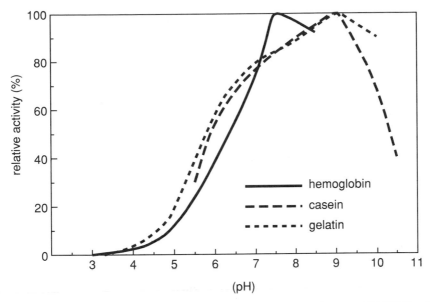

Figure 3.90   Effect of pH on the activity of pancreatic protease (Corolase PP [RM]) with various substrates. Activities determined with a hemoglobin substrate by a modified Anson method; with gelatin by measuring the drop in viscosity; and with casein by the PU method.

sheep by pulverizing, squeezing, and subsequent salt precipitation of the expressed fluids. In these soluble preparations, the proteins were fully activated, whereas a part of the lipase activity was destroyed.

A further means of production involves the by-products of insulin manufacture. Here the pulp of ground glands is extracted at acidic pH, yielding dried preparations that are relatively rich in trypsin, whereas the chymotrypsin is partly denatured. Such preparations represent good starting material for the manufacture of pure trypsin.

See also Figures 3.90–3.92.

Application: A large amount of pancreatin is used in the pharmaceutical industry in digestive-aid preparations. The accompanying lipase activity is particularly important here. Soluble products are preferred in leather manufacturing for the tanning of hides. They are also used in the manufacture of protein hydrolysates and for the liquefaction of fish protein and meat residues.

Products and Manufacturers: Soluble preparations; Corolase PP [RM]; Novotrypsin [NO]; Pancreatin: [KC].

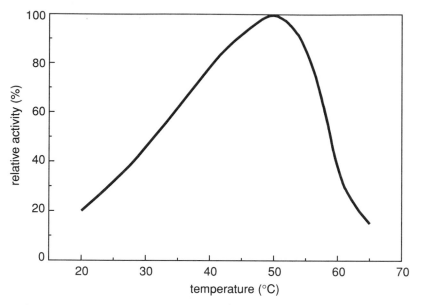

Figure 3.91   Effect of temperature on the activity of pancreatic protease [RM] (pH 8.0; reaction time 10 min).

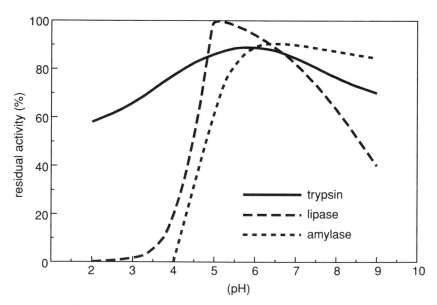

Figure 3.92   Effect of pH on the enzyme stability in a pancreatin preparation [NO] (reaction time 4 h; 25°C).

REFERENCES

Ammon, R., and Dirscherl, W., in *Fermente, Hormone, Vitamine,* 3rd edition, Vol. 1, Georg Thieme, Stuttgart, 1959, p. 246.

Anson, M. L., *J. Gen. Physiol.* **20,** 663 (1937).

Blow, D. M., in Boyer, P. D. (ed.), *The Enzymes,* 3rd edition, Vol. 3, Academic Press, New York, 1971.

Cheeseman, G. C., "Rennet and cheesemaking," in Birch, G. G., Blakebrough, N., and Parker, K. J. (eds.); *Enzymes and Food Processing,* Appl. Science Publ., New York, 1981, p. 195.

Delange, R. J., and Smith, E. L., in Boyer, P. D. (ed.), *The Enzymes,* 3rd edition, Vol. 3, Academic Press, New York, 1971, p. 81.

Desnuelle, P., Neurath, H., and Ottesen, M., "Structure, function, relationships of proteolytic enzymes," *Proc. Int. Symp. IUB,* Munksgaard Verlag, Kopenhagen, 1969.

Fruton, J. S., in Boyer, P. D. (ed.), *The Enzymes,* 3rd edition, Vol. 3, Academic Press, New York, 1971, p. 119.

Hankinson, J., and Palmer, S., *J. Dairy Sci.* **25,** 277 (1942).

Hartley, B. S., and Shotton, D. M., in Boyer, P. D. (ed.), *The Enzymes,* 3rd edition, Vol. 3, Academic Press, New York, 1971, p. 323.

Keil, B., in Boyer, P. D. (ed.), *The Enzymes,* 3rd edition, Vol. 3, Academic Press, New York, 1971, p. 250.

Kunitz, M., *J. Gen. Physiol.* **29,** 149 (1946).

Lipscomb, W. M., *Accounts Chem. Res.* **3,** 81 (1970).

Northtrop, H. J., in *Nord-Weidenhagen* vol. **1,** 667 (1940).

Quiocho, F. A., and Lipscomb, W. N., *Advances Protein Chem.* **25,** 1 (1971).

Radhakrishnan, T. M., Walsh, K. H., and Neurath, H., *Biochemistry* **8,** 4020 (1969).

Rovary, R., Poilroux, D., and Desnuelle, P., *Biochim. Biophys. Acta* **23,** 608 (1957).

Shotton, D. M., and Hartley, B. S., *Nature* **225,** 802 (1970).

Smillie, L. B., and Hartley, B. S., *Biochem. J.* **101,** 232 (1966).

Vallee, B. L., and Neurath, H., *JACS* **76,** 5006 (1959).

3.2.4 Microbial Proteases: Brief Overview

Of all the proteases, those produced microbiologically have the largest commercial importance. Proteases with a value of about $254.6 million were produced in 1987; 75% of these proteases were used in the manufacture of detergents, and about 10% were used in cheesemaking. Other applications are in tanneries and numerous food processes. Proteases are formed by a large number of different microorganisms, although only a few are useful for the production of commercial preparations. Only those organisms that produce enzymes in very high yield and release the proteases into the culture medium are suitable.

As was mentioned in the introduction, proteases are classified according to their structure or the properties of the active site. The most important microbial proteases belong to the serine-, metallo-, and carboxyl proteases. In industrial products, the classification relates to the pH optimum of the respective enzyme. There are acidic, neutral, and alkaline proteases, with some overlapping in definition. The serine proteases generally have optimum activity at alkaline pH levels, although "neutral" serine proteases have also been described.

In addition, there are classifications in the literature derived from various applications. There are references to "gelatinases," "keratinases," and "caseinases." For example, the protease from *Streptomyces fradiae* (Novel et al., 1963) was named *keratinase*. It is indeed very active against keratin, as are other proteases, but it also cleaves casein, gelatin, and hemoglobin. This case clearly demonstrates the difference between analytic determination and practical application. It was originally hoped that this enzyme could be used for dewooling, but it was later shown that other bacterial proteases, when used in combination with reducing chemicals, are much better suited for this purpose.

The collagenase from *Clostridium histolyticum* owes its name to the fact that it can only cleave collagen. The collagenase-specific substrate (Z–Gly–Pro–Gly–Gly–Pro–Ala) is also cleaved by so-called unspecific collagenases (Nordwich and Jahn, 1968). Compared to trypsin, the microbial enzymes have, in general, a significantly broader activity; specifically, they cleave proteins much less specifically (Fig. 3.93 and Table 3.33).

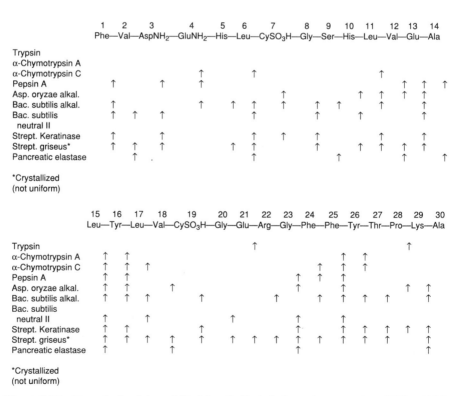

Figure 3.93 Proteolysis of the oxidized insulin B-chain by various proteases (Uhlig, 1970).

TABLE 3.33 Properties of Bacterial Proteases Compared to Trypsin

Enzyme	Molecular Weight (daltons)	Inhibitor (Activator)	Effective pH Range (80%)	Activity/ mg Protein
Subtilisin Carlsberg	26,000	DFP[a]	8.5–11	2,200
B. subtilis, alkaline protease	25,000	DFP	8.5–11	2,300
Neutral bacterial protease (Tsuru I)	44,700	EDTA[b] (Zn, Ca)	6.5–7.5	13,600
Neutral bacterial protease (Tsuru II)	34,000	EDTA	7.5–8.0	12,500
A. oryzae, alkaline protease	35,000	DFP	7.5–9.5	3,500
Trypsin	24,000	DFP	7.0–9.0	1,700

[a]DFP, isofluorophate.
[b]Ethylenediaminetetraacetic acid.

The specific activities of the crystallized enzymes were compared using the Kunitz test. Casein was hydrolyzed at the pH optimum for 10 min at 30°C. Measurements were made at 280 nm. The technical proteases are produced by bacteria and fungi. The different types of enzymes, the producing microorganisms, and examples of the respective commercial products are shown in Table 3.34.

TABLE 3.34 Technical Enzymes, Microbial Sources, and Manufacturers

Enzyme Source	Commercial Product/[Producer]
Acid carboxyl proteases	
Rhizopus sp.	Sumizyme RP [SN]; Newlase [AM]
A. oryzae	Corolase PS [RM]; Veron PS [RM]
	Sumizyme LP [SN]; Sanzyme [SA]
A. niger	Sumizyme FP [SN]; Proctase [MJ]
A. saitoi	Seishin Pharm. Co., Noda, Japan [SP]
Mucor miehei	Rennilase [NO]; Fromase [GB]; Marzyme [SOE]; Morcurd Plus [NO]
Mucor pusillus	Meito Rennet [MSJ]; Noury Lab [MSJ]
Metalloproteases	
B. amyloliquefaciens (B. subtilis)	Neutrase [NO]; Corolase N [RM]; Veron P [RM]; Bioprase [NG]; Rhozyme [GR]; Orientase [UA]
B. thermoproteolyticus	Thermoase [DA]
Serine proteases	
A. sojae	Corolase PN [RM]
A. melleus	Sumizyme MP [SN], [AM]
B. licheniformis	Alkalase [NO]; Optimase [SO]; Maxatase P [GB]
Bacillus sp.	Esperase [NO]; Savinase [NO]
B. griseus	Pronase [KK]

Table 3.34 is not comprehensive, in that other enzyme preparations are commercially available that contain more than one enzyme. In addition, all the source organisms for the commercial products listed have not been cited.

3.2.4.1   Serine Proteases from Bacteria
References: Aunstrup (1980), Ward (1983).

The extraction of a serine protease from *B. subtilis* and its purification and crystallization are described by Güntelberg and Ottesen (1952). The organism was later classified as *B. lichenformis.* The enzyme became known under the name "Subtilisin Carlsberg" and has been produced since 1960 by the Novo Company on an industrial scale. The enzyme gained great commercial significance following its introduction as an active component in detergents.

In 1959, a detergent containing alkaline bacterial protease (Bio 40) was developed by Jaag (1968) at the Snyder Company in Switzerland. Prior to this, attempts had been made to use pancreatic proteases for this application by Otto Röhm. However, the pancreatic proteases were not sufficiently stable at high pH levels in the presence of tripolyphosphate and anionic detergents. After the introduction of the protease from *B. licheniformis,* which is stable in alkali even in the presence of complexing agents, a period of rapid development of technical enzymes followed.

Today, the protease-containing detergents are used in many countries. Other enzymes, such as cellulases and lipases, are also being added in increasing measure.

The serine proteases include a number of very specific proteases from different organisms that, until now, were used only for structural determinations such as amino acid sequencing of proteins. One of these is the serine protease from *Staphylococcus aureus.* This enzyme cleaves peptide linkages C-terminal from Glu and Asp (Glu-X, Asp-X) (Endoprotease Glu-C, [BöM]). The Endoprotease Lys-C [BöM] from *Lysobacter enzymogenes* specifically cleaves peptides with lysine linkages.

The molecular weights of serine proteases range from 15,000 to ~30,000 daltons, and the isoelectric points are at approximately pH 9.0. The following gives the properties and the most significant constants of the three most commercially important subtilisin enzymes.

Subtilisin Carlsberg (EC 3.4.21.14; CAS 9014-01-1) (Organism: B. licheniformis)

PROPERTIES   The molecular weight is 27,277 daltons, calculated from the 274 amino acid residues of the single-chain molecule. It contains no cysteine and thus no disulfide bridges. The isoelectric point is at pH 9.4.

Activity: Optimum pH 8–9. (See also Figs. 3.94 and 3.95.)

Stability: Inactivated under pH 5.0 and above pH 11. Calcium is not necessary for activation or for stability. The enzyme remains stable in the presence of complexing agents.

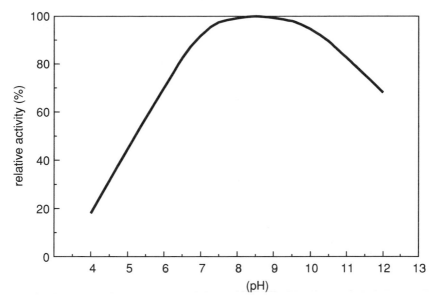

Figure 3.94   Effect of pH on the activity of subtilisin Carlsberg [hemoglobin, denatured with urea, phosphate buffer, 25°C; Aunstrup (1980)].

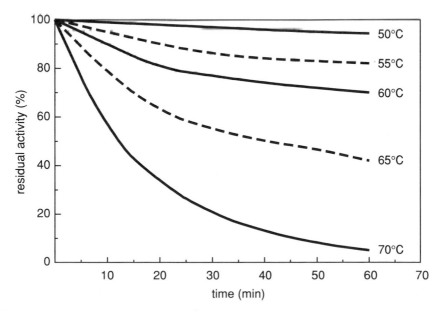

Figure 3.95   Effect of temperature on the stability of subtilisin Carlsberg (pH 8.5, Tris-maleate buffer; Aunstrup, 1980).

Specificity: Largely nonspecific; cleaves most peptide linkages; prefers those with aromatic amino acids and many esters. Up to 40% of the peptide linkages in casein are cleaved.

Inhibitors: Diisopropyl fluorophosphate (DPF), phenylmethylsulfonyl fluoride (PMSF).

Subtilisin BPN   Haghihara in 1954 first isolated and crystallized this enzyme from a commercial preparation of the bacterial protease "Nagarse" (NG). In 1960, Ottesen and Spector isolated an identical enzyme (Subtilisin Novo) from the bacterial protease Novo. The producing strain was identified as *B. amyloliquefaciens.* Subtilisin BPN has 275 amino acid residues in a chain that does not contain cysteine. The enzyme has an isoelectric point at pH 9.1 and has a high homology (identical amino acid sequence) to Subtilisin Carlsberg. Only 58 of the 274 peptide linkages are different. The stability of Subtilisin BPN is slightly more dependent on calcium ions at high temperatures and high pH values than is Subtilisin Carlsberg. In addition, it shows different specificity against esters.

Proteases from alkalophilic Bacillus sp.
References: Aunstrup et al. (1972), Horikoshi and Ikeda (1977).

These enzymes are also serine proteases with a molecular weight of 20,000–30,000 daltons and an isoelectric point at approximately pH 11. The characteristics are similar to the above-mentioned subtilisins with the exception of a higher activity in the alkaline pH range.

Activity: 80% of the maximum activity between pH 8 and 12.
Stability: pH 6–12.

Endoprotease from Streptomyces griseus   This pure enzyme, isolated from the commercial product "pronase" [KK], is used in research for total hydrolysis of proteins (Yoshida et al., 1988).

PROPERTIES

Molecular Weight: 20,000 daltons (gel filtration), 22,000 daltons [sodium dodecylsulfate–polyacrylamide gel electrophoresis (SDS–PAGE)].
Isoelectric Point: At pH 8.4.
Activity: Activation optimum at pH 8.4.
Specificity: Preferably cleaves peptide C-terminal of Glu and Asp.
Inhibitors: DPF, no inhibition by EDTA, *p*-chloromercuribenzoate, or SH blockers.

Jurasek et al. (1969) found that in many respects, the enzyme is similar to trypsin. It has a similar amino acid sequence, a similar active site, and is sensitive to trypsin inhibitors. They found the enzyme cleaves preferentially at Lys and Arg in the oxidized B-chain of insulin.

Technical Preparations   The most important application for serine proteases is in the detergent industry, and for this area most preparations are produced as granulates. The activity of the granulates lies between 0.1 and 2 Anson units. In the new liquid detergents, the enzymes are delivered in a stabilized liquid form.

3.2.4.2   Serine Proteases from Fungi (EC 3.4.21.15)   Technical fungal proteases contain a mixture of serine proteases and proteases active in neutral and acidic pH ranges. The enzyme known as *Aspergillus* protease B is produced by the *Aspergillus* strains: *A. flavus, A. sydowi, A. candidus, A. oryzae, A. sojae,* and *A. melleus.* The enzyme is usually obtained with the solid-bed culture process and has a specificity similar to that of the subtilisins. This statement has to be treated with caution because the enzymes from various *Aspergillus* strains differ strongly in their characteristics. Incomplete purification and the tendency of proteases to autolyze can be attributed to inconsistencies in some of the literature.

Protease from A. candidus   This protease has an isoelectric point at pH 4.9. The pH optimum with casein lies at pH 11–11.5 (30°C). The maximum activity at 47°C is reached at pH 7.0. Enzyme stability is at pH 5.9–9. Ions have no influence on activity, but calcium stabilizes against heat denaturation. The inhibition through DPF and potato inhibitor points to a serine protease. There is no inhibition by SH reagents (Ohara and Nasuno, 1972).

Protease from A. flavus   This enzyme resembles in many of its properties the enzyme from *A. candidus.* It can be incubated without loss of activity in 8 M urea at 20°C but is rapidly inactivated by 4 M guanidine + HCl. It is stable between pH 5 and 10.

Protease from A. melleus   This enzyme mixture is virtually identical to that from *A. oryzae* or *A. sojae.* Differences, however, are apparent in practical applications; for example, in the cleavage rates of different proteins.

Protease from A. oryzae   One of the most fundamental studies on the protease complex from *A. oryzae* was done by Bergkvist (1963/1; 1963/2), who isolated the enzyme from a submersion culture by precipitating it with tannin. After purifying on DEAE cellulose, 28% of the protease activity was found in three fractions. These were separated according to their pH optima into three subfractions: alkaline (1), neutral (2), and acidic (3).

Protease 1:   ~29% of the activity with casein at pH 6; 66% activity with casein at pH 7.4.

Protease 2:   ~24% with casein at pH 6; 35% with casein at pH 7.4.

Protease 3:   ~47% yield with the test at pH 6 and 35% at pH 7.4. The pH curves were recorded using gelatin as substrate. The drop in viscosity of a 5% gelatin solution in 5 min at 37°C was monitored (Fig. 3.96).

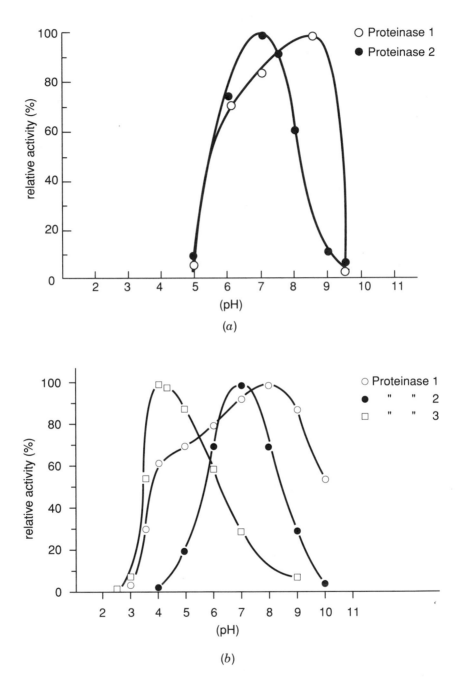

Figure 3.96   Effect of pH on the activities of three proteases from *A. oryzae*; (*a*) caseinase activity; (*b*) hemoglobinase activity; (*c*) gelatinase activity (viscosimeter). (Bergkvist, 1963/2).

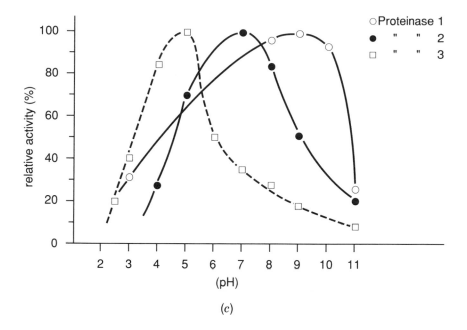

(c)

Figure 3.96 *Continued*

In 1950, Crewther and Lennox were able to crystallize a protease from Takadi-astase (Takamine). The enzyme was named *Aspergillopeptidase A* (EC 3.4.4.17). Because of its differences from protease 1 of Bergkvist, the classification of this protease is still uncertain. Table 3.35, from Bergkvist (1963/1), shows some charac-teristics of the three enzymes.

Protease from Tritirachium album Limber   This enzyme, which is also extensively used for protein hydrolysis in research, is known as "Protease K" [MK]. The "K" refers to its ability to hydrolyze keratin. The primary and tertiary structures of this enzyme are known (Jany et al., 1986). The active site is similar to that in subtilisin.

PROPERTIES

Molecular Weight: 28,500 daltons, polypeptide chain with 277 amino acids. Contains two disulfide bridges and one free cysteine group.

Specificity: The enzyme cleaves peptide linkages C-terminal to aliphatic, hy-drophobic, and aromatic amino acids. Native proteins are rapidly hydrolyzed and similarly, enzymes are also rapidly inactivated by unspecific degradation or limited proteolysis.

Specific Substrates: Glutaryl–Ala–Gly–Phe–pNA or glutaryl–Ala–Ala–pNA.

TABLE 3.35   Properties of Purified Proteases from *A. oryzae* (Bergvist, 1963/1)

Property	Protease		
	1	2	3
pH optima for			
Casein	8.2	6.8	—
Denatured hemoglobin	7.6	6.8	4.3
Gelatin	9.0–9.5	6.3	4.5
TAME[a]	8.5	—	—
Temperature optima (°C)	50	50	45
pH-stability	5–8.5	4.5–10.5	3.0–6.3
Inhibitors			
Na-laurylsulfonate	+	−	+
Laurylamine	+	+	+
L-Cysteine	−	+	−
EDTA[b]	−	+	−
6-Aminocaproic acid	−	−	−
Soybean inhibitor	−	−	−
Ascorbic acid	−	+	+−

[a]N-α-p-tosyl-L-arginine methylester hydrochloride.
[b]Ethylenediaminetetraacetic acid.

Technical Preparations   Technical fungal proteases that contain primarily alkaline proteases (see Fig. 3.97) are used to a limited extent in different branches of the food industry. A special application is in leather manufacturing where a constant enzyme complex composition is of great importance. When properly applied, the enzyme works well in a number of processes without damaging the leather (see Section 5.9.3).

   Products and Manufacturers: Corolase PN [RM], Sumizyme MP [SN], Rhozyme P 11 [RH].

3.2.4.3   *Metalloproteases: Neutral Proteases (EC 3.4.24.4; CAS 9068-59-1)*   The neutral proteases from bacteria and fungi that are used in many applications belong to this group. These enzymes distinguish themselves in that their activity is dependent on one or more metal atoms at the active site. The activation optimum lies at pH 7.0. These enzymes generally have zinc as the metal ion. They cleave peptide linkages relatively unspecifically but prefer amino acids with hydrophobic side chains.

   Amides and esters are not cleaved, and this characteristic can serve as a means of distinguishing them from the serine proteases. Model substrates eluded discovery for a long time. Ward (1983) mentioned a substrate, furylacrolylglycyl leucinamide (FAGLA), which is cleaved only by neutral but not serine proteases.

   The most important technical neutral proteases are produced from the following

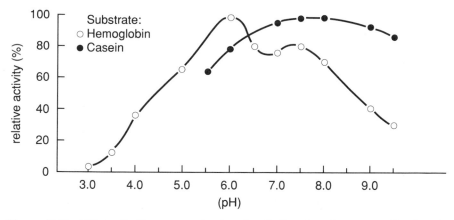

Figure 3.97 Effect of pH on a predominantly alkaline fungal protease (Corolase PN [RM]).

organisms: *B. amyloliquefaciens, B. subtilis* (it can be assumed that, in many technical preparations, the enzyme was not obtained from *B. subtilis* but from *B. amyloliquefaciens*), *B. stearothermophilus, A. oryzae, A. niger.*

Neutral Protease from B. amyloliquefaciens

PROPERTIES

Molecular Weight: 37,000 daltons.

Structure: One zinc atom and two calcium atoms per mole of enzyme.

Stability: Optimal stability lies at pH 4.8–6.6. Calcium and protein stabilize the enzyme. Strong complexing agents (EDTA) inhibit it completely. If calcium ions are removed during the purification process, for example, during precipitation with sulfates, the stability is seriously reduced. Therefore, it is recommended that the enzyme is used in the presence of Ca^{2+} ions. Dancer and Mandelstam (1975) showed that neutral protease was already degraded during fermentation by alkaline serine protease. By inhibiting the latter with PMSF in the medium, larger fractions of neutral protease are protected.

Inhibitors: Heavy-metal salts such as Hg^{2+}, Ag^{2+}, Fe^{3+}, EDTA, and phosphate.

Specificity: Cleavage at amino end of Phe, Leu, and Val.

See also Figures 3.98 and 3.99.

Neutral Protease from B. thermoproteolyticus   This protease is known as "Thermolysin" or as "Thermoase" [DA], which are the pure and technical enzyme preparations, respectively.

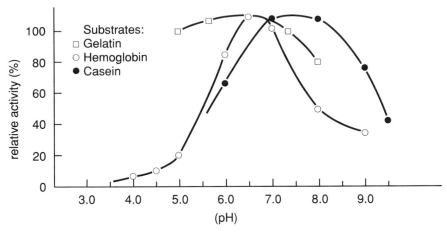

Figure 3.98   Effect of pH on proteolytic activity with casein, hemoglobin, and gelatin substrates. (Corolase N [RM].

PROPERTIES

Molecular Weight: 34,000 daltons.

Structure: One zinc atom and four calcium atoms per mole. Zinc is necessary for activity and calcium for stability (Matthews et al., 1972).

Activity: 90% of optimum activity between pH 7 and 9. At pH 7.5 the optimum temperature is 80°C.

Stability: Of the optimum stability, 90% is between pH 5.5 and 7.5 at 40°C.

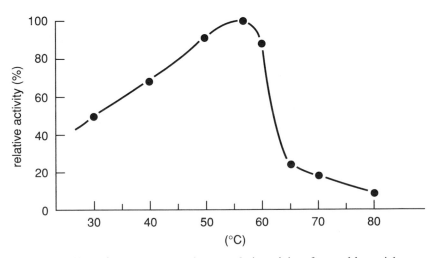

Figure 3.99   Effect of temperature on the proteolytic activity of neutral bacterial protease.

When an aqueous solution at pH 7.2 is heated to 84°C for 15 min, 50% residual activity remains. For the enzyme from *B. liquefaciens,* this point is reached at 59°C under the same conditions (Aunstrup, 1980).

Neutral Protease from A. oryzae   The enzymes isolated by Bergkvist (1963/1) (see Section 3.2.4.2) contain a neutral protease that is inhibited by EDTA and cysteine. Two neutral proteases from *A. oryzae* with different heat stability were found by Nakadai et al. (1973).

Protease 1:   pH optimum pH 7.0. Stability: pH 5.5–12. Inactivation above 50°C.

Protease 2:   pH optimum pH 5–6. Stability: 70% residual activity remains on heating for 10 min at 90°C.

In addition, two types of neutral proteases are produced from *A. sojae* that do not cleave esters or amides (Sekine, 1972). These enzymes act specifically on peptide linkages containing amino acids with hydrophobic side chains and on various synthetic peptides containing Phe, Leu, Tyr, and other amino acids.

Technical Preparations

NEUTRAL BACTERIAL PROTEASE   Neutral bacterial proteases in powder form show activities between 0.5 and 3 Anson units. The following side activities have been observed: α-amylase, alkaline protease, and β-glucanases. Preparations formulated with sugar or salts are offered for special applications in food and leather manufacturing.

The preparations are very stable during dry storage conditions. At a storage temperature of 20°C, activity decreases by less than 10% per year. Liquid preparations with 0.5–1.0 Anson units are stabilized with glycerol or, for commercial purposes, with ethylene glycol.

Products and Manufacturers: Corolase N [RM], Neutrase [NO], GODO-BNP [GO], HT Proteolytic [SOE].

NEUTRAL FUNGAL PROTEASE   In technical proteases, the neutral protease is only one component among various proteases, and thus activity ranges from pH 5 to 8.5 depending on the source of the preparation.

Products and Manufacturers: Panazyme [AM], Veron PS [RM], Fungal Protease P [SO], Maxatase [GB], [SA], [SN], [UA].

3.2.4.4   Acid Proteases: Carboxyl Proteases   Acid proteases in industry are those in which the activity depends on the presence of a carboxyl group, such as in aspartic acid, at the active site. Enzymes are active in the pH range 2–6. In the systematic classification, acidic proteases are those with an activity equal or similar to pepsin.

These enzymes are produced by several fungi. Those from the *Aspergillus* and *Rhizopus* strains are important in industrial processing. Acidic proteases are not inhibited by alkylating agents, such as DPF, chelating agents (EDTA), and SH-group blocking reagents.

This group of proteases also includes the microbiologically produced milk coagulating enzymes that have, to some extent, replaced rennet in cheese production. These proteases are obtained from *Mucor pusillus, Mucor miehei,* and *Endothia parasitica.* In terms of value, these enzymes are the most important of the acid proteases. Other applications are in the manufacture of baked goods, the hydrolysis of proteins, and the clarification of wine and beer.

Acid Protease from A. saitoi (EC 3.4.23.6; CAS 9025-49-4)   This enzyme is known as *Aspergillopeptidase A.* It was first isolated and characterized by Yoshida (1956). According to Ichishima (1970), the enzyme can be produced either in surface cultures or by submerse fermentation.

PROPERTIES

> *Molecular Weight:* 34,000–35,000 daltons.
>
> *Structure:* Single-stranded peptide chain with 283–289 amino acids, with two cysteine residues that form a disulfide bridge.
>
> *Activity:* The pH optimum lies between pH 2.5 and 3.0 with casein and soybeans. The temperature optimum lies at approximately 30°C.
>
> *Stability Range:* pH 2–5.
>
> *Specificity:* Preferably cleaved are Glu, Asp, and Leu peptides, as well as Leu–Val, Ala–Leu, and Glu–Arg linkages.

Protease from A. niger var. Macrosporus   Two acidic proteases are produced from this fungal strain (Koaze et al., 1970). Acid stability and pH optimum are different for each enzyme. The data in Table 3.36 were obtained from a publication by the Meiji Company (1972).

Acid Protease from A. oryzae   Takadiastase, mentioned earlier, contains at least one acid protease. The characteristics (pH-activity profile and temperature stability) of this enzyme are similar to the acid protease component described by Bergkvist (1963/2).

TABLE 3.36   Properties of Two Acid Proteases from *A. niger* var. *macrosporus*

Property	Enzyme A	Enzyme B
Molecular weight (daltons)	20,000–21,000	40,000
pI	1.5–2	~4
Activity optimum at pH	2	2.5–3
Temperature optimum (with casein substrate)		
At pH 1.5	60°C	<30°C
At pH 2.0	65°C	40°C
At pH 2.6	70°C	55°C
Gelatin liquefaction	Strong at pH 2	Weak at pH 2
Collagen digestion	Very strong	Strong

Although several isoenzymes can be isolated during the fractionation of these enzymes, only two different acid proteases from *A. oryzae* seem to exist. The isozymes are presumably formed by autolytic processes. One of the acidic proteases has a pH optimum at pH 4.0–4.5 with hemoglobin as substrate. It activates trypsinogen and has a limited esterase activity (Matsubara and Feder, 1971). The second protease has an activity optimum from pH 4.8–5.2 and is less heat-stable than the first one. It is the main component of the technical preparation Rhozyme A4. Both enzyme types can be isolated from preparations produced in surface cultures.

> *Products and Manufacturers:* Proctase [MJ], Sumyzyme FP [SN], Molsine [Tanabe], Corolase PS [RM].

Acid Protease from Rhizopus sp. (EC 3.4.23.4)
Reference: Matsubara and Feder (1971)

This enzyme is produced by different *Rhizopus* strains. *R. oligosporus* forms two proteases with optimum activity at pH 3.0 and 5.5. The *R. chinensis* protease has a molecular weight of 35,000 daltons and consists of a single peptide chain with 324 amino acid residues. No histidine residues are present. The pH optimum lies between 2.9 and 3.3 and the optimum stability, between pH 3 and 6. The protease has a broader specificity than pepsin or rennin (Tsuru et al., 1970).

3.2.4.5   Microbiological Rennet Preparations
References: Green (1977), Turk and Vitale (1981)

Microbiologically produced rennet has been available for the last 25 years. Whether natural rennet should be replaced was, at one time, an issue of much discussion. The introduction of a replacement was supported by the fact that pure rennet was becoming rare and expensive and the production of cheese after World War II had increased significantly worldwide. Over the last 15 years, the new products have been improved. Now they are accepted by the dairy industry and have found their place next to natural rennet. In addition, there are ethical and religious reasons for renouncing animal rennet. Consequently, various plant proteases were also used for milk coagulation. Because these proteases do not have the specificity of natural rennet, problems occurred in the production of cheese.

These days, microbiological rennet preparations are prepared from cultures of *Mucor pusillus, Mucor miehei,* and *Endothia parasitica* (see Table 3.37).

Rennet from Mucor pusillus   The *Mucor* strain for the production of the first satisfactory rennet preparation was found in 1964 by Arima (1964/1), an important Japanese researcher in applied biotechnology. Because of the developmental work of Arima, the enzyme is still obtained today in very high yields from surface cultures. After extraction of the solid, mycelia-penetrated culture medium, the enzyme is purified and a solid product is obtained with ethanol precipitation or freeze drying (Arima et al., 1964/2; 1964/3).

TABLE 3.37   Producers and Sources of Some Commercial Microbial Rennet Preparations

Enzyme (Tradename)	Producer	Source
CHY-MAX	Pfizer Inc. [PF]	*E. coli* K-12
EMPORASE	Sanofi Bio-Industries Inc. [SBI]	*M. pusillus lindt*
FROMASE	Gist Brocades N.V. [GB]	*M. miehei*
HANNILASE	Chr. Hansen's Laboratories A/S [CH]	*M. niehei*
MARZYME	Solvay Enzymes Inc. [SOE]	*M. miehei*
Meito-RENNET	Meito Sangyo Co., Ltd. [MSJ]	*M. pusillus lindt*
MORCURD PLUS	Novo Nordisk A/S [NO]	*M. miehei*
RENNILASE	Novo Nordisk A/S [NO]	*M. miehei*

PROPERTIES

Molecular Weight: 30,000 daltons.

Structure: Contains two cysteines without formation of a disulfide bridge.

Activity: pH optimum with hemoglobin pH 4; with casein, pH 4.5.

Stability: Optimum from pH 3 to 6.

Rennet from Mucor miehei   The strain isolated by Aunstrup is quite suitable for rennet production in submersed cultures.

PROPERTIES

Molecular Weight: 38,000 daltons.

Structure: Single-stranded peptide chain, contains ~6% carbohydrate.

The enzyme is very similar to the enzyme in *M. pusillus,* although it differs significantly with respect to the effect of calcium on coagulation. At the same calcium concentration, the relative coagulation capacity of the *M. pusillus* rennet is about 50% greater.

Rennet from Endothia parasitica
Reference: Whitaker (1970)

The enzyme found by Sardinas (1972) gives a less specific protease action than the enzyme from *Mucor* sp., as shown by the cleavage spectrum of the oxidized insulin B-chain. Its application is limited as it readily produces bitter peptides.

PROPERTIES

Molecular Weight: 34,000–39,000 daltons. Single-stranded peptide chain, without a carbohydrate fraction.

Isoelectric Point: At pH 5.5.

Optimum Activity: At pH 2.0–4.0. Denatured hemoglobin, pH 2.0.

Stability Optimum: pH 4.0–4.5.

3.2.4.6 Exopeptidases from Aspergillus sp.

3.2.4.6 Exopeptidases from Aspergillus sp. Fungi and yeasts, apart from producing a complex spectrum of endoproteases, also produce a series of exopeptidases. These include some very specific carboxypeptidases and aminopeptidases.

An aminopeptidase from *A. niger* was isolated from the mycelia of a surface culture by Uhlig et al. (1965), purified, and characterized. The enzyme, which was stabilized by Co^{2+} ions, was produced free of protease activity. Enzymatic activity was inhibited by complexing agents, but not by iodoacetamide or PMSF. The molecular weight was 60,000 daltons as determined by gel chromatography. Noteworthy of these peptidases is their ability to degrade native proteins, such as egg albumen–lysozyme and native ribonuclease, from the amino end. The preparation (aminopeptidase-*O*) was used to sequence the amino-terminal sequence of lysozyme (Jolles et al., 1968). An industrial application was not apparent at that time.

Today, there is interest in the use of peptidase preparations for the debittering of protein hydrolysates, in cheese ripening, and in the preparation of amino acid mixtures from proteins.

Plainer and Sprössler (1989) reported on the peptidase activity of various industrial protease preparations during their work on the debittering of protein hydrolysates. As the peptidase activity varied with each production batch, the authors, as can be seen in Table 3.38, listed broad activity ranges.

TABLE 3.38 Peptidase Activities of Technical Protease Preparations from Various Sources (Plainer and Sprössler, 1989)

Enzyme Source	Tradename	Relative Activity							Carboxy-peptidase B
		Amino Peptidase							
		Ala*	Gly*	Leu*	Phe*	Pro*	Tyr*	Arg*	
Papaya	Corolase L10	—	—	—	—	—	—	2	—
Pancreas	Corolase PP	3	2	2	3	4	4	—	3
B. subtilis	Corolase N	—	—	1	—	—	—	—	—
A. sojae	Corolase PN	2	1	5	2	2	1	3	1
A. oryzae	Corolase 7092	4	3	6	2	2	2	2	—
A. oryzae	Röhm 40-88	3	2	5	1	2	2	1	—
A. oryzae	Röhm 51-86	2	1	5	1	1	1	—	—

[a]As the L-amino acid *p*-nitroanilides:

Level	Activity Range (units/mg)
1	$0–10^2$
2	$10^2–10^3$
3	$10^3–10^4$
4	$10^4–10^5$
5	$10^5–10^6$
6	$10^6–10^7$

REFERENCES

Amano, Nagoya. Company information.

Arima, K., U.S. Patent 3,151,039 (1964).

Arima, K., Yu, J., and Iwaski, S., in Perlmann, G. E., and Lorand, L. (eds.), *Methods in Enzymology,* Vol. 19 Academic Press, New York, 1964, p. 446.

Aunstrup, K., Outtrup, H., Andersen, O., and Dambmann, C., in Terui, G. (ed.), *Fermentation Technology Today,* Kyoto, Japan, 1972, p. 229.

Aunstrup, K., in Rose, A. H. (ed.), *Economic Microbiology,* Vol. 5, 1980, p. 50.

Bergkvist, R., *Acta Chem. Scand.* **17,** 1521 (1963/1).

Bergkvist, R., *Acta Chem. Scand.* **17,** 1541 (1963/2).

Crewther, W. G., and Lennox, F. G., *Nature* **165,** 680 (1950).

Dancer, B. N., and Mandelstam, J. J., *J. Bacteriology* **121,** 406 (1975).

Green, M. L., *J. Dairy Sci.* **44,** 159 (1977).

Güntelberg, A. V., and Ottesen, M., *Nature* **170,** 802 (1952).

Haghihara, B., *Ann. Rep. Scien. Works, Faculty of Science of Osaka University* **2,** 35 (1954).

Horikoshi, K., and Ikeda, Y., U.S. Patent 2,182,638 (1977).

Ichishima, E., *Methods in Enzymology* **19,** 379 (1970).

Jaag, R., Thesis Europ. Hochschulschr. Reihe VIII Chemie/Abt. B Biochemie, Verlag Herbert Lang, Bern, 1968.

Jany, K. D., Lederer, G., and Mayer, B., *FEBS Lett.* **199,** 139 (1986).

Jolles, J., Jolles, P., Uhlig, H., and Lehmann, K., *Hoppe-Seyler's Z. Physiol. Chem.* **350,** 139 (1968).

Jurasek, L., Fackre, D., and Smillie, L. B., *Biochem. Biophys. Res. Comm.* **37,** 99 (1969).

Koaze, Y., Goi, H., and Hara, T., U.S. Patent 3,492,204 (1970).

Matsubara, H., and Feder, J., in Boyer, P. D. (ed.), *The Enzymes,* 3rd edition, Vol. 3, Academic Press, New York, 1971.

Matthews, B. W., Colman, P. M., Jansonius, J. N., Titani, K., Walsh, K. A., and Neurath, H., *Nature and New Biol.* **238,** 41 (1972).

Nakadai, T., Natsuno, S., and Iguchi, N., *Agric. Biolog. Chem.* **37,** 2703 (1973).

Nordwich, A., and Jahn, W. F., *Europ. J. of Biochem.* **3,** 519 (1968).

Noval, J. J., Nickerson, W. J., and Robinson, R. S., *Biochim. Biophys. Acta* **77,** 73 (1963).

Ohara, T., and Nasuno, S., *Agr. Biol. Chem.* 36 (10) 1797 (1972).

Otteson, M., and Spector, A., *Comptes Rendues des Traveaux du Laboratoire Carlsberg* **32** 63 (1960).

Plainer, H., and Sprössler, B., Int. Conf. Biotechnology and Food, 1989, University of Hohenheim—Stuttgart, Poster.

Sardinas, J. L., *Adv. Appl. Microbiol.* **15,** 39 (1972).

Sekine, H., *Agric. Biol. Chem.* **36,** 207 (1972).

Tsuru, D., Nattori, H., Tsuji, H., and Fukumoto, J., *J. Biochem. Tokyo* **67,** 415 (1970).

Turk, V., and Vitale, L., *Proteinases and Their Inhibitors: Structure, Function and Applied Aspects,* Pergammon Press, Oxford, 1981.

Uhlig, H., Lehmann, K., Salmon, S., Jolles, J., and Jolles, P., *Biochemische Zeitschrift* **342** 553 (1965).

Uhlig, H., *Seifen, Öle, Fette, Wachse* **96**(3), 55 (1970).

Ward, O.P., in Fogarty, W. M. (ed.): *Microbial Enzymes and Biotechnology,* Applied Science Publishers, New York, 1983, p. 278.

Whitaker, D. R., in Perlman, G., and Lorand, L. (eds.), *Methods in Enzymology,* Vol. 19, Academic Press, New York, 1970, p. 436.

Yoshida, F., *Bull. Agr. Chem. Soc. of Japan* **20**, 252 (1956).

Yoshida, N., Tsuruyama, S., Nagata, K., and Hirayama, K., *J. Biochem.* **104**, 451 (1988).

3.3 ESTER CLEAVAGE: FAT HYDROLYSIS

References: Brockerhoff and Jensen (1974), Borgström and Brockmann (1984), Antonian (1988)

3.3.1 Brief Overview

3.3.1.1 Fats and Oils   The fats and oils that occur widely in the plant and animal kingdoms are compounds (esters) of free fatty acids with glycerol. They are known as lipids or lipoids. They play important roles in human nutrition and affect the taste of foods. They are also important as carriers of aroma compounds and influence the richness of taste and texture of foods, such as meat or baked goods. Fats and oils are used in a number of applications (see Table 3.39) as commercial raw materials without any modifications. On hydrolysis, they are converted into glycerol and fatty acids, which are basic chemicals for the chemical industry. Soap is obtained by saponifying the fats with alkali.

The conversion of natural fats and oils to those with special functional properties was previously achieved only by chemical or physical methods. As enzymes are able to catalyze not only the degradation of fat but also, under specific conditions, the synthesis of fats and esters, many new possibilities have been opened up for lipid chemistry. Keywords here are enzymatic esterification, transesterification, and the stereospecific cleavage of racemates by esterases and lipases.

The lipolytic enzymes have a physiological importance in the degradation and formation of lipids in organisms. Lipases are formed mainly by the pancreatic glands, while various esterases are produced in the liver. These enzymes are found in blood, muscle tissue, and milk. The milk lipases are especially significant in the formation of aroma substances in milk products and cheese.

TABLE 3.39   Industrial Use of Lipids and Oils (Dechema, 1988)

Raw materials for laundry detergents, softeners for synthetics	~40%
Rubber and textiles	~22%
Paints and coatings	~13%
Food additives	~11%
Lubricants, flotation, metal protectants, etc.	~11%

Lipases occur in many plant cells and seeds, such as *Rhicinus,* grains, and beans; they are also generated by fungi, yeasts, and bacteria. Until the late 1960s, lipase preparations in medicine were limited to use as digestive aids. Now, with the new technologies mentioned, these applications have been considerably extended. The development of new enzyme technologies, which is driven by new lipase preparations from microorganisms, is fully under way. Also worth mentioning is the current introduction of lipases as active ingredients in detergents.

3.3.1.2 Esterases and Lipases   Esterases contain a number of enzymes. These are classified according to the chemical nature of the respective acid components. Generally, esterases (EC 3.1.1.1) cleave carboxylic acids from esters according to the formula

$$R{-}CO{-}O{-}R^1 + H_2O \longrightarrow R{-}COOH + R^1{-}OH$$

For this large group of enzymes, the term *aliesterases* is still frequently used. Simple carboxylic esters, especially those of acetic acid, are also cleaved by proteases. Reference is often made to the esterase activity of protease. Trypsin, chymotrypsin, and papain all show relatively nonspecific esterase activity. The pectinesterases described in Section 3.1.6.2 also belong to this group. Besides these general esterases, there are a number of specific esterases, which are, in part, already used commercially. For example, the phospholipases are used for the specific cleavage of lecithin.

If the alcohol component of the ester is a glycerol residue, the acid component is a fatty acid, and the triglyceride is water-insoluble, then the effective enzyme is a lipase (EC 3.1.1.3). There are still some problems of definition, but lipases can be described as enzymes that hydrolyze esters at the oil–water interface in an insoluble or heterogeneous system (Jensen, 1971).

3.3.2   Lipases

3.3.2.1 Nomenclature   Glycerides are triester of glycerol with two primary groups and one secondary ester group. The two primary ester groups can be distinguished sterically. The carbon atoms of glycerol are labeled 1, 2, and 3 from top to bottom. In the Fischer projection, the secondary alcohol group is placed on the left. The triglycerides are represented in a simplified manner in Figure 3.100. The fatty acids are labeled by the first letter of their names (Holmann, 1966) or by the number of their carbon atoms. The number of double bonds present in a given fatty acid provides added classification.

3.3.2.2 Substrates   Plant and animal fats and oils are substrates for the lipases. For the analysis of lipases, purified or synthetic triglycerides, such as triolein or tributyrin, are used. The latter is easier to use in routine analysis as it forms relatively stable emulsions with water or buffer solution without the use of stabilizers. It should be noted that tributyrin is also cleaved by esterases that are not real lipases according to the preceding definition.

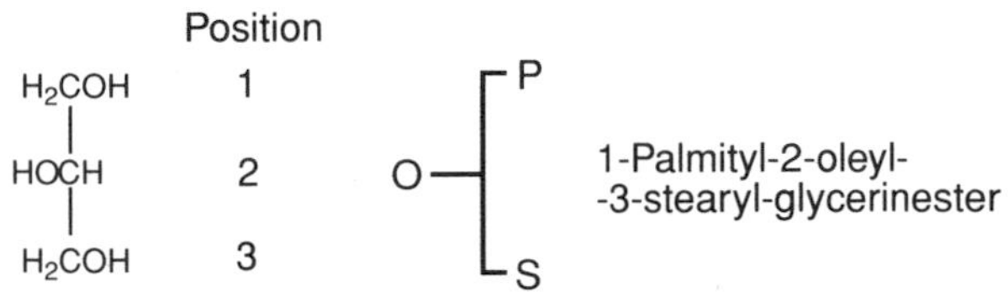

Position

H₂COH   1   ┌─ P
 │
HOCH   2   O ─┤   1-Palmityl-2-oleyl-
 │   -3-stearyl-glycerinester
H₂COH   3   └─ S

16:0 = P = Palmitic acid
16:0 = S = Stearic acid
18:1 = O = Oleic acid
18:2 = L = Linoleic acid
18:3 = Ln = Linolenic acid

Figure 3.100   Acylglycerol nomenclature (Holmann, 1966).

3.3.2.3 Factors Influencing Lipase Activity   Lipid hydrolysis depends on different parameters such as pH, temperature, water content, and the phase boundary area. It is, therefore, difficult to compare the results of different authors on this subject. The pH optimum of most lipase lies between pH 7.5 and 9. Different microbial lipases are active between pH 5.5 and 8.5 and milk lipases, between pH 4.1 and 6.5. The lipase activity depends on the temperature. Lipase activity has been detected in frozen foods at $-30°C$.

Salts in different concentrations influence lipase activity. NaCl is needed to activate pig pancreas lipase. The presence of calcium ions is absolutely necessary for the hydrolysis of long-chain triglycerides so that the free fatty acids produced can be removed in the ionic form as their insoluble calcium salts.

Sodium taurocholate stimulates the lipolysis of triolein in a weakly alkaline medium but inhibits it in the weakly acidic range. In general, bile acids inhibit lipolysis but can, in low concentrations, stabilize the pancreatic lipases.

The colipase, a protein with a molecular weight of 10,000 daltons seems to be essential for lipolysis in the intestines of vertebrates. One possible function of this enzyme is to neutralize the bile acid inhibition of lipid cleavage. Characteristics of the pancreatic lipases are their affinity for the oil–water interface and the high rate of hydrolysis. The lipase activity depends strongly on the inside surface of an oil–water emulsion (Sarda and Desnuelle, 1958). Extreme care in the production of emulsions has to be exercised to obtain reproducible hydrolysis rates. Stable emulsions from olive oil can be produced by vigorous stirring or by homogenizing the oil with an aqueous solution of gum arabic, carboxymethylcellulose, or polyvinyl alcohol.

3.3.2.4 Specificity   Many lipases are unspecific and cleave various natural lipids and oils, such as olive oil, soybean oil, coconut oil, butterfat, and pork and beef fat as well as synthetic substrates, with very similar conversion rates. Nevertheless, two kinds of specificities (MacCrae, 1983) can be distinguished:

1. Position specificity or regiospecificity.
2. Fatty acid specificity, depending on the structure of the fatty acid residues.

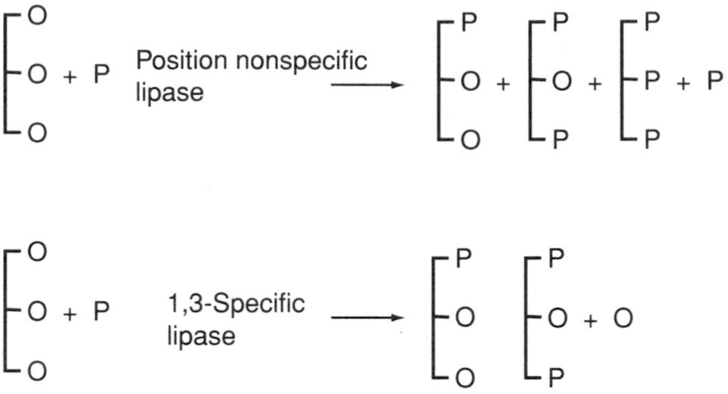

$$\begin{bmatrix} O \\ O \\ O \end{bmatrix} + P \quad \xrightarrow[\text{lipase}]{\text{Position nonspecific}} \quad \begin{bmatrix} P \\ O \\ O \end{bmatrix} + \begin{bmatrix} P \\ O \\ P \end{bmatrix} + \begin{bmatrix} P \\ P \\ P \end{bmatrix} + P$$

$$\begin{bmatrix} O \\ O \\ O \end{bmatrix} + P \quad \xrightarrow[\text{lipase}]{\text{1,3-Specific}} \quad \begin{bmatrix} P \\ O \\ O \end{bmatrix} \begin{bmatrix} P \\ O \\ P \end{bmatrix} + O$$

Figure 3.101   Method for determining the 1,3 regiospecificity of lipases (Nielsen, 1985).

Regiospecificity   Two groups of lipases can be distinguished by referring to their regiospecificity. This specificity can be determined through lipase-catalyzed trans-esterification of triolein (OOO) with palmitic acid (P).

The formation of tripalmitin (PPP) shows that the position specificity is not exhibited while the absence of PPP points to 1,3-regiospecificity. A method for determining the 1,3-regiospecificity of lipases is shown in Figure 3.101.

1. Nonspecific lipases that cleave triglycerides into free fatty acids produce intermediate 1,2- or 2,3-diglycerides. The first reaction (I) is very fast, while the next step (II), namely, the hydrolysis into 2-monoglycerides, proceeds very slowly. The last step (III), the hydrolysis to free glycerol and fatty acid, is also a slow reaction (Fig. 3.102). The 1,2- or 2,3-diglycerides and the 2-monoglycerides are chemically unstable; these glycerides are transformed into 1,3-diglycerides or 1-monoglycerides, respectively, through an intramolecular acyl migration. For technical use, the most important representatives of this group are the lipases from *Candida cylindracea* and *Humicola lanuginosa*. Bühler and Wandrey (1987) obtained 95% cleavage of soybean and olive oil with a lipase preparation from *Candida cylindracea* under technically feasible conditions.

2. The 1,3-specific lipases, including the pancreatic lipase, can completely hydrolyze certain triglycerides. This 1,3 specificity occurs relatively frequently in the microbial lipases, such as in the lipases from *A. niger*, from *R. delemar* (Okumara et al., 1976), from different *Mucor* species and from *Pseudomonas* sp. According to the investigations of Slotboom et al. (1970), microbial lipases, particularly those from *R. arrhizus*, show stereospecificity with respect to positions 1 and 3 of the glycerol residue.

Fatty Acid Specificity   It is common to investigate fatty acid specificity using the alkyl esters of the fatty acids as substrate. However, the speed of the reaction—which is very slow if triglycerides are used as substrate—must be considered. The

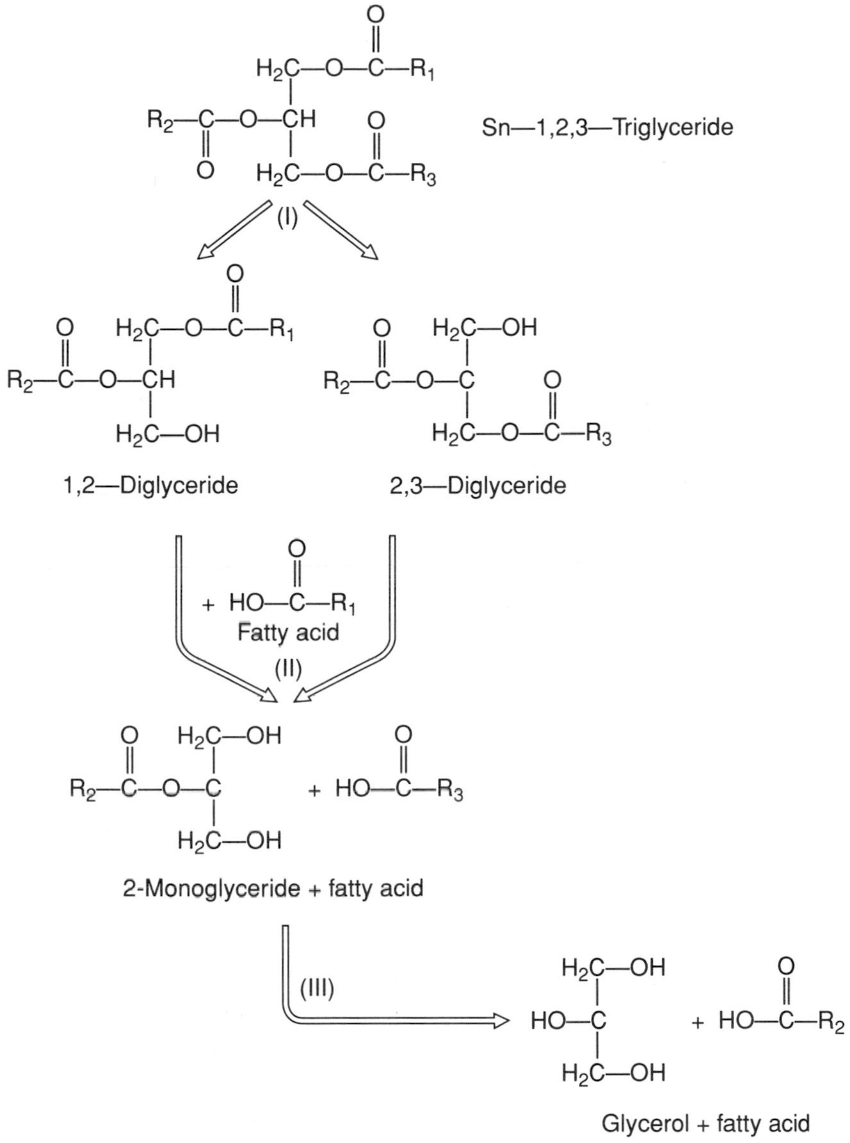

Figure 3.102   Reaction steps in triglyceride hydrolysis.

problem with triglycerides is that some are liquid at normal reaction temperature, whereas those with long-chain fatty acid residues are solids. The lipase from *Geotrichum candidum* shows a remarkable fatty acid specificity. This lipase preferably liberates long-chain fatty acids with a cis double bond in the 9 position from triglycerides. Thus, the methylesters from oleic acid, linoleic acid, and linolenic acid are hydrolyzed relatively fast, while the esters of the saturated fatty acids such as palmitate and stearate are attacked very slowly. The triglycerides—triolein (cis-d_9-18:1), trilinolin (cis-$d_{9,12}$-18:2) and trilinolen (cis-$d_{9,12,15}$-18:3) are cleaved relatively rapidly. Tributyrin, trimyristin (14:0), and tristearin (18:0) are poor substrates (Jensen and Pitas, 1976).

Many microbial lipases show only a low specificity in the hydrolysis of natural oils and fats. Fish oils and butterfat are the exception. Differences in the specificity of different *Aspergillus lipases* can be observed; some prefer to cleave short-chain fatty acids, whereas others prefer the long-chain fatty acids.

3.3.2.5 Animal Lipases

Pancreatic Lipase   As early as 1923, Willstätter and Memmen isolated a lipase from pig pancreas, although complete purification of the enzyme was not accomplished until many years later by Verger et al. (1969). Problems were encountered as a result of the instability of the enzyme during the purification steps. Stability is improved if the activity of the pancreas protease can be inhibited during the purification procedure. The specific activity with respect to defatted pancreas powder was improved by a factor of 30.

PROPERTIES

> *Molecular Weight:* Hog pancreatic lipase, 45,000 daltons (Sephadex), 50,000 daltons (UC). Two species exist: LA, a more acidic protein; and LB, a more alkaline protein. Both exhibit the same specific activity. Beef pancreas lipase, 48,000 daltons.
>
> *Isoelectric Point:* At pH 4.9–5.0.
>
> *Structure:* The enzyme molecule contains six disulfide bridges and two free SH groups. One of the SH groups appears in the vicinity of the active site, but does not seem to be necessary for the catalytic activity. SH reagents in low concentrations do not influence the activity. Because of the inhibitory action of organophosphates, it was concluded that the enzyme was of a serine type. As the activity is also inhibited or destroyed by photooxidation, it was concluded that a histidine participates in the catalytic activity.

Pancreatic lipase requires NaCl in order to be active (7 mM for optimal activity).

Even though many substances influence the lipase activity, no cofactors are necessary. The role of calcium, the influence of bile acids, and the colipase have already been mentioned.

LIPASE ACTION   According to the model by Brockerhoff (1973), the pancreatic lipase, on approaching a "supersubstrate," such as the surface of an oil droplet, is bound in the correct orientation for the reaction. The hydrophobic area of the enzyme is linked to the hydrophobic substrate. The active site lies close to this area but is also accessible to hydrophilic regions because water is necessary for the hydrolysis of the intermediate acylenzyme (Fig. 3.103).

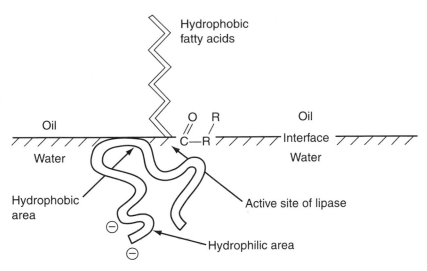

Figure 3.103 Orientation of pancreatic lipase at an oil–water interface [model according to Brockerhoff (1973)].

Specificity: Analysis of the lipolysates from glycerol-1-oleate-2,3-dipalmitate and from 1,2-dipalmitate-3-oleate showed that the two primary positions, 1 and 3, are cleaved equally. Pancreatic lipase does not cleave stereospecifically. The enzyme is regiospecific for the primary ester position (1,3). If the glycerides are present in the emulsified form, the short-chain fatty acids (tributryn) are hydrolyzed faster than the long-chain ones. A pronounced structure specificity does not exist.

TECHNICAL PREPARATION Pancreatin preparations are used in technical as well as medical applications. The lipase activity is standardized according to the analytic procedures of the FIP or the NF. There are pancreatins with a six- to eightfold lipase NF activity, or 30,000–35,000 FIP units.

One preparation with an industrial application, 7023 C [RM], contains ~8000 FIP units. In all preparations, the lipase is available only in a partially soluble state. A significant part is bound to the fibers of the pancreatic tissue and is insoluble until contact is made with the substrate; thus, the enzyme acts as if it were immobilized. Such preparations are ideally suited for conversions in minimally aqueous media.

Manufacturers: [BR], [KC], [RM].

Pregastric Lipases or Lipase Esterases
Reference: Nelson et al. (1977)

These enzymes are obtained from the parotid glands of calves, goats, and lambs by extraction with glycerol or salt solutions. Molecular weights lie between 150,000 and 172,000 daltons. The optimal activity range stretches from pH 4.8 to 5.5 at 32–42°C. The optimal activator is 0.5 M NaCl.

The regiospecificity targets preferentially the Sn-3-position of the glycerol. In the case of butterfat, the short fatty acids C_4–C_8 are primarily cleaned off.

The lipase from pregastric calf glands produced by the Amano Company [AM] exhibits the following fatty acid specificity, with the activity determined with a tributyrin standard at 100: triacetin = 2; tricaproin = 113; tricapryl = 58; tricaprin = 49; trilaurin = 13; trimyristin = 1; tripalmitin = 9; tristearin = 0. *Aspergillus* and *Mucor* lipases show similar specificities. Comparison with the pancreas and fungal lipases reveals pronounced differences in the specific hydrolysis of mono- and dibutyrins, relative to activity with tributyrin at 100 (Richardson and Nelson, 1967; Table 3.40).

Activity: Tributyrin, tricaproin, β-naphthylbutyrate, or milk fat, emulsified in 10% gum arabic, are used as substrates. The pH optimum of these esterases lies within pH 4.5–5, at 37°C to a maximum of 40°C. Optimal stability is pH 5.0–7.0. In the United States, these preparations are used to enhance aroma and flavor in the production of Italian cheeses, such as provolone or romano. The activity of gastric lipases obtained from calf stomachs differ from pregastric lipases (Richardson et al., 1971). A combination of the two enzymes is used in the manufacture of provolone and cheddar cheese.

Products and Manufacturers: From animal organs: Italase [SOE], Kapalase [SOE], Picantase [GB], Lipase PGE [AM]. From microorganisms: Palatase [NO], Lipolact [RM].

3.3.2.6 Microbial Lipase Preparations

Lipases are produced by numerous bacteria and fungi. Until about 15 years ago these enzymes were known only as preparations for scientific research. Later, Japanese producers started to offer lipases from *Rhizopus* or *Aspergillus* to the pharmaceutical industry as substitutes for pancreatin. With the use of this new technology, several new lipase preparations from different microorganisms have been developed. The characteristics of some of these enzymes, now produced on an industrial scale, are discussed in the following section.

Lipases from Aspergillus sp.   Several lipases from *A. niger* with different specificities are known. An enzyme isolated by Fukumoto et al. (1963) following eightfold enrichment and recrystallization has been well described. The enzyme has an optimal activity at pH 5.6 at ~25°C and hydrolyzes triglycerides without regiospecificity.

TABLE 3.40   Relative Fatty Acid Specificity of Lipases from Different Sources

Enzyme Source	Tributyrin	Dibutyrin	Monobutyrin
Pregastric esterase	1.00	0.04	0
Pancreatic lipase	1.00	0.74	0.95
Fungal lipase	1.00	0.26	0.15

TABLE 3.41   Characteristics of Two Lipase Isozymes from *Aspergillus niger*

Characteristic	Lipase I	Lipase II
Molecular weight (daltons)	31,000	19,000
pI	4.0	3.5
pH optimum	5–6	5–6
Substrate specificity	Long-chain fatty acids	Short-chain fatty acids
Hg^{2+}	No influence	Inactivates
Iodoacetamide	Inhibits	No influence

Two isozymes from *A. niger* were isolated and characterized by Hofelmann et al. (1985). Some typical properties are given in Table 3.41.

See also Figures 3.104 and 3.105.

During butterfat hydrolysis, the authors discovered differences in aroma development and aroma spectra between the two enzymes. Other *A. niger* lipases with molecular weights between 22,000 and 25,000 daltons have been described by various authors. The classification of technical preparations is, however, difficult.

Products and Manufacturers: Lipolact [RM], Lipase AP [AM], Lipozyme [NO].

Lipases from Mucor sp.   *Mucor miehei* produces a lipase with 1,3 regiospecificity, which cleaves off short- and long-chain fatty acids; 70% of the activity is reached

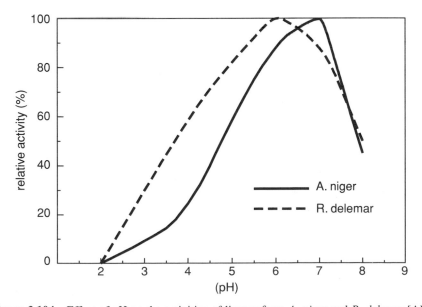

Figure 3.104   Effect of pH on the activities of lipases from *A. niger* and *R. delemar* [AM].

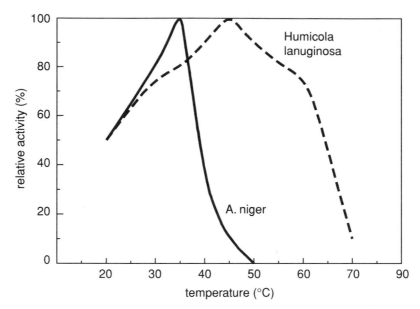

Figure 3.105   Effect of temperature on the activities of lipases from *A. niger* (lipase AP [AM]) and *Humicola lanuginosa* [NO].

between pH 7 and 8.2. The optimal temperature is 40°C. Up to 93.7% of olive oil can be hydrolyzed.

This lipase is a good catalyst for transesterification in the 1,3 position. *Mucor javanicus* (molecular weight 21,000 daltons, 2.6% carbohydrate) is another strain used for the production of lipase products.

Manufacturers: [AM], [NO].

Lipase from Candida cylindracea

Molecular Weight: 120,000 daltons, isoelectric point at pH 4.2, 4.2% carbohydrate.

Structure: The amino acid composition was determined by Tomizuka et al. (1966) following crystallization of the enzyme.

Activity: Range pH 5–8.5. Optimum (90%) pH 7–8, optimum temperature ~45°C.

Specificity: Regiospecificity does not exist; short-chain to long-chain fatty acids are hydrolyzed at comparable speeds.

Product and Manufacturers: Lipase AY: [AM], [SA].

Lipases from Rhizopus sp.   A series of lipases from *Rhizopus* has been described. These include *R. arrhuzus*, *R. javanicus*, *R. niveus*, *R. delemar*, and *R. rhizopodiformis*. Of these, the lipase from *R. arrhizus* has been characterized in most detail. The other enzymes show a 1,3 specificity, the activity optimal lies within pH 5.0–7.0, and the optimum temperature range is 30–45°C.

Lipase from R. arrhizus

Molecular Weight: 43,000 daltons, isoelectric point at pH 6.3.

Structure: Autolysis of the purified lipase gives a glycopeptide and a carbohy-drate-free protein with molecular weights of 8500 and 32,000 daltons, respectively. The latter shows full lipase activity (Semeriva et al., 1969). The gly-copeptide contains 13–14% mannose and 2 mol of hexosamine per mole.

Activity: The pH activity range and the specificity are similar to that of the pancreatic lipases. Fatty acids are cleaved off the primary ester groups. There is no stereospecificity with respect to the sn-1 or sn-2 positions.

Lipase from R. delemar   The lipase preparation purified by Mohsen et al. (1986) had a molecular weight of 48,720 daltons. For activity, see Figure 3.104.

Product and Manufacturer: Lipase D [AM].

Lipases from Penicillium sp.   The most important lipase source is *P. roquefortii,* which forms at least two different exocellular enzymes, both of which are 1,3-specific. Both lipases cleave, in order of decreasing rate, tributyrin, tricapryl, tricaprin, trimyristin, and triolein. The two lipases differ with respect to optimum activity. The acidic lipase has optimum activity at pH 5.5; the alkaline lipase, at pH 8.0.

A *Penicillium* lipase has been described in which synthesis reactions, such as esterification, shows a strong specificity for the sn-2 position. However, with respect to hydrolysis, the enzyme is nonspecific, which is presumably related to the different reaction conditions.

Manufacturer: [AM].

Lipase from Pseudomonas sp.   *Pseudomonas aeruginosa,* the enzyme purified by Stuer et al. (1986), has a molecular weight of ~29,000 daltons and isoelectric point at pH 5.8. The *Pseudomonas* lipases are recommended for application in detergents because of their activity and stability in alkaline pH levels.

Manufacturer: [AM].

Lipase from a Transformed Fungal Strain   According to the manufacturer's information, the donor strain was an alkalophilic fungal strain. The lipase gene was inserted into the *Aspergillus oryzae* genome. A particular property of this enzyme is its high activity at pH 10–pH 11, as well as an adequate stability at 50–55°C in this pH range. This lipase is manufactured for use in detergents.

Manufacturer: [NO].

Lipase from Humicola lanuginosa   Omar et al. (1987) described an enzyme from *Humicola lanuginosa* with enhanced temperature stability that was suitable for the hydrolytic as well as the esterification reaction. The need for a higher temperature stability is due to the fact that hard fats such as tallow are solid at 25–30°C and are difficult to hydrolyze, and that syntheses involving palmitic or stearic acid should be performed above their melting-point temperatures.

The optimum conditions for hydrolysis are at 45–50°C, at pH 7.0 (in buffer) and a water content of at least 60%. For esterification, the optimum is at 45°C, pH 7.0, and 8% water.

The enzyme is relatively specific for C_{12} to C_{18} fatty acids. No synthesis is observed for C_2 to C_8 fatty acids. The best solvents for synthesis reactions are n-octanol, n-decanol, and isoamyl alcohol.

3.3.3 Phospholipases

Reference: Brockerhoff and Jensen (1974)

3.3.3.1 Lecithin
Reference: Schäfer (1988)

The phospholipid, lecithin, is indispensable in food processing technology. It is found in many plants and animal organs and can be obtained from soybean, sunflower, rapeseed, and corn oils. Gobley, in the nineteenth century, isolated lecithin from egg yolk. The name is derived from the Greek word for egg yolk, *lekitos.*

The lecithin used today in industry and food processing is obtained from raw soybean oil. Lecithin swells if water is added at high temperature and can be separated from the oil phase by centrifugation. The raw lecithin contains, besides 60–70% polar lipids, approximately 27–37% soybean oil and 0.5–1.5% water. Phospholipids are insoluble in acetone; therefore, oil-free powder products can be produced from raw lecithin by means of acetone precipitation.

3.3.3.2 Structure of Lecithins and Enzymatic Attack   Phospholipids are cleaved in different ways by two groups of enzymes. These enzymes are lipases (phospholipase A_1 and A_2) and lecithin–phosphodiesterases (phospholipase C and D). (See Fig. 3.106.)

Phospholipase A_1 (EC 3.1.1.32; CAS 9043-29-2)   This enzyme is C-1-(α-)-specific and is found in the venoms of snakes, scorpions, and bees, as well as in the pancreas and in microorganisms. It requires Ca^{2+} for activity. At one time it was thought that the action of these venoms was traceable to the phospholipase. However, in the case of rattlesnake venom, the enzyme-free toxin, crotoxin, was later separated from the enzyme. Alpha- or sn-2-lysolecithins that are formed by the action of phospholipase A_1 are capable of lysing erythrocytes.

Phospholipase A_2 (EC 3.1.1.4; CAS 9001-84-7)   This enzyme occurs in pancreatic glands as its inactive precursor, pro-a-phospholipase A_2. It is activated by trypsin, hydrolysis of an arginine–alanine linkage, and the release of a heptapeptide.

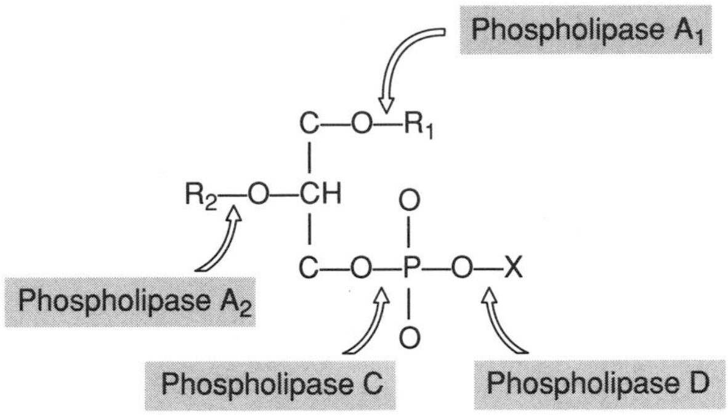

Figure 3.106 Structure of lecithin and the specificity of phospholipases: (1) when R_1 and R_2 are fatty acids, the compound is known as sn-1,2-diacetyl glycerin or generally, lecithin; (2) when R_1 or R_2 is an H atom, then the compound is known as monoacylglycerin phospholipid or lysophospholipid, or simply lysolecithin; (3) the most important phosphate esterifying alcohol groups (X) are $X = H$ = phosphatidic acid (PA), $X = -CH_2-CH_2-NH_2$ = phosphatidylethanolamine (PE), $X = -CH_2-CH_2-N(CH_3)_2-CH_3$ = phosphatidyl choline (PC), and $X = -CH_2-CH(NH_2)-COO^-$ = phosphatidyl serine (PS).

It is isolated as a technical enzyme for the production of lysolecithin from porcine pancreatic glands.

PROPERTIES, PRODUCT, AND MANUFACTURER   The enzyme has a molecular weight of 14,000 daltons. It consists of a peptide chain of 123 amino acids, which are linked by six disulfide bridges (De Haas, et al., 1970). The isoelectric point is at pH 6.3.

Activity: Calcium is the cofactor for this metalloenzyme. The pH-optimum at 40°C lies between pH 8 and 9, the optimal temperature at pH 8.0 is about 50°C.

Stability: See Figure 3.107.

Product and Manufacturer: Lecitase [NO].

Phospholipase C   This enzyme occurs in *Clostridium* sp., especially *Clostridium perfringens* (*C. welchii*). The molecular weight is between 20,000 and 24,000 daltons, and the pH optimum is around neutral. Other enzymes with equal specificities are isolated from *A. oryzae, B. cereus,* and *Pseudomonas* sp., as well as from animal tissue.

Phospholipase D   This enzyme occurs in cabbage and carrot leaves (Hanahan and Chaikoff, 1948) and is also produced by *Corynebacteria.* It requires calcium for its activity. The plant enzyme has a pH optimum of pH 5.6–5.8, the microbial enzyme from pH 7–8. The latter has a molecular weight of about 90,000 daltons.

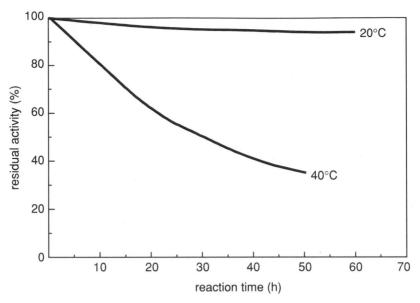

Figure 3.107   Effect of temperature on the stability of phospholipase A-2 [NO] (enzyme concentration: 1 mg/ml; 0.2 M Tris HCl, pH 8.0; 0.01 M $CaCl_2 \cdot 2H_2O$).

Phytase (EC 3.1.3.16; CAS 9001-89-2), Phytate-6-phosphatase
Reference: Schwimmer (1981/1)

The hydrolytic release of phytic acid from myoinositol hexaphosphate ester was the first phosphatase-catalyzed reaction known. Because of the nutritional significance of phytase, especially in dietary fiber-rich materials, some aspects of the enzyme will be detailed here.

Phytate occurs in plant materials, the endosperms of different cereals (particularly wheat bran), and also in immature potatoes, where it is bound to starch. It serves as a phosphate reserve for the seeds and grains. Corn contains about 1%; sesame, about 5% phosphorous. The calcium–magnesium salt is called *phytin*. The highly ionized orthophosphate groups form salts with cations and complexes with proteins. The poor availability of mineral salts, especially calcium and magnesium, of certain protein hydrolyzates could be attributable to the formation of protein–phytate–mineral salt complexes. Other substances, such as fibrous bulk fillers, oxalates, and phenolic substances, may also be responsible for the poor availability. Nevertheless, the presence of phytate seems to be a factor in the antinutritive properties of beans and bran products. In World War II, there was a calcium deficiency illness in children, which was associated with phytate intake (McCance and Widdowson, 1942). This deficiency was compensated for with bread that was produced from debranned wheat that is low in phytate, and by addition of calcium.

Erdmann and Frobes (1977) showed that in rats, piglets, and chickens, the availability of phosphate, calcium, iron, and zinc was limited. Also in human nutrition, diets containing large amounts of unfermented cereals with high phytin contents can be a problem (Schwimmer, 1981/2).

Phytase occurs in cereals, in the blood of amphibians and birds, in the intestinal membranes of rats, and in innumerable microorganisms. The phytase from wheat bran is inhibited by an excess substrate and by phosphate ions. Phosphatides do not affect the enzyme. The phytase from wheat bran shows optimal activity from pH 7 to 7.5. This enzyme can be activated by autolytic treatment of wheat bran (6 h, 50°C, pH 4.5). In breads made from flour to which such pretreated bran has been added, calcium, magnesium, and phosphate are more available than without such pretreatment. In bread production without such phytate-rich materials—that is, without an excess of bran—the small amount of phytate disappears as a result of the action of the native wheat enzymes and enzymes in the yeast.

The claim that phytate is an antinutritive substance is questionable because the intestines of most monogastrics, including that of humans, contain phytase. This activity is usually weak, but is induced by phytate. Calcium inhibits phytase activity. Processes have been developed, for example, with beans, which reduce phytate by autolytic processes. Incubation for ~24 h and a temperature of ~50°C is required. The phytate content of beans is also lowered by germination. A phytase, which was obtained by Rojas and Scott (1969) from *Aspergillus ficuum,* was used for the hydrolysis of phytate in linseeds.

Product and Manufacturer: A technical phytase is now available [AO], [GB].

REFERENCES

Antonian, E., *Lipids* **23**(12), 1101 (1988).

Borgström, B., and Brockmann, H. L., *Lipases,* Elsevier, New York, 1984.

Brockerhoff, H., *Chem. Phys. Lipids* **10**, 215 (1973).

Brockerhoff, H., and Jensen, R. G., *Lipolytic Enzymes,* Academic Press, New York, 1974.

Bühler, M., and Wandrey, C., *Fett Wissensch. Technol.* **89**(4), 156 (1987).

De Haas, G. H., Slotboom, A. J., Bonsen, P. P. M., Nieuwenhuizen, W., van Deenen, L. L. M., Maroux, S. Dlouha, V., and Desnuelle, P., *Biochim. Biophys. Acta* **221**, 54 (1970).

Erdmann, J. W., and Frobes, R. M., *Food Prod. Dev.* **11**(10), 46 (1977).

Fukumoto, J., Iwai, M., and Tsujisaka, Y, *J. Gen. Microbiol.* **9**, 353 (1963).

Hanahan, D. J., and Chaikoff, I. L., *J. Biol. Chem.* **172**, 191 (1948).

Hofelmann, M., Hartmann, J., Zink, A., and Schreier, P., *J. Food Sci.* **50**, 1721 (1985).

Holmann, R. T., *Prog. Chem. Fats Other Lipids* **9**, 1 (1966).

Jensen, R. G., *Prog. Chem. Fats Other Lipids* **11**, 347 (1971).

Jensen, R. G., and Pitas, R. E., in Paolette, R., Porcellati, G., and Jacini, G. (eds.), *Lipids,* Vol 1, Raven Press, New York, 1976, p. 141.

McCance, R. A, and Widdowson, E. M., *J. Physiol.* **101**, 304 (1942).

Mohsen, S. M., Allan, A. M., and El-Azhary, T., *Egypt. J. Food Sci.* **14**(1), 147 (1986).

Nelson, J. H., Jensen, R. G., and Pitas, R. E., *J. Dairy Sci.* **60**(3), 327 (1977).

Nielsen, T., *Fette, Seifen, Anstrichmittel* **87**(1), 15 (1985).

Omar, I. C., Nishio, N., and Nagai, S., *Agric. Biol. Chem.* **8**, 2153, 2154 (1987).

Richardson, H. G., Nelson, J. H., and Fraham, M. G., *J. Dairy Sci.* **54**(5), 643 (1971).

Richardson, H. G., and Nelson, J. H. *J. Dairy Sci.* **50**, 1061, (1967).

Rojas, S. W., and Scott, M. L., *Poultry Sci.* **48**, 819 (1969).

Sarda, L., and Desnuelle, P., *Biochim. Biophys. Acta* **30**, 513 (1958).

Schäfer, W., *Getreide, Mehl und Brot* **42**, 26 (1988).

Schwimmer, S., *Source Book of Food Enzymology*, p. 656, Avi Publishers, Westport, Connecticut, (1981).

Semeriva, M., Benzonana, G., and Desnuelle, P., *Biochim. Biophys. Acta* **191**, 598 (1969).

Slotboom, A. J., De Haas, G. H., Bonsen, P. P. M., Burbach-Westerhuis, G. J., and van Deenen, L. L. M., *Chem. Physics Lipids* **4**, 15, (1970).

Stuer, W., Jaeger, K. E. and Winkler, U. K., *J. Bacteriol.* **168**(3), 1070 (1986).

Tomizuka, N., Ota, Y., and Yamada, K., *Agric. Biol. Chem.* **30**, 1090 (1966).

Verger, R., de Haas, G. H., Sarda, L., and Desnuelle, P., *Biochim. Biophys. Acta* **188**, 272 (1969).

Willstätter, R., and Memmen, E., *Hoppe-Seyler's Ztschr. Physiol. Chem.* **129**, 1 (1923).

3.4   OXIDOREDUCTASES

References: Scott (1975), Bright (1974)

Oxygen causes changes in food. Such oxidative processes are mainly catalyzed by enzymes. There are many frequently encountered examples such as the browning of fruit or vegetables caused by polyphenoloxidase, which is also apparent in the darkening of mushrooms or potatoes when tissue is injured. Another oxidase, lipoxygenase, which occurs in soybeans, oxidizes unsaturated fatty acids. A related oxidation can occur with carotene, and this is used for bleaching flour in the manufacture of baked goods. Enzymatically catalyzed oxidation plays a major role in the formation of aroma compounds. These flavor compounds are changed by oxidation. Also, in addition to the above-mentioned rancidity due to lipolysis, oxidative rancidity can occur in milk fat. Oxidases can destroy valuable vitamins, such as ascorbic acid.

Enzymes of this group are widely distributed in plants, microorganisms, and animal tissue. Glucose oxidase is liberated by the pharynx gland of the honeybee, which leads to its presence in honey. Peroxidase that—in the presence of hydrogen peroxide—produces strongly colored products from phenolic substrates is prepared from horseradish.

To eliminate the damaging effects of the oxidoreductases in food, it is possible to inactivate these enzymes by heat or to completely eliminate the oxygen from the food. This can be done with glucose oxidase. In spite of their potential in food processing, only a few enzymes of this group have presently been produced in commercial quantities. These are glucose oxidase and catalase. Catalase serves to degrade residual hydrogen peroxide when this substance is used for cold sterilization of milk.

Scott (1975) has written a good review on the oxidation reaction. He has worked on the characterization, manufacture, and the application of oxidoreductases since

1950. There are a variety of reactions that are catalyzed by oxidases. In the classic reaction (1), two atoms of oxygen are transferred to a substrate:

$$A + O_2 \longrightarrow AO_2 \tag{1}$$

Example: catechol-1,2-oxygenase

As an example for the action of lipoxygenase, Scott (1975) uses the oxidation of pyrocatechol with two oxygen atoms to yield muconic acid and oxidized double bonds in fatty acids.

According to an alternative mechanism, the substrate is oxidized by the transfer of two hydrogen atoms to molecular oxygen, thereby producing hydrogen peroxide:

$$AH_2 + O_2 \longrightarrow A + H_2O_2 \tag{2}$$

Example: glucose oxidase

Glucose + $O_2 \longrightarrow$ gluconic acid + H_2O_2

As an example of this reaction (2), the reaction of glucose with glucose oxidase is shown. However, hydrogen peroxide is produced only if the enzyme is highly purified. Commercial glucose oxidase contains catalase, which immediately decomposes any hydrogen peroxide formed to water and oxygen.

$$AH_2 + \tfrac{1}{2}O_2 \longrightarrow A + H_2O \tag{3}$$

Example: ascorbic acid oxidase

Ascorbic acid + $\tfrac{1}{2}O_2 \longrightarrow$ dehydroascorbic acid + H_2O

The oxidation of ascorbic acid [reaction (3)] is important for dough preparation in bread production. It is assumed that dehydroascorbic acid can oxidize the free SH groups of gluten to S—S bridges. This would explain the role of ascorbic acid in strengthening gluten, an important process in baking technology.

Another oxidation mechanism occurs when the oxygen atom, necessary for oxidation, comes from H_2O rather than O_2:

$$A + H_2O + B \longrightarrow AH_2 + BO \tag{4}$$

Example: glucose dehydrogenase

Glucose + H_2O + NAD $\longrightarrow$ gluconic acid + NADH

The oxidation of glucose by a dehydrogenase is shown as an example.

3.4.1 Oxidases

3.4.1.1 Fungal Oxidases

Glucose oxidase (GO, GO_d, GO_x; β-D-glucose:oxygen-1-oxidoreductase) (EC 1.1.3.4; CAS 9001-37-0)

Other Names: Notatin, glucose oxyhydrase.

GO catalyzes the oxidation of β-D-glucose to β-D-gluconic acid and hydrogen peroxide. The lactone hydrolyzes spontaneously, not enzymatically, to D-gluconic acid. This reaction is accelerated by the action of D-gluconolactone hydrolase, which is incorporated in commercial GO preparations. Maximov (1904) was the first to discover GO activity in a dry powder from the mycelia of *A. niger.* Müller (1928) was the first to relate this oxidative action to an enzyme. It was assumed that the yellow color of the purified enzyme originated from a prosthetic group that was bound to the enzyme. Keilin and Hartree (1948) confirmed the theory that the oxidation was caused by flavin–adenine–dinucleotide (FAD). The idea that GO acts as an antibiotic (Notatin) was dropped when it was discovered that the antimicrobial activity resulted from the formation of hydrogen peroxide.

The activity of hydrolytic enzymes such as amylases and proteases relates to the special arrangement of amino acids at the enzymes' active site, whereas the activity of the oxidoreductase is linked to the presence of a special group of substances, the coenzymes of which are capable of transporting hydrogen. Hydrogen is taken up in a reversible reaction and released again, for example, to oxygen. A portion of the so-called FAD, referred to above, should illustrate this (Fig. 3.108).

The first commercial product entered the market in 1952. Today, technical preparations are commonly produced from *A. niger,* or, as in Japan, from *Penicillium amagasakiense* (Kusai et al., 1960). Earlier, other *Penicillium* strains such as *P. notatum* (Keilin and Hartree, 1948) or *P. glaucum* were used. GO is an intracellular enzyme that is not released to the medium during culturing but remains within the cell. This does not seem to be the case for *Penicillium.* However, the rapid release of the enzyme during fungal culture was traced back to autolysis of the mycelia.

Therefore, the enzyme has to be freed from cells during the manufacturing process. This is accomplished by the mechanical destruction of the cell walls or by the introduction of autolytic processes, supported by exogenous enzymes. The solubilized enzymes are then purified, usually by filtration, ultrafiltration, or similar, then concentrated and stabilized or dried.

PROPERTIES   The enzyme from *A. niger* has a molecular weight of 160,000 daltons and contains two FAD groups. This enzyme consists of two subunits that are connected by disulfide bridges. GO from *Penicillium* consists of four subunits: peptide chains with a molecular weight of ~45,000 daltons. Two pairs of subunits each form dimers linked by disulfide bridges. The two dimers are associated to a tetramer.

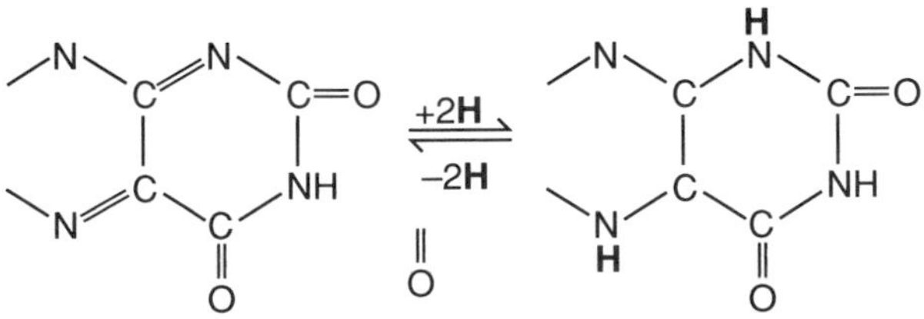

Figure 3.108   Hydrogen transport via flavin adenine dinucleotide (FAD).

TABLE 3.42   Comparison of Glucose Oxidase from Different Sources (Scott, 1975)

Enzyme Source	Molecular Weight (daltons)	K_m	iP
P. amagasakiense	154,000	1.15×10^{-2}	4.35
P. notatum	138,000–152,000	0.96×10^{-2}	—
A. niger	192,000	1.1×10^{-2}	4.5

The enzymes from *Aspergillus* and *Penicillium* differ in amino acid composition but have similar properties (Table 3.42).

Activity: The pH optimum for the enzyme from *A. niger* lies at pH 5.5. The pH activity ranges from pH 2.6 to 6.5. The pH range depends closely on temperature and other reaction conditions (Fig. 3.109).

Stability: GO from *A. niger* contains ~16% carbohydrate. If this fraction is removed by oxidation, the activity is preserved while temperature stability is lost. Scott (1975) determined stability in a practical application by removing oxygen from acidic drinks. He used Cola at 30°C and pH 2.6 and grape juice at pH 3.2 (Fig. 3.110).

Enzyme: Fermcozyme [SB].

Inactivation: The enzyme is 90% degraded by heating at 80°C for 2 min, 70°C for 3 min or 65°C for 10 min (Scott, 1980).

Products and Manufacturers: Deoxin [NG]; Oxygo, Ovazyme [FR]; [SOE], [GB], [BÖM], [AM]; [SOEG].

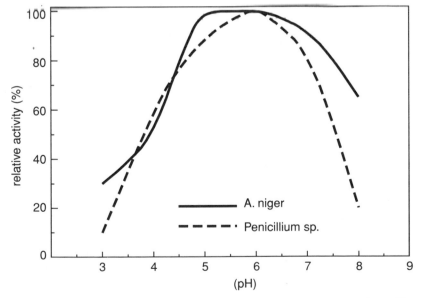

Figure 3.109 Effect of pH on activity of glucose oxidase from *A. niger* and *Penicillium* sp.

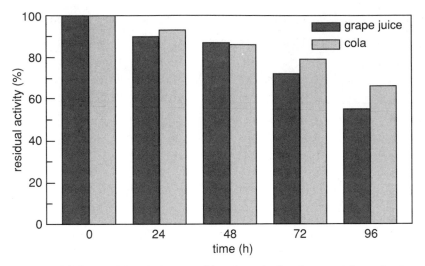

Figure 3.110   Stability of glucose oxidase from *A. niger* in grape juice and cola.

Catalase (EC 1.11.1.6; CAS 9001-05-2)
References: Scott (1975), Ammon and Dirscherl (1959)

Tenard (1918) realized that animal as well as plant cells degrade hydrogen peroxide by formating oxygen. Jakobsen (1892) showed that the reaction was caused by an enzyme. Loew (1901) suggested the name *catalase* for the enzyme. For a long time, the action of catalase was known only for its degradation of hydroperoxide. Keilin and Hartree (1936) showed that catalase could also act as a peroxidase. Thus, catalase can be regarded as a special peroxidase that oxidizes hydrogen peroxide to oxygen, with a simultaneous reduction of a second molecule of hydrogen peroxide to water. With the degradation of peroxide, the oxidation of certain alcohols, or nitrites or phenols, can proceed.

Catalase is isolated from beef liver, *A. niger,* and *Micrococcus lysodeikticus.*

PROPERTIES, PRODUCTS, AND MANUFACTURERS

Molecular Weight: Liver enzyme, 250,000 daltons.

Cofactor: Four ferric–protohemin molecules per mole of enzyme.

Activity: pH optimum: Liver enzyme pH 7, activity range pH 5–8, *A. niger* pH 6, activity range pH 3–9; *Ml lysodeikticus* pH 6, activity range pH 5.5–8. (See also Fig. 3.111.)

Temperature: The *A. niger* enzyme can tolerate 65°C for a few minutes (Scott, 1960). The enzyme from *M. lysodeikticus* acts optimally at pH 7.0 and 30°C; it is stable at pH 7.0 for 15 min at 50–55°C.

Analysis

   BU: One Baker unit (BU) is the amount of catalase (fungal catalase) that will

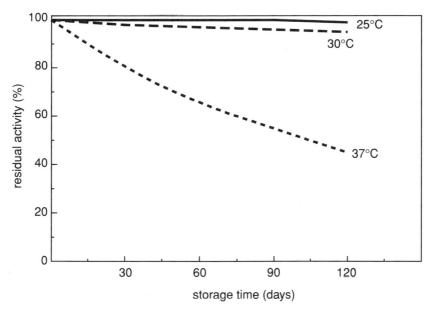

Figure 3.111 Stability of catalase from *Micrococcus* (Catalase U5L [UA]).

decompose 264 mg of hydrogen peroxide under the reaction conditions defined by Scott and Hammer (1960).

CU: One catalase unit (CU; Japanese method [UA], [AM]) is the amount of catalase that degrades 1 mM of hydrogen peroxide in 1 min at 30°C (iodine thiosulfate titration according to Stellmach, 1988).

KU: One Keil unit (KU) is the amount of catalase (liver catalase) necessary to decompose one g of 100% hydrogen peroxide in 10 min at 25°C and pH 7 in an inert atmosphere of CO_2 or N_2 (Keil, 1954; Scott and Hammer, 1960).

Technical Products: Stabilized liquid preparations, with 1000 BU/ml; technical powder preparations with 2000 to 8000 BU/g, or 100 KU/ml. Reagent quality: ~40,000 units/g.

Manufacturers: [UA], [FR], [AM], [SOE].

For sulfhydryloxidase, see Section 5.6.

REFERENCES

Ammon, R., Dircherl, W., in Fermente. Hormone, Vitamine, Georg Thieme, Stuttgart, 1959, p. 516.

Bright, H., "Food related enzymes," *Symp. Adv. Chem. Ser.* 305 (1974).

Jakobsen, K., Hoppe-Seyler's Zschr. Physiol. Chem. No. 340 (1892).

Keilin, D., and Hartree, E. F., *Proc. Roy. Soc.* Ser. B., **119,** 141 (1936); Ser. B, **121,** 173 (1936).

Keilin, D., and Hartree, E. F., *Biochem. J.* **42,** 221 (1948).

Kusai, K., Sekuzu, I., Hagihara, B., Okunuki, K., Yamauchi, S., and Nakai, N., *Biochim. Biophys. Acta* **40,** 555 (1960).

Loew, O., U.S. Dept. Agric. Report No. 68 (1), U.S. GPO, Washington, DC, 1901.

Maximov, N. A., *Ber. Dtsch. Bot. Ges.* **22,** 225, 488 (1904).

Müller, D., *Biochem. Ischr.* **199,** 136 (1892, 1928).

Müller, D., *Ergeb. Enzymforsch.* **5,** 259 (1936).

Scott, D., *Enzymologia* **22,** 223 (1960).

Scott, D., "Oxidoreductases," in Reed, G., (ed.), *Enzymes in Food Processing,* Academic Press, New York, 1975, p. 222.

Scott, D., "Industrial enzymes," in *Kirk-Othmer Encylcopedia of Chemical Technology,* 3rd ed., Wiley, New York, 1980, 202.

Scott, D., and Hammer, F., *Enzymologia* **22,** 194, 229 (1960).

Stellmach, B., *Bestimmungsmethoden der Enzyme für die Pharmazie,* Lebensmitteltechnik. Biochemie und Medizin, Steinkopff, Darmstadt, 1988.

Tenard, L. J., *Ann. Chim. Phys.* **9,** 314 (1918).

3.5 GLUCOSE ISOMERASE (EC 5.3.1.5; CAS 9055-00-9)

References: Bucke (1981, 1983)

One of the most important new developments in the area of starch conversion to sugar syrup was the introduction of glucose isomerase. Previously it had been possible to produce glucose syrup and dry glucose from starch; it now became possible to produce a mixture of glucose and fructose with the help of this new enzyme. Glucose isomerase catalyzes the isomerization of glucose to fructose. The mixture is the same as that obtained by the inversion of sucrose when using invertase. It is this form of artificial honey that has a sweetness equal to that of sucrose. This is attributable to the fructose portion; on a weight basis, fructose is 1.2–1.8 times sweeter than sucrose. Fructose, which can be extracted from the mixture, has extra significance as dietetic sugar. The sugar syrup, which is also called *isomerose,* is generally obtained from corn and contains about 42% fructose. By concentrating, a "high-fructose corn syrup" (HFCS) containing 55% fructose is obtained. In the United States and Canada it is used mainly in beverages. Currently about 4 million tons of HFCS are manufactured worldwide.

This unusual enzyme was discovered by Marshall and Kooi (1957). They found that an enzyme from *Pseudomonas* that converted xylose to xylulose was also able to convert glucose to fructose. Since the reaction was very slow, feverish activity followed in the 1960s in search of suitable enzymes. Various microorganisms that produced the enzyme were discovered that could be used commercially. However, the process, which had to compete with invert-sugar production, was still too expensive. Only with the introduction of immobilized enzyme systems and the associated continuous recyclization did the process become economically feasible. The great commercial value of this process has helped to encourage much research and development in the area of immobilization technology.

3.5.1 Glucose Isomerase from Microorganisms

3.5.1.1 Glucose Isomerases from Bacteria

Glucose Isomerase from Lactobacillus sp. Kent and Emmery (1973) showed that a xylose isomerase can be induced in *Lactobacillus* and can efficiently isomerize xylose or glucose equally. This enzyme requires magnesium ions and cobalt ions for the isomerization of xylose and glucose, respectively. Because the enzyme is not very stable, the organism is not used for the commercial production of glucose isomerase.

Glucose Isomerase from Streptomyces sp. Tsumura and Sato (1965) showed that *S. phaechromogenes* also formed a xylose or xylan-induced xylose isomerase that was able to isomerize glucose equally well. Magnesium ions were necessary for both activities, and heat stability was improved by cobalt ions. However, the activity optimum of this enzyme was between pH 9.3 and 9.5, which was too high for a commercial process.

Takasaki et al. (1969) were able to obtain a glucose isomerase from *S. olivochromogenes* that showed a pH optimum range from pH 8 to 8.5. CPC International [CPC] developed a commercial product from this strain that did not require cobalt. The isomerase from *S. griseofuscus* (GODO) has a molecular weight of 180,000 daltons. This enzyme contains four cobalt atoms per molecule and is also activated by magnesium ions. It is stable over the pH range of pH 5.0–11.0 for 12 h at 30°C. Maximum temperature is 80°C. The isomerase product of Clinton Corn Processing from *S. albus* is dependent on manganese ions. The molecular weight is 165,000 daltons. The molecule consists of four subunits.

Various *Bacillus* species also form xylose isomerase, which isomerises glucose, as well as ribose, in the presence of cobalt ions.

Glucose Isomerase from Bacillus coagulans The enzyme from *B. coagulans*, produced by Novo [NO], is active up to 90°C, but optimal stability is between 60 and 65°C. Optimal activity is between pH 8 and 8.5; irreversible denaturation occurs below pH 5. Danno (1971) reported on the role of cobalt ions with this enzyme. It appears that they are not active as cofactors at the active site, but rather effect conformational changes of the enzyme such as significantly potentiating its activity to isomerize glucose.

Glucose Isomerase from Arthrobacter sp. This enzyme from *Arthrobacter* requires cobalt ions for neither activity nor for stability.

The enzyme from *Actinoplanes missouriensis* appears to be a genuine glucose isomerase because it is formed in the absence of xylose or xylan in the culture medium. The enzyme requires magnesium and cobalt ions for full activity. In immobilized enzymes, the cobalt problem is solved by immobilizing both the enzyme and cobalt. Currently, glucose isomerase is used technically in its immobilized state in solid-bed reactors. Immobilized enzymes are described in Chapter 4.

Products and Manufacturers: See Table 3.43.

TABLE 3.43   Various Microorganisms Used in Glucose Isomerase Production

Organism	Producer	Tradename
Streptomyces griseofuscus	Godo Shusei Co. Ltd. [GO]	GODO-GI
S. olivaceus	Solvay Enzymes Inc. [SOE]	Taka-Sweet
A. rubiginosus	Cultor Ltd. [CL]	Spezyme GI
	Solvay Enzyme GmbH [SO]	Optisweet
S. olivochromogenes	CPC International/Div. Enzyme Biosystems Ltd. [CPC]	G-zyme G 993
S. violacenoniger	Roquette Freres S. A. [RF]	
S. phaeochromogenes	Nagase Co. Ltd. [NG]	Swetase
Bacillus coagulans	Novo Nordisk A/S [NO]	Sweetzyme
Actinoplanes missouriensis	Gist-Brocades N.V. [GB]	Maxazyme GI— immobilized
Arthrobacter sp.	ICI Biological Products [ICI]	Glucose isomerase

REFERENCES

Bucke, C., "Enzymes in fructose manufacture," in Birch, G. G., Blakebrough, N., and Parker, K. J. (eds.), *Enzymes and Food Processing,* Appl. Science Publ. Ltd., London, 1981.

Bucke, C., "Glucose-transforming enzymes," in Fogarty, W. M., (ed.), *Microbial Enzymes and Biotechnology,* Appl. Science Publ. Ltd., London, 1983, p. 93.

Danno, G., *Agric. Biol. Chem.* **35,** 997 (1971).

Kent, C. A., and Emmery, A. N., *J. Appl. Chem. Biotechn.* **23,** 689 (1973).

Marshall, R. O., and Kooi, E. R., *Science* **125,** 648 (1957).

Takasaki, Y., Kosugi, Y., and Kanbatashi, A., *Agric. Biol. Chem.* **33,** 1527 (1969).

Tsumura, N., and Sato, R., *Agric. Biol. Chem.* **29,** 1123 (1965).

4 Carrier-Bound Enzymes: Methods of Immobilization

References: Godfrey (1983), Lam and Malikin (1994), Gorton et al. (1994), Adlercreutz (1993)

4.1 PRINCIPLES OF COUPLING TO CARRIER

The definition of immobilized enzymes issued by the European Federation of Biotechnology (1983) includes "all enzymes that can be recycled." In the 1970s, the advantage of the coupled versus the noncoupled, or free, enzyme was thought to be a great innovative step in enzyme technology. These expectations were only partially realized. Presently, only two large-scale processes exist that employ immobilized enzymes: glucose isomerization with immobilized glucose isomerase and the splitting of Penicillin G with immobilized penicillin amidase. However, a number of processes still await large-scale industrial application. Their implementation is not so much an issue of inadequate technical knowhow, but rather of market demand, market acceptance of the products, and regulatory laws.

Basic scientific or commercial interests can lead to enzyme immobilization. In this text, the latter are emphasized. Deciding criteria are as follows:

1. For the facility:
   a. High efficiency, using concentrated enzyme beds, thereby reducing (thermal) stress on the product.
   b. More automation is possible; however, this is associated with higher investment costs, specifically, new investment.
2. For the process:
   a. Improved process control.
   b. Continuous operation; however, this may result in new problems such as microbial contamination.
3. For the product:
   a. Maintaining constant quality.
   b. Product not contaminated with enzyme; therefore, no requirement for enzyme inactivation. Possibly better sensory quality. High-molecular-weight or turbid substrates are not suitable for immobilized enzymes.

Immobilized enzymes basically compete with native enzymes or with conventional chemical processes. A careful evaluation of all factors mentioned under 1–3 (above) is required. Enzyme costs themselves play a deciding role. Immobilization of cheap bulk enzymes has little value. The carrier and immobilization costs, along with the loss of enzymatic activity during immobilization, would not be recovered.

Furthermore, for many enzymes, continuous operation does not permit running at optimum temperatures but rather requires temperatures at least 10°C lower. This represents an added disadvantage to using the free enzyme. Nevertheless, immobilization is, in principle, profitable:

- For expensive enzymes (e.g., penicillin acylase).
- For low-molecular-weight substrates (e.g., sugar, amino acids).
- For controlling those processes where microbial contamination may occur. Here, a sterile substrate or reaction temperature of >60°C will be helpful.
- Given the need for an absolutely enzyme-free product (e.g., allergen-free diets).

In addition, the important limitation due to the molecular size of a given substrate must be considered. The use of immobilized enzymes for many enzymatic processes such as the hydrolyses of starch, pectin, proteins, and cellulose is futile. Here, the limited diffusion of these large molecules would, at best, effect a chromatographic preselection of the substrate (Plainer and Sprössler, 1987); the hydrolysis of a high-molecular-weight substrate would occur only on the surface of the carrier because the immobilized enzymes, trapped in the carrier center, would not be reached because of the length of the diffusion paths.

A concise and timely review of this technology with emphasis on analytic applications of immobilized enzyme reactor systems has been published by Lam and Malikin (1994).

4.2   IMMOBILIZATION METHODS

At the meeting on "Enzyme Engineering" in Henniker, New Hampshire (USA) in 1971, the possibility of enzyme immobilization was initially discussed. The four classes of immobilization established then are still relevant today but have acquired different degrees of importance.

The diagram in Figure 4.1 also includes special cases such as "immobilized cells" (group 3), specifically, enzymes embedded in a matrix. Combinations are also possible, such as the adsorption of an enzyme at a carrier surface (Fig. 4.1, group 2), with subsequent covalent cross-linking between enzyme molecules (Fig. 4.1, group 1), a frequently used and effective method of immobilization.

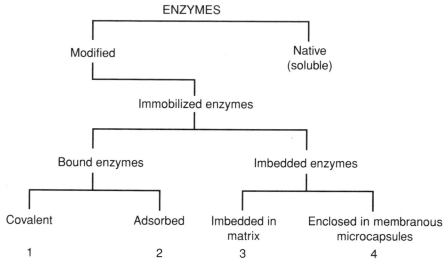

Figure 4.1   Nomenclature for immobilized enzymes.

4.2.1   Enzyme Carriers

Most industrial processes involving immobilized enzymes are conducted in fixed-bed reactors. The following carrier requirements have been developed:

- Particle size > 0.1 mm, usually 0.5–1 mm.
- Interior surface > 50 m^2/g.
- Average pore diameter $= 10$–50 nm.
- Pore volume > 1 ml/g, usually 2–3 ml/g.
- Mechanical stability is required; gels cannot be used.

Carrier materials having these morphological specifications require synthesis because most materials of natural origin cannot meet these requirements.

4.2.2   Covalent Coupling

Enzymes are immobilized by various means. One method is the covalent coupling of the enzyme to a reactive support or matrix. The covalent binding should be accomplished through functional groups in the enzyme that are not involved in the enzyme's catalytic activity. Generally, the binding is achieved by the nucleophilic groups of the enzyme reacting with activated functional groups of the support material. Aminohydroxyl and thiol groups on the enzyme may participate in the linkage. Cysteine, lysine, tyrosine, and histidine residues may be considered the most active

(Kennedy et al., 1990). Such intermolecular bonds may result in an extended useful life of the enzyme.

As a rule, the covalent binding shown in group 1 of Figure 4.1 is preferred. The introduction of a functional group on the carrier is required, by way of either copolymerizing with functional monomers or a polymer-type reaction (Fig. 4.2).

Another often overlooked requirement for good enzyme coupling is the adsorption of the enzyme prior to the coupling reaction. This includes hydrophobic or

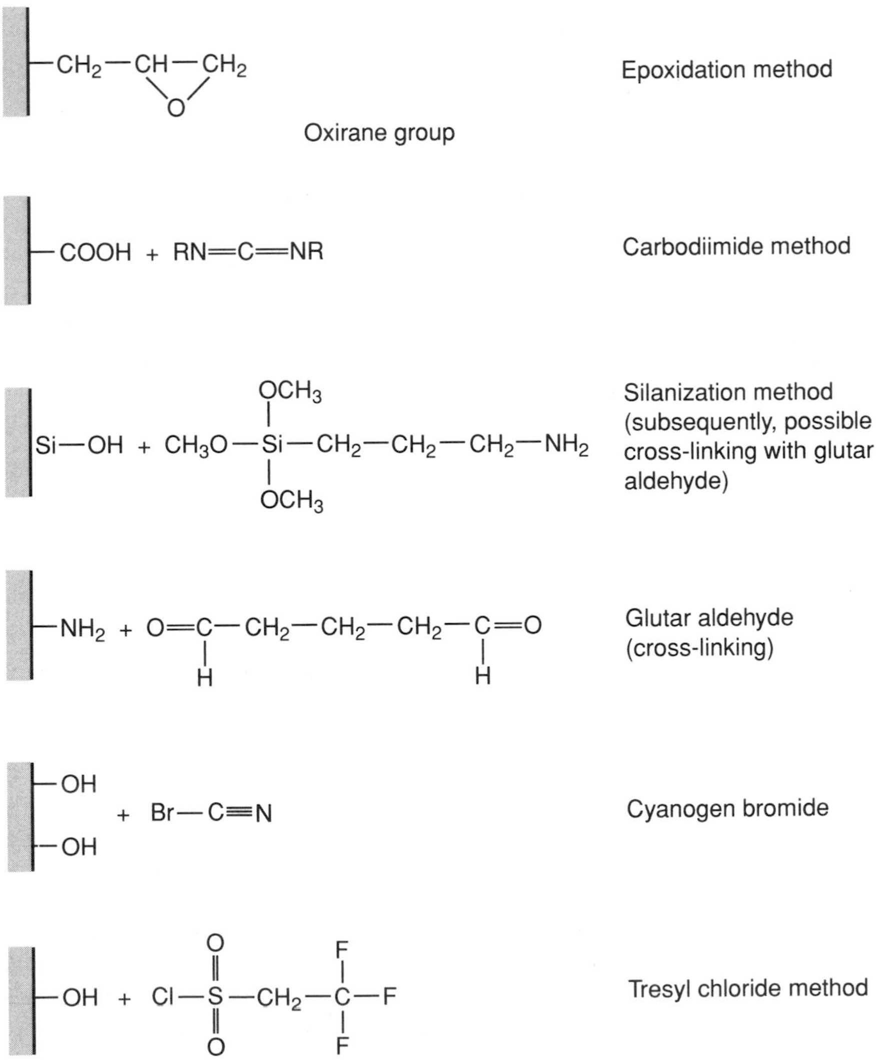

Figure 4.2   Activating the carriers for binding enzymes.

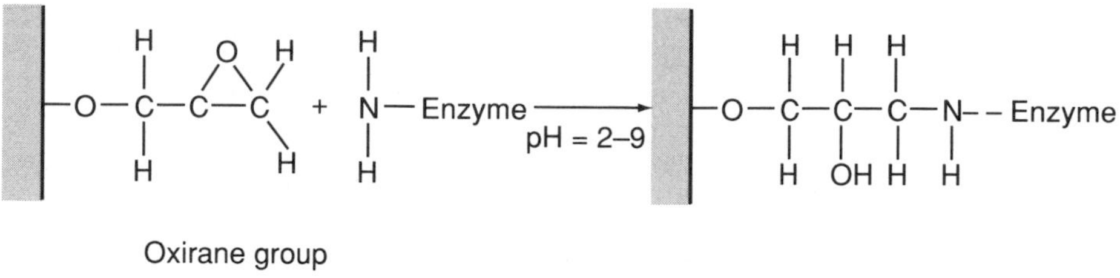

Oxirane group

Figure 4.3 Oxirane binding of enzyme.

ionic adsorption. This is shown for epoxide coupling. Eupergit C (Röhm Pharma, Weiterstadt, Germany) hydrophobically adsorbs the enzyme in the presence of a high electrolyte concentration (Krämer et al., 1985, 1991). Its accumulation on the carrier's inner surface permits the reaction with the oxirane group over a wide pH range (Fig. 4.3).

Usually, the enzyme's lysine α-amino groups as well as its free sulfhydryl groups are coupled, predominantly in a cross-linked manner.

Epoxide coupling is used commercially for the immobilization of penicillin amidase (e.g., Eupergit PCA, Röhm Pharma).

4.2.3 Ionotropic Gel Formation

Imbedding the enzymes in a matrix of natural or synthetic polymers is quite a simple process (Fig. 4.4). Anionic polymers [e.g., Na-alginate, carrageenan, carboxymethylcellulose (CMC)] or cationic polymers (e.g., chitosan) are cross-linked with their counterions (calcium or polyphosphate).

Unfortunately, this procedure has some drawbacks. The Ca^{2+}-ion-mediated linkage, for example, easily dissolves on salt addition. The gel spheres are soft and ill-suited as fillers in fixed-bed reactors. Klein and Wagner (1983) report a number of ways to circumvent such difficulties; one method involves reinforcing the matrix with epoxide resins.

Despite the disadvantages, the ionotropic gelling process has proved itself as the method of choice for immobilizing whole cells. For example, yeast cells, immobilized in alginate, are used in alcohol production; or *E. coli* cells, embedded in carrageenan, are used for producing aspartic or malic acids (see Table 4.2).

4.2.4 Cross-Linking (Lysed) Cells

This simple method is now the accepted way of immobilizing glucose isomerase; it does not require enzyme isolation. By volume, this method has become the most important one. The main problem resides in the limited mechanical stability of the immobilized system; this can partially be resolved by concomitantly including structural support substances (gelatin, polymers). Various alternatives are illustrated in Table 4.1.

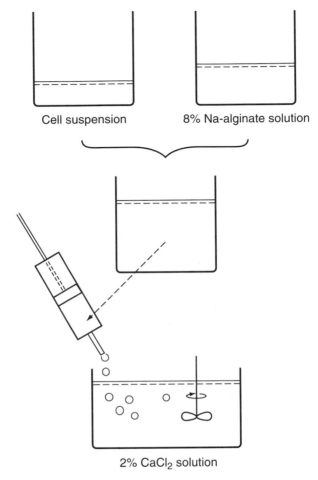

Cell suspension   8% Na-alginate solution

2% CaCl$_2$ solution

Figure 4.4 Laboratory method for making ionotropic gels.

TABLE 4.1 Immobilization of Glucose Isomerase by Cross-Linking

Supplier	Country[a]	Immobilization Method
Novo	DK	*Bacillus coagulans* cells, homogenate, cross-linked with glutyraldehyde (GA), extruded
Gist Brocades	NL	*Actinoplanes missouriensis* cells mixed with gelatine, cross-linked with GA
Miles Labs	USA	*Streptomyces olivaceus* cells, cross-linked with polyamine + GA, extruded
ICI	UK	*Arthrobacter globiformis* cells, coagulated with cationic + anionic polymers
Nagase	J	*Streptomyces phaechromogenes* enzyme, GA cross-linked with intact but dead cells

[a]*Countries:* DK—Denmark; NL—the Netherlands; USA—United States; UK—United Kingdom; J—Japan.

4.2.5 Enzyme–Membrane Reactors (EMRs)

Immobilized enzymes (e.g., glucose isomerase) are used primarily in fixed-bed reactors; for penicillin amidase, a batch reactor can be used as well. The membrane
reactor represents a special case of enzyme immobilization. It basically involves
physically separating the enzyme from the substrate or product with a semiperme-
able membrane; generally, it is an ultrafiltration membrane with an exclusion limit
of 10,000 daltons. The main advantages of EMRs are the high degree of permeability and the control of microbial contamination. The most important EMR designs
are illustrated in Figure 4.5.

1. One method permits product passage through the membranes, while retaining
the enzymes and allowing sufficient substrate retention time in the reaction vessel.
Commercial examples are L-amino acid production by splitting the racemate of
acetylated amino acids with L-aminoacylase (Wandrey and Flaschel, 1979) or stereo-
selective reductive amination of α-keto acids with amino acid dehydrogenase and
formate dehydrogenase + NAD (Leuchtenberger et al., 1984). In this case, the problem with the low-molecular-weight coenzyme NAD was solved elegantly by coupling NAD to the higher-molecular-weight polyethylene glycol, which increased its
size and thus prevented its passage through the membrane.

2. Another method permits only the limited passage of substrate and product
through an ultrafiltration membrane; the enzyme circulates within the closed loop
on the catalyst side. This system shows a close relationship to carrier-bound enzymes. The membrane pores can be compared to carrier pores; the diffusion paths,
however, are far shorter because the enzyme can immediately interact with the substrate on the catalyst side. With sufficiently large and thin membrane surfaces, the
system's diffusion capability is relatively unlimited. AKZO has developed a steam-
sterilizable dialysis membrane reactor for this system (Czermak et al., 1988).

3. A third method permits immobilization of the enzyme within the membrane.
The reaction takes place while substrate passes through the membrane and is not
diffusion-limited. Obtaining higher production yields becomes a problem even un-

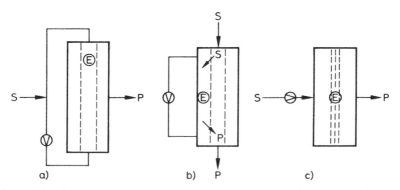

Figure 4.5   Enzyme–membrane reactors; S = substrate; E = enzyme; P = product.

der moderate reaction kinetics. This can be circumvented by coupling many membranes in series. This method has also found technical application, such as in the Amerace filters made of PVC/silica (Schmidt-Kastner and Gölker, 1986) for lactose hydrolysis. Problems arise only with solutions containing particulates or having high viscosity.

4.3 CURRENT TECHNICAL APPLICATIONS

Reference: Tischer (1992)

Until 1979, high expectations for immobilized enzymes still existed. Immobilizing strategies were suggested for practically all known enzyme reactions (List and Knechtel, 1980). However, five years later, Poulsen (1984) delivered a realistic appraisal of the possibilities for immobilized enzyme applications. At that time, only two immobilized enzymes were successful in large-scale commercial settings. Today, nothing much has changed. Other than the two "highlights"—glucose isomerase and penicillin amidase, both exclusively used in immobilized systems—only very few commercial applications (amino acid acylase, aspartase, fumarase, nitrilase) exist, predominantly in Japan, that can compete with processes employing free enzymes or chemical methods. There is no shortage of fully developed methods for enzyme immobilization. However, for economical reasons they cannot come to fruition. In the past, many projects were started without prior market analysis. Table 4.2 lists and evaluates the most important technical applications. The immobilization methods listed refer to Figure 4.1 and are identified by their group number in Table 4.2.

4.4 SELECTED EXAMPLES

4.4.1 Immobilized Glucose Isomerase

About 1500 tons of immobilized glucose isomerase (value ~$58.5 million) are required to produce 6–7 million t/yr (metric tons per year) of isosyrup or high-fructose corn syrup (HFCS) internationally (see Table 4.2). Three factors were responsible for this success:

1. In comparison to the 1960s, the price of unrefined sugar increased 10–12-fold by the mid-1970s. Consequently, there was a need to explore sugar resources other than sucrose.
2. Glucose isomerase is too expensive for use in its native form.
3. The enzyme has sufficient thermostability to permit continuous operation at 60°C, which will eliminate microbial problems.

Basically, three immobilization methods exist:

TABLE 4.2 Industrial Application of Immobilized Enzymes (Status in 1988)

Immobilized Enzymes	Production Scale[a]	End Product (t/yr)	Immobilization Method	Group[b]	Most Important Suppliers (Country)[c]
Amino acid acylase	L	1000 t Phe, Met, Try, Val	DEAE sepharose	(2)	Tanabe (J)
			Membrane reactor	(4)	Degussa (G)
			Eupergit	(1)	Röhm (G)
Amyloglucosidase	S	5000 t glucose syrup	Charcoal	(2)	Tate & Lyle (UK)
Aspartase	L	1000 t	Cells in carrageenan	(3)	Tanabe (J)
Fumarase	S	Malic acid	Cells in carrageenan (first immobilized cells)	(3)	Tanabe (J)
Glucose Isomerase	I	6–7 million t HFCS	See Table 4.3		Table 4.3
Hydantoinase	S	D-Phenylglycine	Polyacrylamide	(3)	Kanegafuchi (I)
Invertase	S	1000 t invert sugar	PMMA	(1)	Röhm (G)
			Corn cob granulate	(1)	Inra (F)
			Ion exchanger	(2)	—(G)
Lactase	S	<10,000 t lactose syrup	Ceramic carrier	(1)	Corning (USA)
			PMMA		Röhm (G)
			—		Valio (I)
			Ion exchanger, SiO_2	(2)	Sumitomo (J)
					Amerace (USA)
					Snam Progetty (I)
Lipase	P	Transesterification, hydrolysis	Nylon fibers	(4)	Novo (DK)
			Ion-exchanger celite	(1)	Unilever (NL)
Nitrilase	L	4000–6000 t acrylamide	Cells in polyacrylamide	(3)	Nitto (J)
RNAse	S	>500 t RNA	Glass, activated cellulose	(1)	VR China, Japan
			Eupergit	(2)	Röhm (G)

211

TABLE 4.2 Industrial Application of Immobilized Enzymes (Status in 1988) *(Continued)*

Immobilized Enzymes	Production Scale[a]	End Product (t/yr)	Immobilization Method	Group[b]	Most Important Suppliers (Country)[c]
Penicillin G amidase	I	4500 t 6-aminopenicillanic acid	Eupergit (e.g.)	(1)	Röhm (G) Beecham (UK) Gist (NL) Bayer (G)
Penicillin V amidase	I	500 t	Polyacrylamide, nylon fibers	(4)	Toyo Jozo (J) Boehringer M (G) Snam Progetti (I) Biochem. Kundel (A) Novo (DK)
Yeast cells	S	Alcohol	Na-alginate	(4)	Kyowa Hakko (J)

[a]See Figure 4.1.

[b]P, pilot scale; S, small installations; L, large installations; I, produced exclusively with immobilized enzymes.

[c]*Countries*: J, UK, USA, DK, NL same as in Table 4.1; G—Germany; I—Italy; A—Austria.

TABLE 4.3   Immobilization Methods for Commercial Glucose Isomerase Products

Immobilization Method	Company
Lysed cells, cross-linked with glutyraldehyde, granulated	Novo
Intact cells, cross-linked with glutyraldehyde, granulated	Miles Lab. Inc., MKC
Intact cells, embedded in gelatin, cross-linked with glutyraldehyde, granulated	Gist-Brocades
Cell-free enzyme bound to ion-exchange resin and granulated	Clinton Corn Products

1. The enzyme, after isolation and purification, is covalently immobilized on an inert, tailor-made carrier.
2. The partially purified enzyme is immobilized on ion-exchange resins.
3. Cross-linking cells that contain the desired enzyme(s).

See also Table 4.3.

The various methods produce immobilized products that can vary significantly in quality, such as activity, mechanical stability, density, and useful life, which effect the resultant productivity. Productivity determines economical feasibility.

In spite of the low productivity of the immobilizates in type 3 of Table 4.4, they seem to hold the greatest market share. But then, of course, they are much cheaper than the highly productive free glucose isomerases (Table 4.4).

The reaction and the process of glucose isomerization are described in Section 3.5.

4.4.2   Immobilized Penicillin Amidase

On separating phenylacetic acid from Penicillin G, 6-aminopenicillanic acid (6-APA) is produced; it serves as the starting material for about 20 semisynthetic penicillins. For these syntheses, immobilized penicillin amidase (also penicillin acylase) is used. The method utilizing Eupergit PCA (Röhm Pharma, Germany) may serve as an example.

TABLE 4.4   Productivity of Various Industrial Glucose Isomerases

Type	Yield (kg HFCS dry weight/kg enzyme)	Supplier (Country)[a]
1. Pure enzyme on inert carrier	20,000	UPO (USA), MKC (G)
2. Crude enzyme on ion-exchange resin	7,000–10,000	Finnsugar (FI), Sanmatsu (J)
3. Cross-linked cells	2,000–4,000	GB (NL), Novo (DK), ICI (USA), Nagase (J), Godo Shusei (J)

[a]*Countries:* USA, G, J, NL, DK, USA, J same as in Tables 4.2 and 4.3; FI—Finland.

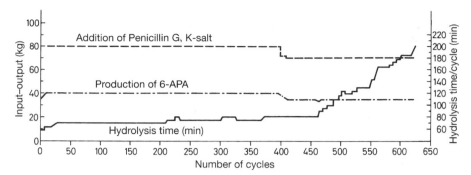

Figure 4.6 Time course of 6-aminopenicillanic acid (6-APA) production employing immobilized penicillin acylase (Eupergit PCA, 620 cycles).

4.4.2.1 Reaction Because this reaction liberates acid, alkali must be added. Ammonium hydroxide has become the agent of choice. Excess titration would inactivate the enzyme. Because of the need for a constant pH (pH 8), only a batch operation can be used. One cycle lasts approximately 100 min (duration of hydrolysis ~70 min, filtration ~30 min). A complete flowchart is shown in Figure 4.6.

Eupergit PCA yields about 250 kg of 6-APA (99% pure) per kilogram of wet carrier. This value has been attained in 620 cycles (Krämer and Goekcek, 1987). Following the 450th cycle, the duration of hydrolysis must be increased stepwise to assure constant quality. Figure 4.6 illustrates the course of this industrial production at the 100-kg scale.

Under certain conditions (acid pH, high substrate concentrations), the reaction can be reversed with penicillin amidase (Kasche, 1985). One example is the ampicillin production, accomplished by the enzymatic coupling of D-phenylglycine methylester to 6-APA.

4.4.3 Immobilized Lactase

Lactase hydrolysis of lactose to glucose and galactose generates a product of increased sweetness and high solubility; it permits the production of a syrup that contains 60–70% sugar and is microbially quite stable (lactose syrup). Additional possibilities exist, namely, the ability to refine the by-product whey, or to develop dairy products that can be used by those suffering from lactose intolerance (a large part of the world population).

Since the late 1970s, considerable effort was expended internationally to develop methods for continuous enzymatic lactose hydrolysis. Special attention was given to the problem of microbial contamination of the reactor caused by substrates such as milk and whey, which are particularly susceptible in this regard. There is no lactose to date that permits a continuous operation at >60°C temperatures that would minimize the danger of microbial growth. Consequently, cleaning and disinfectant cycles are required, usually employing H_2O_2, or quaternary ammonium salts (benzalkonium chloride). Milk as a substrate must be sterilized prior to its passage through the column (Marconi and Morisi, 1979; Snamprogetti SpA, Italy, 1979); the reactor temperature of 5°C retards microbial growth. Usually the substrate used is acidic whey or the ultrafiltered permeate; because of their acid pH of 4–5, they are less likely to become contaminated.

Important technical applications are listed in the "Lactase" row in Table 4.2. The Corning Glass Company (USA) should be mentioned because it has attempted, with considerable publicity, to install lactose hydrolysis as a large-scale commercial operation. Like many other proponents, the company has not met the goal. Only Nutrisearch (USA) operates a large lactose hydrolysis facility, employing the Corning process with an acidic ultrafiltered permeate for yeast production.

One must ask why this process could not succeed commercially in spite of many technically refined features. Several reasons are evident:

1. Liquid sugar can be obtained more economically from starch.
2. The possibility of improving the process with hydrolyzed whey is insignificant.
3. Whey utilization today, via spray drying, ends in the animal feed industry. Wastewater problems are not an issue.
4. The high investment costs required for lactose hydrolysis suppress producer interest.

Also, "whey syrup" or "lactose syrup" are end products that lack acceptance by the consumer, and are, therefore, of little interest to the producer.

The Plexazym LA 1 system (Röhm GmbH, Germany) may serve as an example of how the problem of lactose hydrolysis can be resolved technically. Productivities reaching 12 t of dry substance of whey (degree of hydrolysis = 80%) can be obtained with Plexazym LA 1 (*A. oryzae* lactase, immobilized on macroporous plexiglass) at 35°C with acidic whey substrate (pH 4–4.5). The continuous operation runs for 200 days with only little loss in activity at an average daily output of 1 t whey/kg Plexazym LA 1. The fixed bed must be aerated once a day, extensively flushed with water, and disinfected with 0.1% solution of benzalkonium chloride (Sprössler and Plainer, 1983). A flowchart is presented in Figure 4.7.

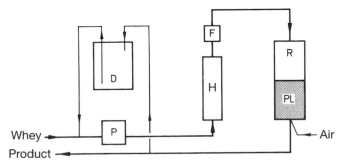

Figure 4.7 Flowchart of Plexazym LA 1 process; P = pump; D = disinfectant; H = heat exchanger; R = reactor, PL = Plexazym LA 1; aeration assists in the cleansing operation; F = filter.

To control microbial growth during acidic whey hydrolysis with Plexazym LA 1, the carrier must be morphologically optimized:

1. The pore diameter is selected to be just large enough to permit the entrance of only low-molecular-weight lactose in the carrier interior, excluding lipid coacervates, casein micelles, and especially microorganisms (Fig. 4.8). The tendency of protein (e.g., lactalbumin) to enter the carrier is strongly dependent on diffusion.
2. The carrier surface is smooth; it consists of perfect spheres with a limited size distribution. This curtails microbial growth.
3. The carrier has no charge and is hydrophilic. There is no adsorption of proteins from the substrate.
4. The high degrees of enzymatic activity related to carrier volume permits a short substrate retention time in the fixed bed and, therefore, high flow velocities (e.g., ~1 cm/s of linear velocity in the exclusion volume at a bed height of 40 cm. Thus, the reactor is continuously cleansing itself by convection; fines can hardly form sediments.
5. Finally, the carrier has such structural stability that intensive cleaning after a hydrolysis cycle (~20 h aeration with mechanical support) causes no attrition.

4.4.4 Immobilized Amyloglucosidase (AMG)

The technically relevant processes have been listed in Table 4.2. Immobilized amyloglucosidase deserves special attention because its immobilization is attractive for a number of reasons:

1. Low-molecular-weight substance (maltodextrin).
2. A large-scale industrial process with the free enzyme has already been estab-

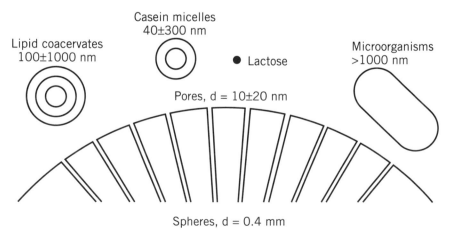

Figure 4.8 Morphology of Plexazym LA 1.

lished. Immobilization should provide economic advantages such as reduction of reaction volume (column instead of tank) or a reduction in reaction time (now 48–72 h for saccharification).

3. Continuous operation is well suited for the subsequent isomerization process with immobilized glucose isomerase.

Until now, the use of immobilized AMG has failed primarily because of low glucose yields. Instead of the 95–96% glucose obtained with free AMG, only a maximum of 94% glucose was produced with immobilized AMG (Rugh et al., 1979). This is due to the limited diffusion in the carrier. As a result of the extended retention time, a thermodynamic equilibrium can be reached inside the carrier. During hydrolysis of 30% maltodextrin, this equilibrium is reached at 1–2% isomaltose. This by-product, together with some maltose and higher-molecular-weight oligosaccharides, reduces the glucose yield to the extent cited earlier.

Alternately, if the thermodynamic equilibrium of the hydrolysis favors only monomers, such as during inversion of sucrose, then the use of immobilized enzymes (such as invertase) will yield very pure end products (Plainer and Sprössler, 1987).

Restricted diffusion in immobilized AMG further causes high-molecular-weight substrate components to pass through the column without being hydrolyzed.

Additional reasons for not using immobilized AMG are the inadequate thermostability of AMG in a continuous operation (a maximum of 55°C can be tolerated) and the relatively inexpensive free enzyme now available.

Immobilized AMG is currently employed only in processing the raffinose stream during glucose isomerization.

Table 4.5 compares the efficacy of immobilized AMG to that of invertase and lactase.

4.4.5 Immobilized Lipase

Reference: Erdmann (1992)

Hydrolysis of natural fats and oils is a large-scale operation in the chemical industry. The conventional technology available for fat hydrolysis is high-temperature steam splitting (250–260°C at 50 bar). With growing ecological concern, increasing energy costs, and a desire to achieve better yields and quality of fatty acids and glycerin, considerable effort has been made in industrial and academic laboratories to split fats and oils enzymatically, using lipase as the catalyst.

4.4.5.1 Economic Considerations A typical steam splitter with an annual capacity of 50,000 metric tons has an energy and operating cost between $0.01 and $0.02 per kilogram of fatty acid. If one would use soluble lipase at current market prices ($200/kg and higher) and at a typical consumption level of 0.1–0.2% (based on the oil used), the cost of lipase would be $0.20/kg of fatty acid. Even with the cheaper crude lipase preparations, a reliable 1985 estimate was $0.09/kg of

TABLE 4.5 Sugar Hydrolizing Enzymes for Industrial Application

Immobilized Enzyme	Production (Röhm)	Objective	Reaction	Parameters	Productivity
Amyloglucosidase	Plexazym AG	Converting dextrins into sugar	Dextrins → glucose	30% Substrate pH 4.5, 40°C	6 t dry weight, syrup with 88% glucose
Invertase	Plexazym IN	Invert sugar production	Sucrose → glucose + fructose	50% Substrate pH 4.5, 40°C	6 t dry weight, syrup with 90% inversion
Lactase	Plexazym LA	Whey utilization	Lactose → glucose + galactose	Acidic whey pH 4.5, 35°C	12 t dry weight, syrup with 80% conversion

coconut-based fatty acids. These economic considerations make recycling the lipase mandatory.

Bühler and Wandrey (1987) reported experiments on continuously recycling soluble lipase in a two-stage stirred-tank reactor system. They found a 98% hydrolysis of soybean oil. Their losses of lipase were 0.62 g/kg of fatty acid produced. Because of the high cost of lipase, methods had to be developed to recycle the enzyme.

4.4.5.2 *Immobilization of Lipases*
In principle, all methods discussed previously can be applied to the immobilization of lipases. Because the target product is a fatty acid, which is a low-price commodity, only those immobilization procedures can be used that retain a high degree of activity after immobilization.

Kroll et al. (1980) reported 67% retained activity of lipase immobilized on Wofatit MC50, using the carbodiimide coupling method. These "yields" seem unusually high. Retained activities as low as 1% have been widely reported in the literature.

The concept of hydrophobic adsorption was introduced in the mid-1970s and has proved to be a definite advantage. Lavayre and Baratti (1982) used iodopropyl-modified Spherosil porous glass for immobilization of *Candida cyclindracea* and reported 103% activity as compared to the soluble enzyme. They even reported activities of 155% when using a crude enzyme. This indicates a selective adsorption of the active enzyme over its companion (inactive) proteins.

Lipases (free and immobilized) catalyze hydrolysis, esterification, and transesterification reactions. Novo Industri (Holte, 1989) has patented a similar immobilization method for the adsorption of *Mucor miehei* lipase on macroporous phenol formaldehyde resins for use in transesterifications or the synthesis of wax esters (98% yield) in fixed-bed reactors. Lever Brothers (Macrae and How, 1988) claim a continuous interesterification process using 1,3-specific immobilized lipase for production of cocoa butter substitute fats.

Similar immobilization methods were reported by Brady et al. (1988). Microporous polyethylene [high-density polyethylene (HDPE)] or polypropylene (PP) powders with high internal surface areas of 100 m^2/g and pore sizes of 0.2–0.5 μm [Accurel, manufactured by AKZO Fibers, Obernburg, Germany; see Castro (1981)] shows high affinity for lipases without chemical pretreatment or coupling reaction. An aqueous buffer solution of *Candida* lipase (Enzeco [EDC]; Maxazyme LP [GB]; or Type VII Lipase [SCC]) was stirred with ethanol-prewetted Accurel powder; the immobilisate showed enzymatic activity of 98% (HDPE) and 107% (PP). Other carriers, less hydrophobic than polyethylene or polypropylene (glass, cellulose, clay, alumina), showed only a low affinity for lipase (9–24%).

Lipase, when immobilized by hydrophobic adsorption, has a number of advantages:

- It is stable during storage in fatty acids.
- It works at an optimum of pH 4 and 42°C.
- High concentrations of glycerin (40%) stabilize the immobilized enzyme and increase its half-life 10-fold.
- The carrier is mechanically stable and chemically inert.

- The purity of fatty acids produced is much better than those produced via steam splitting.
- Very high-purity (waterwhite) glycerin can be produced in concentrations as high as 40% as compared to the 10–15% concentrations with steam splitters.
- Enzyme losses in fatty acid or glycerin can be minimized.
- The conditions during hydrolysis are extremely mild and permit splitting of unsaturated fats without excess losses due to polymerization.

4.4.5.3   Use of Immobilized Lipase on Pilot Scale   In 1981, Miyoshi Oil and Fat Company reported that it had replaced its fatty acid production method for powdered soaps with an enzymatic process (Miyoshi Oil and Fat Co., 1983). While still using soluble lipase, the energy savings were 99% ($850,000/yr) for a monthly production of 1000 metric tons. The degree of hydrolysis, however, was only 92% and not suitable for high-purity fatty acids.

Brady et al. (1988) of AKZO Chemicals and Tsukishima Kikai Company Ltd. (1987) studied and initiated the use of immobilized lipase for cleaning natural fats and oils. A typical reaction scheme is shown in Figure 4.9.

In both cases continuously stirred tank reactors (CSTR) were used instead of fixed beds. The immobilized enzyme is stable under reaction conditions (agitation), and no attrition has been reported. The oils and fresh water flow in a countercurrent mode and the final concentrations of glycerin reach 30%, which yields additional

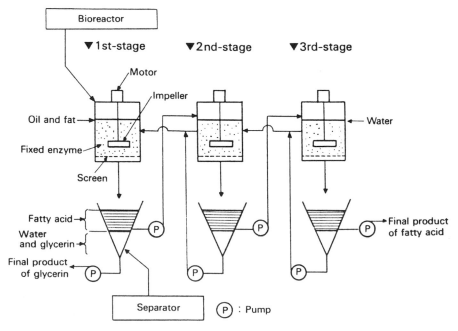

Figure 4.9   Conceptual diagram of fatty acid production using countercurrent bioreactor. (*Source: Bioprocessing Technology,* June 1987, p. 3.)

energy savings when concentrating the glycerin because less water has to be evaporated. The productivity of well-agitated CSTRs is 10-fold higher than that measured in fixed-bed reactors, known for their limited interface between the two immiscible phases. AKZO reports 1100 kg of fatty acid per kilogram of immobilized enzyme (IME) for the CSTR and surprisingly high half-lives (5680 h). Depending on the formulation, the IME can contain between 5 and 20% *Candida* lipase on the Accurel carrier. While AKZO's pilot reactor had been operated for 237 consecutive days and hydrolysis as high as 98% could be achieved with olive or soybean oil, the Tsukishima Kikai Company, Ltd. could obtain only 90% hydrolysis of beef tallow in 600-h runs with *Pseudomonas*-derived lipase.

Thus, enzymatic hydrolysis of fats and oils seems to be possible on an industrial scale if the substrates are carefully selected. The advantages of enzymatic hydrolysis are the mild conditions, which return high yields and purities of the target (unsaturated and hydroxy) fatty acids and energy savings. Successful esterifications are dependent primarily on the enzyme–substrate specificities (selectivities), parameters that are unattainable with general chemical catalysis.

4.5 OUTLOOK: TECHNICAL APPLICATIONS OF IMMOBILIZED ENZYMES

Today's immobilization technology is flexible, and has currently advanced to the point where virtually every immobilization problem that may arise can have a quick and effective solution. However, there is no universal method; each enzyme has to be considered separately.

Currently, the array of industrially utilized enzymes consists (with the exception of glucose isomerase) primarily of hydrolases. New conversions to immobilized enzyme processes are unlikely to occur at present. The picture looks different, however, for enzyme systems requiring cofactors [Hartmeyer (1985) refers to them as "second-generation biocatalysts"] and for synthetases (anabolic enzymes). Innovations are due here. An additional promising field is that of enzyme reactions in organic solvents, namely, largely selective and stereospecific synthesis.

Some examples (Buchholz, 1989) are as follows:

- Synthesis of oligosaccharides for dietetic sugars.
- Selective conversion of sucrose for the same purpose [palatinose is already produced this way; see Munir (1988)].
- Selective oxidation of sucrose at the 3-position.
- Peptide syntheses.
- Antibiotic synthesis using reversible reactions.
- Immobilized enzyme use in wastewater treatment.

Despite the present sophisticated state of the art in immobilization technology, its development will continue. Technical applications of immobilized enzymes will profit from the consistently high number of scientific publications.

REFERENCES

Adlercreutz, P., in Nagodawithana, T., and Reed, G. (eds.), *Enzymes in Food Processing,* 3rd ed., Academic Press, New York, 1993, p. 103.

Brady, C. D., Metcalfe, L. D., Slaboszewski, D. R., and Frank, D., "Hydrolysis of fats," U.S. Patent 4,629,742 (1986).

Brady, C. D., Metcalfe, L. D., Slaboszewski, D. R., and Frank, D., *J. Am. Oil Chem. Soc.* **65,** 917 (1988).

Buchholz, K., *Chem. Ing. Tech.* **61,** 611 (1989).

Bühler, M., and Wandrey, C., *Fat Sci. Technol.* **89,** 598 (1987).

Castro, A. J., "Method for making microporous products," U.S. Patent 4,247,498 (1981).

Czermak, P., König, A., Tretzel, J., Reimerdes, E. H., and Bauer, W., *Forum Mikrobiol* **11,** 368 (1988).

Erdmann, H., in Finn, R. K., Präve P., Schlingmann, M., Crueger, W., Esser, K., Thauer, R., and Wagner, F. (eds.), *Biotechnology—Focus 3, Fundamentals, Applications, Information,* Hanser, Munich, 1992, p. 339.

Godfrey, T., in Godfrey, T., and Reichelt, J. (eds.), *Industrial Enzymology: The Application of Enzymes in Industry,* The Nature Press, New York, 1983, p. 437.

Gorton, L., Marko-Varga, G., Domínguez, E., and Emnéus, J., in Lam, S., and Malikin, G. (eds.), *Analytical Applications of Immobilized Enzyme Reactors,* Blackie Academic & Professional (imprint of Chapman & Hall), New York, 1994, p. 1.

Hartmeyer, W., *Naturwissenschaften* **72,** 310 (1985).

Holte, P. E., "Immobilized *Mucor miehei* lipase for transesterification," U.S. Patent 4,798,793 (1989).

Kasche, V., *Biotechnol. Lett.* **7,** 877 (1985).

Kennedy, J. F., White, C. A., and Melo, E. H. M., *Chim. Oggi* **5,** 21 (1990).

Klein, J., and Wagner, F., in Chibata, I., and Wingard, L. B., eds., *Applied Biochemistry and Bioengineering,* Vol. 4, *Immobilized Microbial Cells,* Academic Press, New York, 1983, p. 12.

Krämer, D. M., and Goekcek, Z., *4th Eur. Congr. Biotechnology,* Vol. II, 1987, p. 341.

Krämer, D. M., Plainer, H., Sprössler, B., Uhlig, H., and Schnee, R., "Verfahren zur Immobilisierung gelöster Eiweisstoffe," German Discl. 3 515 252 (1985).

Krämer, D. M., Plainer, H., Sprössler, B., Uhlig, H., and Schnee, R., "Verfahren zur Immobilisierung gelöster Eiweisstoffe," Eur. Patent 200 107 (1991).

Kroll, J., Hassanien, F. R., Glapinska, E, and Franzke, C., *Die Nahrung* **24,** 215 (1980).

Lam, S., and Malikin, G. (eds.), *Analytical Applications of Immobilized Enzyme Reactors,* Blackie Academic & Professional (imprint of Chapman & Hall), New York, 1994, p. 276.

Lavayre, J., and Baratti, J., *Biotechnol. Bioeng.* XXIV, 1007 (1982).

Leuchtenberger, W., Karrenbauer, M., and Plöcker, U., Forum Mikrobiol. (special issue: *Biotechnology*) **7,** 40 (1984).

List, D., and Knechtel, W., *Industr. Obs Gemüseverwertung* **65,** 415 (1980).

Macrae, A. R., and How, P., "Rearrangement process," U.S. Patent 4,719,178 (1988).

Marconi, W., and Morisi, F., in Wingard, L. B., Katchalski-Katzir, E., and Goldstein, L. (eds.), *Applied Biochemistry and Bioengineering,* Vol. 2, *Enzyme Technology,* Academic Press, New York, 1979, p. 219.

Miyoshi Oil and Fat Co., in "Why growth is slow for industrial enzymes," *Chem. Week* **133,** 34 (1983).

Munir, M., "Sugar substitute intermediate isomaltulose production by enzymatic conversion of pure sucrose solutions using immobilized cells of isomaltulose-producing organisms," German Patent 3 038 219 (1988).

Plainer, H., and Sprössler, B. G., *Proc. Internatl. Congr. Biochem. Eng.,* Stuttgart, 1986, 368.

Plainer, H., and Sprössler, B. G., *Forum Mikrobiol.* **5,** 161 (1987).

Poulsen, P. B., *Biotech. Genetic Eng. Rev.* **1,** 121 (1984).

Rugh, S., Nielsen, T., and Poulsen, P. B., *Die Stärke* 31, 333 (1979).

Schmidt-Kastner, G., and Gölker, C., *GBF Monographie* **9,** 201 (1986).

Snamprogetti, SpA., technical information, 1979.

Sprössler, B., and Plainer, H., *Food Technol.* **37,** 93 (1983).

Tischer, F., in Finn, R. K., Präve, P., Schlingmann, M., Crueger, W., Esser, K., Thauer, R., and Wagner, F. (eds.), *Biotechnology—Focus 3, Fundamentals, Applications, Information,* Hanser, New York, 1992, p. 237.

Tsukishima Kikai Co., Ltd., in *Bioprocessing Technol.* **9**(4), 3 (1987).

Wandrey, C., and Flaschel, E., *Adv. Biochem. Eng.* **3,** 147 (1979).

5 Application of Technical Enzyme Preparations

Enzymes chosen for a specific process must be selected to assure their suitability for the given technical and commercial application. To facilitate such a selection, the enzymes and their characteristics were listed and described earlier in this book (see Chapter 3; also Chapter 2, Fig. 2.1).

Some of the prerequisites for a particular application were described in the chapter on general characteristics of enzyme preparations (see Chapter 2). To summarize the various reaction parameters:

- The pH must be adjusted to meet enzyme characteristics; or, when the process pH cannot be changed, the appropriate enzyme must be selected.
- Reaction velocity can be influenced by the reaction temperature. Operating at temperatures just below inactivation reactions can be significantly accelerated. Enzymatic activity, however, can still occur under refrigeration.
- If, for technical reasons, the reaction time is fixed, the reaction can be governed by the temperature or by the amount of enzyme employed.

Figure 5.1 illustrates the interaction of reaction parameters in practice. Table 5.1 lists some of the major technical enzyme applications.

5.1 ENZYMES IN THE STARCH AND SUGAR INDUSTRIES

References: Hebeda (1983, 1992, 1993), Schenck and Hebeda (1992), Woods and Swinton (1995)

Starch itself, as a natural product, is relatively insignificant in everyday use. At one time, starching shirts and shirt collars (i.e., the use of starch in laundering) had widespread application. With time, this practice has lost importance, and now when starch is mentioned, it is usually listed as an additive in puddings and sauces, and more recently, as a basic commodity for the chemical industry. Generally, most people are unaware of the incredibly broad spectrum of industrial processes involving the utilization of starch and starch products. A few examples of processing and use can be cited: the isolation, modification, and digestion of purified starch for the food industry (usually for desserts, sauces, dressings, fillings, soups, ready-to-serve dishes, gum and jelly products); for the production of flours, baked goods, breakfast

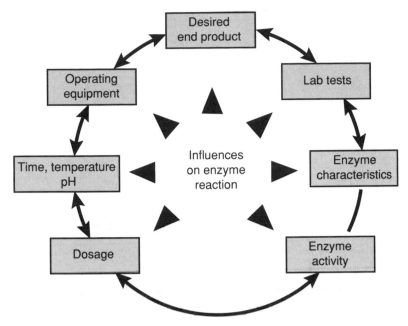

Figure 5.1   Interaction between reaction parameters ([RM] technical data).

cereals, beer, and other alcoholic beverages; and in the manufacture of paper, textiles, adhesives, and for other technical applications. Recently, starch has become an important raw material for the chemical industry.

Table 5.2 shows the significance of starch as a basic commodity. Starch, second only to cellulose, is the most abundant polysaccharide in nature, and the most important in industrial utilization. It represents 0.05% of the total biomass. Only about 2% of the total available starch, present primarily as starch granules isolated from the plant biomass, is purified and processed further.

Important field crops for starch production are listed in Table 5.3.

Several of the most interesting features of starch are its insolubility at room temperature, its ability to solubilize when heated, and the forming of a gel, depending on the concentration, when its "pasting point" is reached. The temperature varies with the type and source of the starch (see Table 5.4).

To understand the following, it is important to note that starch modification or conversion, whether performed chemically or enzymatically, can generally be accomplished only in the gelatinized state and that the desired product properties can be achieved only under these conditions.

It should also be pointed out that the starch granules generated during the technical extraction processes will contain, in spite of costly separation and washing steps, not only amylose and amylopectin but also a series of companion compounds that are not carbohydrates yet are constitutive parts of the granules. These include lipids and phospholipids, fatty acids, proteins, celluloses, and hemicelluloses (pentosans, β-glucans), phosphoric acid (especially in potatoes), minerals, and, of

TABLE 5.1 Survey of Technical Enzyme Applications

Process or Product	Enzyme Preparations
Alcohol	α-Amylases, β-amylases, amyloglucosidase, β-glucanase, proteases, cellulases
Amino acids	Amino acid acylase (immobilized), proteases with peptidases
Animal feed	Proteases, amylases, hemicellulases, β-glucanases
Beer	α-Amylases, β-glucanases, amyloglucosidase, proteases
Berries	Pectinases
Bread	α-Amylases, β-amylase, amyloglucosidase, fungal proteases, pentosanases
Bread preservation	Bacterial α-amylase, branching enzyme, β-glucanases, pentosanases, β-amylase
Cheese ripening	Lipases, proteases, peptidases
Cheesemaking	Animal and microbial rennet, lysozyme
Chicken feed	β-Glucanases
Chitin breakdown	Chitinase
Cider	Clarification: pectin esterase
Citrus juices	Lowering viscosity or clarification: pectinases Eliminating bitterness: naringinase
Citrus slicing	Cellulases
Cleaning lipid traps	Lipases, proteases
Coffee	Extracts: β-glucanases
Confections	Invertase
Cookies, crackers	α-Amylases, proteases
Dential hygiene	Dextranase, mutase, glucanases, peroxidases
Detergents	Proteases, amylases
Dextran breakdown	Sugar industry: dextranase
Dextrins	From starch: α-amylases
Egg products	Preservation: glucose oxidase, catalase Whippability: lipases
Fish curing	Pancreatin
Fish protein	Lowering viscosity of water extracts: proteases
Flavors	From fruits: pectinases, hemicellulases, cellulases From milk products: lipases, proteases In baked goods: lipases
Flax processing	Pectinases, macerases
Flour processing	At the mill: fungal α-amylase Malt flour: proteases, lipoxygenase
Fructose production	From inulin: inulinase From glucose: glucose isomerase
Fruit juices	Pectinase, hemicellulases, cellulases, amylases, amyloglucosidase, arabanase
Fruit, vegetable pulps	Macerases
Gelatin hydrolysis	Proteases
Glucose production	From starch: α-amylases, amyloglucosidase
Glucuronic acid production	Glucose oxidase, catalase
Invert-sugar production	Invertase
Lactose	Milk hydrolysis: yeast lactase

TABLE 5.1 Survey of Technical Enzyme Applications (*Continued*)

Process or Product	Enzyme Preparations
Laundry detergents	Proteases, cellulases, lipases, amylases
Leather manufacture	Proteases, carbohydrases, lipases
Lipid hydrolysis	Lipases
Lipid transesterification	Lipases
Lipid separation	Proteases
Maceration	Plants: pectinases, hemicellulases, amylases
Maize gluten	Isolation: cellulases, β-glucanases
Maltose syrup	α- and β-amylases
Meat tenderizing	Proteases, especially papain
Meat trimmings	Proteases with peptidases
Milk	Cold sterilization: H_2O_2/catalase
Oils	Olive and citrus oil production: pectinases
	Hydrolysis, transesterification: lipases
Phytic acid, phytate	Hydrolytic liberation of phosphate and calcium: phytase
Plant pigments	Anthocyanin production—pectinases
Removing bitterness	From citrus juices
	From protein hydrolysates: peptidases
Removing sizing	Textiles: α-amylases, cellulases
	Films: bacterial proteases
Rubber	Proteases
Silage	Grasses: cellulases, β-glucanases
Soft drinks	Stabilization: glucose oxidase
Soybean hydrolysis	Proteases, α-glucosidase
Tea	Extraction: tannase
Wafers	Bacterial proteases
Wastewater	Proteases, cellulases, complex enzyme preparations
Whey processing	Lactose hydrolysis: lactases
	Whey protein modification: proteases
Winemaking	Maceration: pectinases
	Clarification: pectinases, proteases, β-glucanases
Yeast hydrolysis	β-glucanases for lysing cell walls

TABLE 5.2 Worldwide Harvest of Starchy Raw Materials (after Tegge, 1985)

Starch	Harvest (t/yr)
Starchy raw material (maize, potatoes, wheat)	1.9 billion
Theoretical starch yield (dry solids)	980 million
0.5% of biomass	200 billion
0.05% of total biomass, (95% cellulose, 5%, starch; lipids and protein)	2000 billion

TABLE 5.3   Important Field Crops for Starch Production (after Tegge, 1985)

Starch			Produced	Processed
Source	Processed	Share (%)	Tons/yr	
Maize	—	70	—	—
Potatoes	—	20	—	—
Wheat	—	5	18–20 mil (=2%)	—
Tapioca	—	4	—	—
Rice	—	1	—	—
—	Saccharified	—	—	9.0 million (47%)
—	Modified	—	—	5.5 million (29%)
—	Native	—	—	3.5 million (18%)
—	(Uncertain)	—	—	1.0 million (6%)

course, water. The amount of such starch "impurities" are normally quite small and will vary with the starch source as shown in Table 5.5. These substances can, nevertheless, play a significant role in determining the success or failure of a large-scale technical operation employing enzymatic starch conversion.

Until recently maize was the most important cereal for starch production because the isolated raw material was best suited for enzymatic or other conversion processes. This, however, meant that European industries have had to import and stockpile expensive dried maize of precisely defined quality. In addition, the grain had to be softened in soaking stations before separation into the technically important fractions was possible. The machinery used for separating maize embryos, pericarps, and silk as well as the maize protein (gluten) includes mills, cyclone separators, jet and other washers, decanters, centrifuges, presses, and dryers. Valuable by-products of corn processing are corn oil, maize gluten, and feed mixes.

TABLE 5.4   Pasting Characteristics of Native Starches (after Tegge, 1984)

Starch		Pasting Range (°C)	Swelling Ability (-fold)
Species	Source		
Potato	Tuber	56–66	>1000
Sago	Pith	—	97
Tapioca	Root	58–70	71
Sweet potato	Root	—	46
Maize	Grain	62–72	24
Sorghum	Grain	69–75	22
Wheat	Grain	52–63	21
Rice	Grain	61–78	19
Waxy maize	Grain	63–72	64
Amylomaize (rich in amylose)	Grain	—	6

TABLE 5.5 Composition of Various Starches (after Radley, 1976)

Component	Maize	Waxy Maize	Wheat	Rice	Potato	Tapioca
Amylose (% of starch)	21–30	0–1	19–25	17–19	18–23	17–19
Amylopectin (% of starch)	70–79	99–100	75–81	81–83	77–82	81–83
Lipids (% of dry solids)	0.4–0.8	0.1–0.2	0.8	0.8	0.1	0.1
Protein (% of dry solids)	0.35	0.2–0.3	0.4–0.8	0.45	0.1	0.1
Ash (% of dry solids)	0.1	0.07	0.15–0.3	0.5	0.2	0.2

The European industries have since learned to use these grains and have switched, on a large scale, to wheat as well as European cultivars of maize. One commercial process for producing wheat starch is shown in Figure 5.2. First, the wheat is dry-milled to flour; then, after adding water and intensive kneading, relatively expensive but effective fractionation operations follow. Products include a *prima-* and a *secunda*-grade starch (distinguished by degree of purity) as well as wheat gluten (primarily as "vital gluten"). The latter is much in demand as a constituent of baked goods and, because of its relatively good market price (despite the usual price fluctuations), assures the profitability of processing wheat.

Obtaining starch from potato or other tubers is quite a different process that involves grinding stations; however, this procedure will not be discussed here because potato starch is unimportant in enzymatic processing. In Third World countries, extraction of tapioca root and broken rice plays a significant role; these starches have a purity satisfactory for enzymatic conversion.

The hydrolysis of the starch molecule—that is, its more or less complete digestion into maltodextrins, maltotriose, maltose, and particularly into glucose with inorganic acids, such as hydrochloric and sulfuric acids—has been known for a long time. It was first attempted by the pharmacist Kirchoff who, during Napoleon's continental blockade, sought to produce gum arabica. Thus, a process formed under duress and once converted to an industrial scale had become, for many middle European countries, the alternative to sugarcane imports from their overseas colonial possessions. From that time on, starch conversion products (primarily syrups) were used as sweeteners for food. However, the acid catalyzed conversion is a crude and poorly controlled process and produces, because of the by-products formed, a very contaminated product by today's standards. Even with costly purification steps, it still does not qualify for many potential applications.

Therefore, considerable interest was created when plant and microbial starch hydrolyzing enzymes became available. The utilization of malt and malt extracts was attempted quite early. But only at the onset of this century, Takamine (1894) was able to produce a fungal amylase on an industrial scale that allowed significant im-

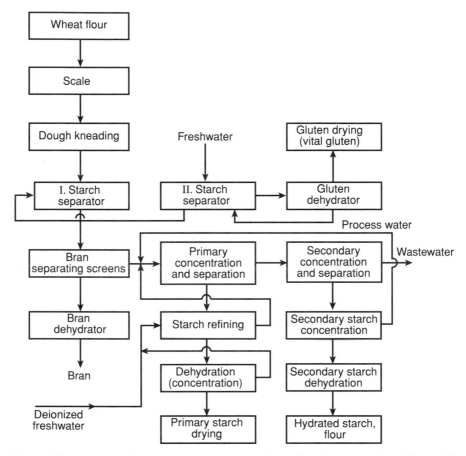

Figure 5.2 Isolation of wheat starch from wheat flour (flowchart after Dorr-Oliver, BV., Amsterdam, the Netherlands, technical information).

provements in starch hydrolysis. The saccharification enzyme, amyloglucosidase (glucoamylase), entered the market in the 1960s and in the mid-1970s, a bacterial α-amylase with adequate thermostability and a glucose isomerase arrived; debranching enzymes such as pullulanase finally became available during the 1980s. All the research and development costs were certainly justified because an extensive spectrum of technical enzymes with very different characteristics are now available that have a strong impact on the manufacture of products as well as on process technology. Table 5.6 lists functions of various amylases. The basic properties of these enzymes, or biocatalysts, are not listed here because they have been defined in earlier chapters.

Although syrups derived from acid-catalyzed conversions still have some commercial importance for reasons of economy, the introduction of enzymatic processes undoubtedly triggered the rapid growth in the starch industry during the past few decades. Significant improvements in process technology and product quality were due to

TABLE 5.6   Functions of Various Amylases (after Richter, 1986)

Enzyme	α-Amylase	β-Amylase	Amyloglucosidase
Site specificity	α-1,4 glucosidic linkage	α-1,4 glucosidic linkage	α-1,4 and α-1,6- glucosidic linkage
Mode of action	Endoamylase α-1,6 bond not hydrolyzed; however, further hydrolysis not inhibited	Exoamylase α-1,6 linkage not hydrolyzed; inhibits further hydrolysis	Exoamylase α-1,6 linkage is hydrolyzed; thus hydrolysis-inhibiting linkage is removed
Reduction in viscosity	Rapid	Slow	Slow
Reduction in iodide–starch color	Rapid	Slow	Slow
Increase in reducing sugars	Slow	Rapid	Rapid
Glucose formation	Slow	None	Rapid
Maltose formation	Slow	Rapid	None
Dextrin formation	Rapid	Slow	Slow

- Lower energy costs for conversion.
- Lower requirements for corrosion-resistant facilities.
- Increased amount of starch conversion (determined as DE value, or dextrose* equivalent); increased product yields (e.g., 1080 kg of D-glucose monohydrate from 1000 kg of dried starch); this represents a 20% increase over the yield obtained with the acid-catalyzed conversion method.
- Reduced discoloration.
- Reduced retrograded products (arising from reverse reactions during the conversion process).
- Significantly reduced expenditure for syrup purification (refining).

To help understand the details of the process technology to follow, Table 5.7 gives a spectrum of products that can be produced by the amylases listed in Table 5.6.

Three basic enzymatic degradative and conversion reactions occur: liquefaction, saccharification, and isomerization.

Figure 5.3 illustrates some of the chemistry involved in starch degradation during liquefaction and saccharification employing the technically most significant enzymes, thermostable α-amylase, and amyloglucosidase.

A flowchart for a large-scale technical process is shown in Figure 5.4. This best illustrates how the respective enzyme characteristics are utilized during the process

*Dextrose is the common or commercial term for D-glucose.

TABLE 5.7   Products of Enzymatic Starch Hydrolysis

Hydrolysate		Produced By
Maltodextrins	$DE^a < 20$	Hydrolysis with acid and/or α-amylase
Glucose syrups		Hydrolysis with acid and/or α-amylase
With low sweetening	DE 20–38	
With normal sweetening	DE 38–48	
With intermediate sweetening	DE 48–58	
Glucose syrups		Hydrolysis with acid and/or α-amylase;
With high sweetening	DE 58–68	saccharification with amyloglucosidase
With highest sweetening	DE 68–98	(glucose, maltose formation)
High-maltose syrups	DE 40 < 50	Hydrolysis with acid and/or α-amylase; saccharification with fungal α-amylase (maltose formation)
High-fructose syrups (HFCS, isosyrup)	DE 97	Hydrolysis with α-amylase; saccharification with amyloglucosidase (glucose formation) and isomerization with glucose isomerase

[a]Dextrose equivalent.

and how they permit a very modern process design with continuous operation, or one requiring special process staging.

The liquefaction phase processes the starch slurry through the hydration and pasting temperatures, reduces its high viscosity, and splits the starch polymer into relatively uniform medium-sized fragments. In practice, this is done by first funneling the end product of starch production, the starch slurry (30–35% solids), into a mixing tank to which cofactors such as Ca^{2+} (as $CaCl_2$) are added to increase the thermal tolerance of bacterial α-amylases (Optiamyl [SOEG] or Tenase [SOE]: maximum 400 ppm Ca^{2+}; Optitherm [SOEG] or Taka-Therm [SOE]; maximum 100 ppm Ca^{2+}), provided the hardness of the water used in the process had not already furnished these optimal conditions (Fig. 5.5).

Further, the pH is adjusted to 6.0; in addition to setting the optimal activity of the individual enzyme (Fig. 5.6), it is important to note that at pH > 6.2, the terminal reducing groups of the starch fragments easily isomerize, and a sizable quantity of maltulose (4-α-glucopyranosyl-D-fructose) can be detected at the conclusion of the starch to glucose conversion process.

Subsequently, all or part of the liquefying enzyme, such as Optitherm-L 420 [SOEG] or Taka-Therm [SOE], is added to the mixing tank or is continuously fed into the succeeding pipeline system. On passing the slurry through one or more jet cookers, which are steam-injection systems, most of the starch granules immediately disintegrate and dissolve on continued heating and mechanical shearing. The temperatures and retention times used in the process rests are inversely proportional; that is, there are processes employing relatively low temperatures (105–107°C) for 5–10 min duration and others using high temperatures of 120–130°C for just a few seconds. Because the latter conditions will inactivate even a thermally stable α-amylase, a sec-

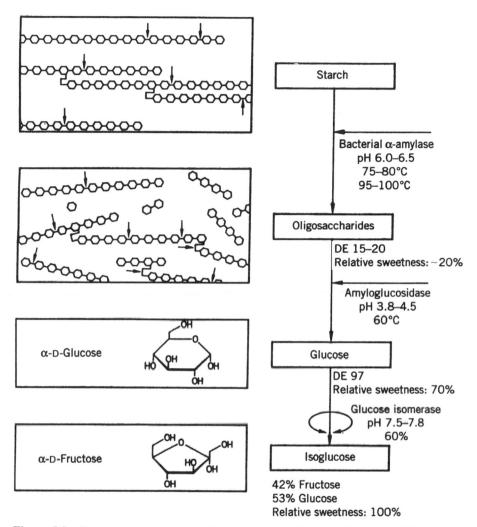

Figure 5.3   Enzyme-catalyzed conversion of starch to glucose and fructose (Richter and Grosser, 1983).

ond addition of Optitherm-L 420 or Taka-Therm into the expansion cooler or at a later stage is required. However, the total amount of liquefaction enzyme required in either case lies between 0.5 and 0.8 liter of Optitherm-L 420 or Taka-Therm per metric ton of dry starch.

Conversion to dextrin is performed in well-insulated pipelines, continuously stirred tank cascades or in simple holding tanks over a 2–3-h period. Customarily, the system is slowly cooled from 100°C to 95 or 90°C. The endpoint of the conversion is determined by either a DE (dextrose equivalent) measurement and/or the absence of an iodine reaction, or by the filterability of the preparation. Opinions as to

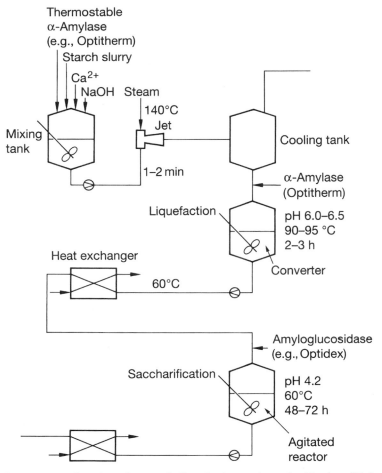

Figure 5.4 Process flowchart for starch liquefaction and saccharification (Richter and Grosser, 1983).

which DE value indicates complete conversion differ widely; some industries consider a DE of 8 or 10 as adequate, whereas others feel that a DE of 15–16 is mandatory. In any case, the DE value as well as jet cooker conditions determine if a syrup after liquefaction and especially after saccharification can be easily and economically filtered, that is, if it can be filtered rapidly without product loss.

Disintegration of the starch granules and the degree of the starch molecules' hydrolysis is, in view of what is known today, very critical because the native composition of starch granules includes, among others, amylose–lipid complexes. When amylose has not been adequately hydrolyzed, these amylose–lipid associations that dissolve only at temperatures above 100°C can reassociate under unfavorable conditions at temperatures below 80°C. In such cases, especially at relatively low saccharification temperatures, turbidity arises from very fine particles of retrograded

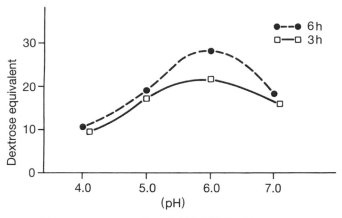

Enzyme concentration: 0.1% (TS) Optitherm-L210
Buffer: 0.1 M Na-phosphate
Substrate: Maltodextrin (10 DE)
            : 100 ppm Ca^{2+}
Temperature: 95 °C

Figure 5.5 Effect of pH on the formation of dextrose equivalents (DE values).

starch; these quickly clog the diatomaceous-earth-coated rotary vacuum filters during syrup clarification.

In addition to the process just described, there are historical methods that employ either total liquefaction or acid hydrolysis with temperatures usually about 150°C prior to the jet cooker stage. For some raw materials such as wheat and potato starches, separating the insoluble components becomes easier; however, it has been repeatedly demonstrated that this does not compensate for the disadvantages of acid hydrolysis described earlier.

When producing maltodextrin products (DE 16–20; roller- or spray-dried), the use of enzymes such as Optiamyl-L [SOEG] or Tenase [SOE] are preferred because of their lower optimal temperature and their concomitantly lower inactivating temperature. The viscosity can be precisely controlled by the enzyme dose (Fig. 5.7); usually 1.0–3.0 liters of Optiamyl-L or Tenase is added to 840 t of dry starch. The enzyme must be rapidly inactivated, usually at 110–120°C, so that the product characteristics such as the ability to absorb water and thicken are retained.

The product of liquefaction or dextrinization is a mixture of oligosaccharides containing small amounts of mono-, di-, and trisaccharides and is, for that reason, a superior substrate for the sweetening process. This process can be carried out with one or several different enzymes, depending on the type of syrup composition and properties desired. Table 5.7 provides a rough classification of glucose syrups, including D-glucose, high-maltose syrups, and high-conversion syrups.

The production of glucose syrups and D-glucose is by and large the most important saccharification process. An amyloglucosidase from *Aspergillus niger* is usu-

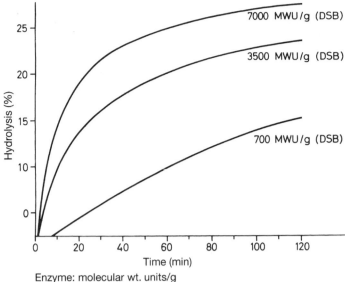

Enzyme: molecular wt. units/g
Substrate: 20% (DSB) Starch
Temperature: 70°C
pH: 6.5
Ca²⁺ 400 ppm
NaCl: 4000 ppm (SOEG)

Figure 5.6   Effect of enzyme concentration on hydrolysis rate. DSB = dry solid base.

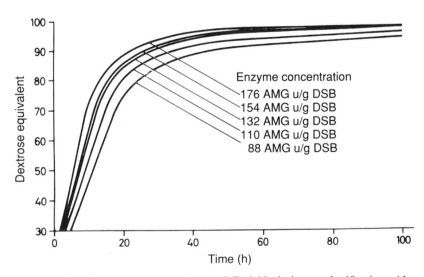

Figure 5.7   Effect of enzyme concentration on DE yields during saccharification with amyloglucosidase.

ally employed with a maximum activity at pH 4.0–4.5 and 60°C. The extent of starch hydrolysis (expressed as DE value) and time required for saccharification are strongly dependent on the substrate concentration and the amount of enzyme. With more enzyme, oligosaccharide hydrolysis will be completed more rapidly; however, this may trigger a potent reverse reaction catalyzed by amyloglucosidase leading to maltose and maltotriose. The proper amyloglucosidase isozyme I:II ratio was long thought to be critical in achieving optimal starch hydrolysis; it has since been shown, however, that a greater percentage of isozyme I will provide limited improvements only with high-molecular-weight substrates (low DE value after hydrolysis). Splitting the α-1,6-linkages in amylopectin fragments is also largely catalyzed by isozyme I; this reaction, however, is not limiting; that is, it will not significantly hamper complete starch hydrolysis down to the glucose unit.

In addition to the main enzyme fractions, other components of amyloglucosidase preparations can influence the course of saccharification. Primarily transglucosidase should be mentioned in that this enzyme can generate panose from two molecules of maltose by splitting off glucose. This trisaccharide with an α-1,6 linkage can, in turn, again be hydrolyzed by amyloglucosidase to isomaltose and glucose. Isomaltose cannot be enzymatically hydrolyzed any further and therefore, systems with high levels of contaminating transglucosidase activity can considerably lower the yield of D-glucose (Shetty and Marshall, 1986). Today, however, one can assume that commercially available amyloglucosidase preparations do not contain transglucosidase in amounts that will affect saccharification yields. This also applies to proteolytic activities of the preparations that, because of the low pH and temperature conditions during saccharification, are generally not thought to be too significant in discoloring syrups.

As seen in Figure 5.4, saccharification is started by cooling the syrup to 60°C with heat exchangers and adjusting the pH to 4.0–4.5 with hydrochloric acid. The saccharification enzyme, such as 0.6–0.7 liter Optidex-L 300 [SOEG] or Diazyme-L [SOE] per metric ton of dry matter, is added either continuously to the pipelines or directly to the saccharification tanks. The latter are well insulated, and the contents may be at rest or may be periodically or continuously stirred. The converters are filled and emptied in batches or are run in continuous operation. To avoid recycling of syrups, which would lower glucose yields, 6–12 conversion tanks are required for continuous operation. The time required to reach optimal DE values (97–98 DE) is approximately 48–72 h. Syrups with 96% glucose or more can be obtained with 2–3% disaccharides (maltose and isomaltose) and with 1–2% higher sugars. Frequently, the pH value drops a little because of microbial contamination and acid formation. In situations where the temperature in the converters cannot be monitored precisely and held at a constant 60°C, a massive contamination with yeasts and molds leading to layers of film cannot be excluded.

In the past few years debranching enzymes such as pullulanases and isoamylases have attained some importance. These enzymes are capable of breaking the α-1,6 bonds within the starch polymer and thus support the action of amyloglucosidase. This means that the saccharification rate will increase slightly with an equal amount of amyloglucosidase, or that with less amyloglucosidase and therefore fewer back reactions, these enzymes produce very good yields and, as experience has shown, the longer reaction times will effect a very uniform saccharification. Debranching enzymes are synthesized by many bacterial species with optimal activities that are usually within pH 5.0–6.0 and 50–55°C. These characteristics make these enzymes relatively unsuitable for use with the saccharification conditions described previously. One exception is the pullulanase isolated from *Bacillus acidopullulyticus,* which is very compatible with amyloglucosidase and therefore is used by itself or in enzyme mixtures by several starch processing industries.

High expectations were held for an immobilized amyloglucosidase in the late 1980s. However, all products offered very quickly showed that the required DE values of 96.0 or higher could not be obtained because of diffusion resistance. The relatively high level of reversion products formed was a consequence of increased enzyme density in the carrier-bound preparations. On the other hand, amyloglucosidase preparations, offered in liquid form, are rather inexpensive, so the technological expenditure necessary for a total continuous operation with its concomitant risks (substrate purity and downtimes) cannot be justified. However, implementation of immobilized amyloglucosidase in the production of other syrup varieties, to be described subsequently, has proved worthwhile in individual cases.

After completing the sweetening process in the manner described, inactivation of the enzyme is not necessary; the same holds for the liquefaction phase of starch conversion. The subsequent syrup purification steps can cause enzyme inactivation and prevent, except with long holding times, the loss of glucose yield due to amyloglucosidase-catalyzed back reactions. Sugar refining starts by precipitating insoluble components such as proteins and lipids on increasing the pH (with sodium carbonate solution) and by removing the particulates with separators or diatomaceous-earth-coated vacuum rotary filters—provided the formation of retrograded starch described earlier does not prevent this.

In addition, the removal of minerals and color is essential and accomplished in most operations by passing the syrup through a combination of strongly acidic and weakly alkaline anion exchange resins which are coupled in series. Treatment with activated charcoal is conducted when there is a high demand for light syrups. The absorbent is added either prior to the first separatory step or at the final ultrapurification step.

Glucose syrups with a high DE value can be stored and transported only at 60°C because they tend to readily crystallize after having been concentrated in the evaporator to 75–78% solids.

To obtain ultrapure α-D-glucose monohydrate, crystallizing vats with cooling mantels and cooling coils are used. Up to 60% of the glucose can be crystallized by gradual cooling to 30°C. The crystallized glucose and mother liquor (hydrol) are separated from each other with sieve centrifuges. The crystallization process itself and subsequent additional purification and drying procedures require a great deal of intuitive skills and are part of the best-kept secrets of this industry.

A whole series of production possibilities exist between the two extremes of glucose and maltodextrin manufacture; these products include glucose syrups with low, normal, intermediate, and high degrees of saccharification tailored by the dose of amyloglucosidase and the duration of saccharification; however, a rapid inactivation of amyloglucosidase is required, by either lowering the pH (pH $<$ 2.5–3), heat treatment (5–10 min at 95–100°C), or a combination of both.

These liquid products seldom become turbid; however, they do crystallize easier than maltodextrins and are more hygroscopic in their dried state.

Maltose-forming enzymes such as β-amylases from barley or barley malt have long been used to provide additional saccharified products for specific applications. However, fungal α-amylase from *Aspergillus niger* has become an interesting alternative for producing high-maltose syrups. Both enzymes are employed under the following conditions.

The initial liquefaction (a low DE value will increase the final maltose content) is reduced to 48–52% solids and adjusted to pH 5.0–5.5. Saccharification commences on addition of 160–240 ml of enzyme such as fungal amylase-L 40000 [SOEG] to 1 t of starch, and proceeds in either batch or continuous operation at 55°C (60°C) for 40–48 h. The liquor with a DE value of 48–52 is composed of

- Glucose 9–12%.
- Maltose 48–52%.
- Maltotriose 10–15%.
- Higher oligomers 21–33%.

This composition is stabilized by the rapid inactivation of the saccharification enzyme (acidification, thermal treatment, or a combination of both).

The advantage of β-amylase derived from higher plants over the fungal α-amylases is that they are less expensive and accomplish their tasks in a significantly shorter time (12–24 h). However, they, too, are unable to hydrolyze the α-1,6 linkage of amylopectin and hydrolysis stops at the β-limit dextrins. A complete hydrolysis of the branched starch molecules with an increased maltose yield of 70% and above can be achieved only by adding microbial pullulanases; these should be adapted as much as possible to the pH and temperature optima of the maltogenic enzymes.

In the manufacture of high-conversion syrups, both amyloglucosidase and fungal α-amylase are used; accordingly, the combined characteristics of glucose and maltose syrups are obtained.

Here, as in the cases described earlier, a broad range of possibilities exist for the individual commercial user.

Normally, the process is started with a substrate of 38–42 DE (derived from acid hydrolysis), which, after concentration to 48–52% solids, is saccharified at pH 5.0–5.2 and 42–60°C with 60 ml fungal amylase-L 40000 [SOEG] and 25 ml Optidex-L 300 [SOEG] or Diazyme-L [SOE]. After 40–48 h, the following approximate syrup composition at DE 63–70 is obtained:

- Glucose 30–35%.
- Maltose 40–45%.
- Maltotriose 8–10%.
- Higher oligomers 20–22%.

To stabilize this composition, rapid enzyme inactivation is required, for example, 2–3 min above 100°C or 20–30 min at 85–90°C.

Refining methods for high-maltose and high-conversion syrups are essentially those used for glucose syrups. In continuous well-controlled operations, the thermal inactivation step can be eliminated and the sweetening enzymes completely removed with activated charcoal.

The methods of starch liquefaction and saccharification discussed so far clearly indicate that syrups can be produced with any combination of oligo-, di-, and monosaccharides, and thus can be tailor-made for a multitude of applications:

1. *Maltodextrins:* Maltodextrins are marketed as white-brown, spray-dried powders; they are slightly sweet, serve to provide bulk, and can be viewed as an easily digestible component of food. They are used primarily in baby food and special diets, in dried soups, as a coffee whitener, and as a flavor carrier.

2. *Glucose Syrups:* Glucose syrups with low or intermediate DE values are utilized not only for their DE-dependent sweetness but also for their nutritional value and special functional properties such as fermentability, viscosity, absorbing and releasing moisture, bulk, foam stabilization, and ability to prevent sugar crystallization which again depends on the DE value. The syrups are of prime importance in the manufacture of sweets of all kinds (especially caramels, jelly confections, chocolates), marmalades and jams, fruit syrups, refreshment beverages, baked goods, fruit juices, ice cream, canned foods, and margarine.

3. *Maltose Syrups:* Maltose syrups have properties similar to those of glucose syrups, but with a higher viscosity and lower hygroscopicity. They have less tendency to brown and crystallize, and are preferentially used in ice cream, hard and soft caramel candies, baked goods, and brewery mashes.

4. *High-Conversion Syrups:* High-conversion syrups are significantly sweeter than maltose syrups and promote browning, moisture uptake, and crystallization. These features make them particularly suitable for marmalades, fruit preserves, sweets in general, sauces, and dressings. Furthermore, they have some importance in making beer and malt liquors.

5. *Glucose Syrups with High DE and Glucose Content:* Glucose crystallizes from high-glucose syrups as its anhydride or monohydrate. Crystalline glucose exists in the α form but when in solution it can mutarotate to its β form with both isomers present at equilibrium. Glucose is by far the most important sugar in nature, and therefore many applications for glucose exist.

Glucose and glucose syrups with high DE values are frequently utilized for their physical and chemical properties. Because of their high nutritional value, they are often substituted for sucrose in manufacturing baked goods (shortbread, biscuits and yeast pastries, breads for toasting, waffles, rusk, and cookies), for beverages, jams and fruit preserves, fish, meat and sausages, and serve as an important ingredient in sweets.

Crystalline glucose is particularly important in the medical and dietetic area. As human blood sugar, it quickly enters the bloodstream and provides energy during heavy physical exertion by athletes. Because glucose can be well tolerated, it is readily administered to patients and babies and can, when fed intravenously, acquire therapeutic functions, including detoxification, adjusting electrolyte and water imbalances, preventing or controlling heart disease, and managing diabetic comas.

Because α-D-glucose can quite easily be converted chemically or by fermentation, this monosaccharide is also used as the raw material for producing sorbitol and mannitol (via reduction), glucuronic acid (via oxidation), vitamin C (microbially), and isoglucose (via enzyme isomerization). Glucose is also involved in other biotechnological production processes and most recently as a reaction participant in the manufacture of plastics.

Glucose isomerization is the latest step in enzymatic starch conversion. The enzyme catalyzing this reaction, glucose isomerase (actually xylose isomerase), is capable of converting about 50% of the D-glucose in a given syrup into D-fructose, the reaction equilibrium of which is as follows:

Isomerization is economically feasible when the fructose yield has reached about 42% (degree of isomerization about 46%); but reaching the complete reaction equilibrium would require an excessive amount of time. The reaction product is called *isoglucose* or *high-fructose syrup* (HFS), and its sweetening power and composition are similar to those of invert sugar. The commercial production began in the mid-1970s and has experienced unbelievable success, particularly in those countries that strongly depend on sugar imports (e.g., United States, Canada, Argentina, Japan, Korea; see Fig. 5.8).

This success, however, was realized only when it became possible to immobilize the enzyme and reduce enzyme costs to an acceptable level (based on the soluble products from first-generation isomerases). Second-generation glucose isomerases, always intracellular enzymes, were developed by using producer organisms as enzyme delivery systems, cross-linking, or fixing the intracellular enzymes with glu-

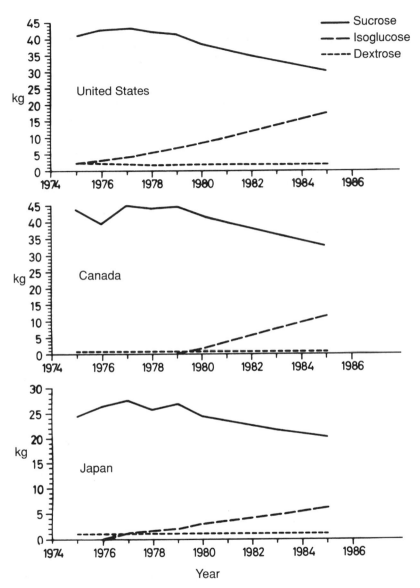

Figure 5.8 Estimated per capita consumption of sweeteners during 1975–1985 (Geyer, 1984).

taraldehyde. Third-generation glucose isomerases are characterized by highly purified enzyme fractions tightly bound to porous, highly inert organic (DEAE cellulose) or inorganic (SiO_2, Al_2O_3) carriers (Weidenbach et al., 1984). (See Fig. 5.9.)

Isomerization with immobilized enzyme products is usually accomplished with a fixed-bed reactor, usually designed as a well-insulated column assembly furnished with a perforated bottom support, inlet, outlet, and observation windows. The substrate is pumped downward through the column. Filling the columns with a second-

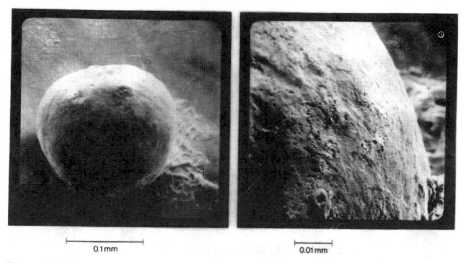

Figure 5.9 Scanning electron micrograph of silicon dioxide carrier (Weidenbach et al., 1984).

generation glucose isomerase is somewhat more labor-intensive than when using the third-generation enzyme preparations because it involves swelling the cells containing the fixed enzyme and carefully washing out all color forming agents.

At a specific substrate flow rate that permits the formation of 42% fructose, a filled column assembly is started at full efficiency (usually measured as the rate of voided volumes). In the course of days and weeks the efficiency drops and the substrate flow has to be reduced—or the temperature increased—to assure constant fructose formation. In most of the enzyme preparations marketed, the activity drops to 50% after 50–55 days (half-life). Continued utilization of the system is economical only for the second half-life (residual activity 20–25%). The total amount of product formed over this period determines productivity and is expressed as kilograms of isoglucose (42% fructose) per kilogram of enzyme or its standard activity. Second-generation glucose isomerases have productivities of 1.5–5 tons of isoglucose per kilogram of enzyme and, therefore, require significantly larger columns (up to 150 cm in diameter and 10 m in height) than do third-generation immobilized enzymes, which permit maximum yields of 22 t of isoglucose per kilogram of enzyme, such as Optisweet 22* [SOEG]. The height of the columns is limited primarily by the pressure gradient between substrate input and output.

To guarantee a sufficiently uniform commercial isoglucose production, several column assemblies (up to eight) are coupled either in parallel or in series and are periodically filled and emptied. Therefore, product formation will follow a sawtoothed pattern, and the unavoidable fluctuations must be compensated with buffer tanks.

Figure 5.10 shows the course of isomerization.

*Optisweet 22 is no longer available from SOEG, formerly MKC—Germany. It is 2.3 × Taka-Sweet 170 from SOE, formerly MS—USA.

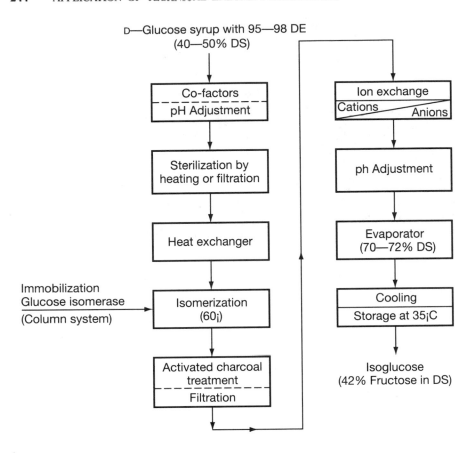

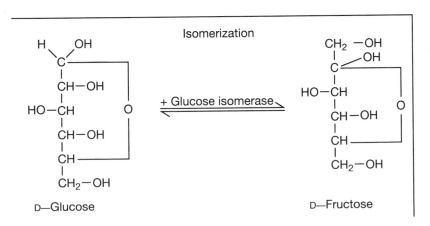

Figure 5.10   Flowchart of the isomerization step with immobilized glucose isomerase (DS, dry solids) (Richter, 1986).

TABLE 5.8 Effect of pH on Productivity of Glucose Isomerase (Optisweet, 22[a])

pH	Productivity (g dry solids/g Optisweet)	Total Reaction Time (hr)	Residual Activity (%)
7.5	22,000	3,200	20
7.8	17,000	2,500	20

Substrate: glucose syrup, 45% dry wt.
Cofactors: 120 ppm Mg^{2+}; 200 ppm SO_2
Degree of isomerization: 46.5%

[a]Optisweet 22, discontinued [SOEG].

Glucose syrup of sufficiently high purity (e.g. <5 ppm Ca^{2+}) with 45–50% solids is adjusted to a pH of 7.5 (the effect of pH on productivity is shown in Table 5.8), and the cofactors Mg^{2+} and SO_2 are added to provide optimal protection of the enzyme from denaturation (Fig. 5.11). After sterilization, the substrate is heated to about 60°C with heat exchangers and discharged into the isomerization columns. The temperature of the charge can be slightly higher because the endothermic reaction in the columns will cause a drop of about 2.5°C by the time it is discharged

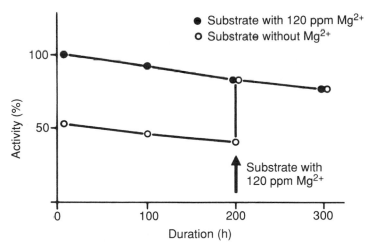

Substrate: Glucose, 45% DS
Cofactor: 200 ppm SO_2
pH: 7.5
Temperature: 60 °C
Isomerization: 46.5%

Figure 5.11 Effect of magnesium ion concentration on Optisweet 22 activity.

from the column. Similar to the syrups described earlier, isoglucose, after lowering the pH to 4.0–4.5 to suppress color formation, is purified with activated charcoal, filtered, demineralized, and reduced in an evaporator to 70–72% solids. In contrast to the glucose syrups, the low-viscosity isoglucose can be stored and transported at 35°C without recrystallization.

Many countries have started to manufacture isoglucose products with more than 42% fructose content, such as HFS 55 (containing 55% fructose). This percentage, however, cannot be met by the technology described here because the reaction equilibrium for glucose isomerases is approximately 50% as mentioned earlier. Therefore, a procedure for the large-scale enrichment of fructose by partition column chromatography and continuously returning the unwanted glucose to the isomerization process has been applied. The higher sweetening power of HFS 55 is particularly advantageous in refrigerated soft drinks and can, if necessary, reduce the caloric content of the sugar added. Isoglucose is further used as an additive in marmalades, jams, canned fruits, fruit juices, and milk products.

The enormous worldwide increase in the production of isoglucose between 1975 and 1985 is shown in Table 5.9. Unfortunately, the possibilities for the use of glucose isomerases within the European Community are very restricted because the sugar lobby was able, as recently as 10 years ago, to effect implementation of a quota system that permits only a few manufacturers to produce relatively small amounts of isoglucose, approximately 2.5×10^5 t/yr, which is less than 5% of the sugarbeet production. Therefore, new starch processing companies operating with new technology have predominantly been built outside the European Community during the last 10 years (eastern Europe, the United States, East Asia, and Third World countries). They have profited from the advances in facility construction and enzyme technology. A whole group of western European businesses play a leading role in the design and construction of such new facilities but have limited opportunity to implement these technologies in their immediate vicinities.

Isolation, purification, and subsequent enzymatic liquefaction and saccharification of starch is the current state of the art in enzymatic processing and is practiced worldwide. The question of whether a high degree of starch purity is always necessary was asked rather early. Professor Tegge, from the German Federal Research Institute of Grain Processing in Detmold, deserves special thanks for coining the term "direct enzymatic saccharification" and for utilizing readily available and inexpensive raw materials such as maize flour, potato peels, broken rice, and millet flour in the enzymatic conversion to glucose syrups. It is important, however, not to impose unrealistic demands on the products so obtained. Further, the enzyme preparations used cannot have contaminating proteolytic activities that will have an unfavorable impact on syrup color.

In an age of conscientious nutrition, natural sweetening of foodstuffs is also considered. Addition of all kinds of enzymatically processed flour to baby food, breakfast cereals, and baked goods has become quite customary. Bacterial α-amylases are also used in potato juice production. Moreover, there is a sizable effort to recover the starch from bakery waste by enzymatic degradation, making recycling old bread and bakery leftovers possible. As shown in Figure 5.12, the process technol-

TABLE 5.9   Estimated Worldwide Isoglucose Production for 1986 and 1990 (1000 t of Solids)

Country	1986[a] 42%[b]	55%[b]	1990 42%[b]	55%[b]
North America				
Canada	140	40	150	50
United States	1587	3175	1855	3438
Total	1727	3215	2005	3488
South America				
Argentina	110	45	120	50
Uruguay	—	15	—	20
Total	110	60	120	70
EC countries				
Belgium	72	—	72	—
France	20	—	20	—
Greece	13	—	13	—
Germany	36	—	36	—
Italy	20.5	—	20.5	—
Netherlands	9	—	9	—
Portugal	10	—	10	—
Spain	83	—	83	—
Great Britain	27.5	—	27.5	—
Total	291	—	291	—
Scandinavia	8	—	—	16
Eastern Europe	120	—	240	30
Africa	10	10	20	50
Asia				
China	10	—	80	80
India	3	—	10	—
Indonesia	30	30	45	45
Japan	300	433	340	525
Malaysia	30	—	75	—
Pakistan	40	—	70	—
Philippines	4	—	5	5
Singapore	3	3	5	5
South Korea	20	130	30	190
Taiwan	2	15	4	20
Thailand	25	—	40	—
Total	467	611	704	865
Australia	10	—	12	—
Worldwide	2743	3896	3392	4519

[a]After Licht, 1986.
[b]Fructose content.

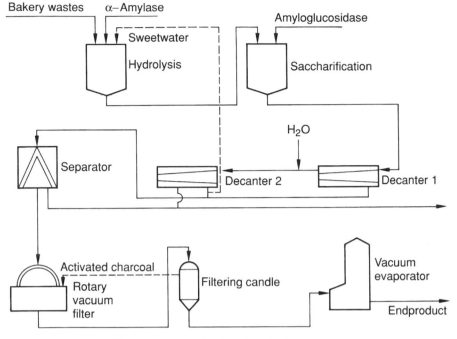

Figure 5.12 Saccharification of bakery wastes.

ogy needed for such recycling is quite elaborate; however, it need not be because often simultaneous hydrolysis and saccharification without subsequent refining suffice to produce suitable additives for doughmaking.

Wastes from potato processing can be converted into animal feed as shown in Figure 5.13, or alternatively, can serve as browning agents or be fermented into alcohol.

Cacao powder also contains starch; the basic properties of many chocolate products can be improved with enzymatic treatment.

Some Third World countries have begun to obtain high-grade sweeteners from broken rice (i.e., inferior grains accumulated during rice harvesting).

Furthermore, direct enzymatic saccharification has long been a well-known process in the fermentation industry. Inexpensive starch-containing raw materials are optimally degraded by enzymes for a specific microbial production strain prior to delivery to the fermenter. Preliminary and prudent steps have been initiated for utilizing amylases and other enzymes in treating wastewater and wastes present in urban clarification plants.

When discussing the enzymatic conversion of starch, the processes used in the textile (see Section 5.11) and paper industries must be mentioned as well. Starch is an important raw material in the paper industry. It is used in bonding paper pulp, in sizing of paper surfaces, and as a thickening agent in paints. Modified starches are almost always used. Modification can be done either chemically, thermochemically, or with bacterial α-amylases such as the α-amylase-LP [SOEG]. The latter process

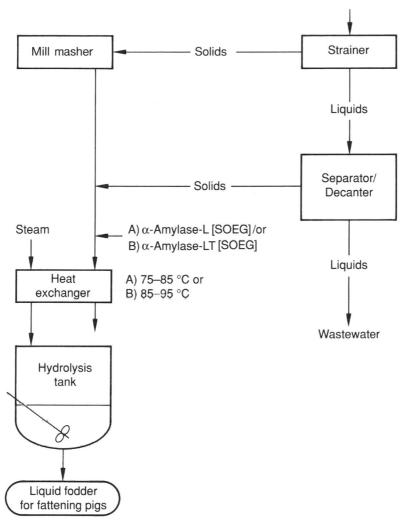

Figure 5.13 Processing wastes from the production of potato chips, french fries, and similar products (Richter and Grosser, 1983).

has since become the most important, and many paper manufacturers have small plants with converters for the enzyme-catalyzed hydrolysis of starch producing the desired viscosity and hygroscopicity.

What is the future of enzymes in the starch industry? Only a few high-performance starch processing enzymes have been added in recent years. The latest really important process improvement was the application of the pullulanases and other debranching activities during saccharification. Recently, serious efforts (Labout, 1985) were undertaken to improve the efficacy of the amyloglucosidase–pullu-

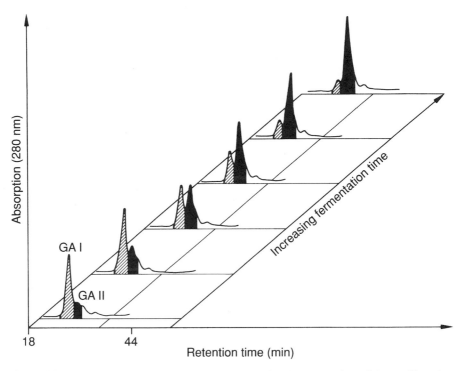

Figure 5.14 Gel-permeation chromatography analyzing the conversion of *Aspergillus niger* amyloglucosidase I into amyloglucosidase II in a pilot fermentation for amyloglucosidase production (2 × BioRad TSK 250, 30 × 0.75 cm, 0.5 ml/min phosphate buffer at pH 6.8) (Konieczny-Janda, 1985).

lanase mixture on saccharification time or yield by optimizing the natural composition of the amyloglucosidase preparations. The ratio of amyloglucosidase isozymes I and II can be adjusted to favor the prevailing desired level of saccharification (Fig. 5.14) either during or after fermentation (DE after hydrolysis, either high or low; Dx after saccharification within a given range; content of di- and trisaccharides). In addition, an acid-stable α-amylase is formed during this process that is particularly well suited to attack the small amounts of high-molecular-weight polysaccharides that are usually present as a consequence of starch degradation. As a rule, almost all amyloglucosidase preparations on the market invariably contain α-amylase activity as well; however, depending on the biological strain and fermentation characteristics, this enzyme is usually present in a relatively low concentration. An increase in acid α-amylase content during *Aspergillus niger* fermentation (in this case a multiple of the standard amount is required) causes the sweetening enzyme to increase glucose yields by an added 0.5–1.0% or to significantly reduce the time required to attain the desired degree of degradation.

Factors dependent on the raw materials play an important role in other interesting new developments. For saccharified wheat starch, the formation of gel-like precipitates make it quite difficult to separate the insoluble components (sediment) in an economically feasible manner. Comparisons of product data from various enzyme producers provided the first clues to the fact that secondary activities in sweetening enzyme preparations can significantly improve the flocculation properties and subsequent filtration behavior of sweetened syrups and prevent foaming and sensory changes in the syrups. It has since been learned that the degradation of companion substances in wheat starch is eminently important in this context. Examples are the enzymatic cleaving of wheat lysophospholecithin (Fig. 5.15), which has excellent emulsification properties and is present in high concentrations; and the cleaving of pentosans, which, when present even at very low concentrations, will increase the syrup's viscosity to such extent that particulates remain suspended. The more recent amyloglucosidase preparations such as Optidex-LW [SOEG] or Diaxyme-L [SOE] contain the required activities of both lysophospholipase and pentosanase in sufficient quantity. Naturally, they can also be added individually in other sweetening processes.

Switching to native maize and wheat varieties led to the development of enzyme cocktails for improving the separation of starch from gluten, or from fibrous material or for the reduction of wastewater generated during wheat processing. These multienzyme mixtures exhibit predominantly cellulase, pentosanase, β-glucanase, and xylanase activities, such as Pentosanase-L [SO; SOE]; functionally, they are considered to be technical aids only.

Surely is it the dream of every process engineer or operations manager in the starch industry, and also of every enzyme technologist, to be able to run hydrolysis, saccharification, and isomerization in a continuous operation under completely standardized conditions. Modern methods of biotechnology suggest that it may be possible to modify the properties of specific enzymes, either chemically or by genetic engineering, to such an extent that they can be tailor-made for specific applications. It would be particularly attractive to be able to balance the pH characteristics of the amylases and isomerases, thus eliminating both the addition of acids and bases and consequently the salts formed during the refining process. A first step in this direction is the effort to employ heat-stable α-amylase at a pH below 6.0,

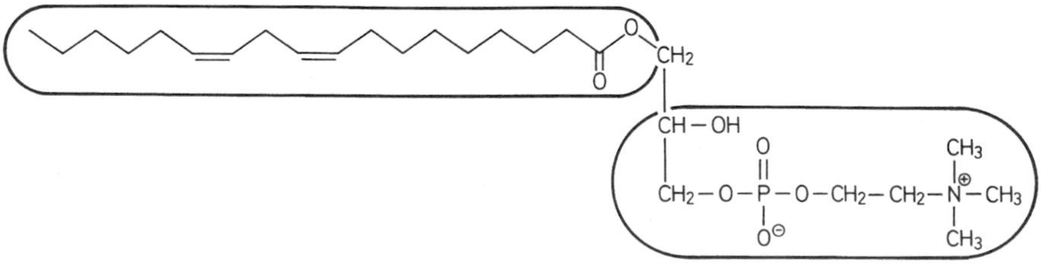

1-Linoleoyl-glycero-3-phosphocholine (Lysolecithin)

Figure 5.15 Model of the structure of lysophospatidylcholine (LPC) (Konieczny-Janda, 1985).

thereby simultaneously and significantly reducing the formation of maltulose (4-α-D-glucopyranosyl-D-fructose).

Currently, there is no method available to shift the pH for isomerization into the neutral range. The upper process temperatures are determined by the pasting point of the starches (60–80°C) and the lower, by the range for microbial contamination (55–60°C). Further, it would be most desirable to remove the rather paradoxical situation of the essential addition of Ca^{2+} to activate α-amylase, and then remove it almost completely prior to isomerization because of its inhibitory effect. For a complete continuous process operation, utilization of an immobilized amyloglucosidase should also be considered. This enzyme, as well as glucose isomerase, should—according to the most recent studies (Geyer, 1991)—be combined in a single-column reactor; loss of activity during prolonged operating times can be compensated for by the stepwise or continuous addition of soluble or carrier-bound enzyme.

5.1.1 Enzymes for Wet Milling of Maize and Cereal Grains

When maize and cereal grains are used for starch processing, the starch granules must be separated from the other kernel components. This is accomplished by the wet-milling process. The kernels are soaked prior to the milling step so that the components are evenly removed. During soaking (30–50 h at 45–50°C) the granules imbibe about 45% water. About 6% of the maize solids are solubilized in the sulfur dioxide–containing steepwater. This steepwater, the first by-product of the process, is used by the industry as an essential fermentation nutrient in other processes. After drawing off the steepwater, the softened kernels enter the milling process started with the addition of cellulase.

The advantages of using enzymes are as follows:

- Increasing starch yield
- Increasing milling capacity
- Reducing steeping time
- Reducing drying costs for gluten by increasing the solids in the separated wet gluten

The enzymes currently employed in this process are cellulases isolated from *Trichoderma* sp. (see Section 3.1.4.4). The preparations used in the process discussed here must be free of amylase activity so that no losses arise during starch refining. Further, the preparations must not show any, or only an extremely low, proteolytic activity, as proteolytic activity would lower the quality of the gluten isolated. Gluten quality is particularly important for wheat gluten. Prior to use, the enzyme preparations should be tested for proteolytic activity with a sufficiently sensitive method. The method measuring the drop in viscosity of a gelatin solution is well suited for such an assay.

Tests to establish the effectiveness of technical cellulose preparations, and their economic advantage must be conducted on an industrial scale. The criteria can include the increase in starch yield and the increase of the wet-gluten dry matter. In maize starch processing, particularly the bonds between gluten and fibrous components must be broken. The success of this separation depends on the amount of water absorbed by the kernel. It must be borne in mind that maize kernel quality and steeping time will affect yield.

An important improvement accomplished with enzymes is the reduction in water content of the gluten protein when obtained by either centrifugation or extraction.

ENZYME DOSING   Generally, liquid enzyme preparations are used at the onset of wet-milling processes. According to Silver (1988), at least 5000 filter paper units of a cellulase from *Trichoderma reesei* are used per ton of steeped maize kernels.

To detect the effects clearly, pilot production runs are initiated with relatively high doses of enzyme; for example, 500 ml of *Trichoderma* cellulase with a specific activity of 2000 C units per milligram of protein (e.g., 500 mg Rohament 7069 [RM]) is added to one ton of starchy raw material.

Depending on the results obtained, the enzyme dose is lowered. Typically, good results are obtained with 40–50% of the enzyme quantity cited above.

An alternative to this application is the addition of the enzyme preparation to a gluten suspension prior to vacuum filter centrifugation. A reaction time of 1 h at 40°C is required in this case to digest the polysaccharides, which significantly hamper filtration.

WHEAT STARCH   For the isolation of wheat starch (see Chapter 5, Fig. 5.2), the use of enzymes has the following advantages:

• Increasing the yield of A-starch
• Improving extraction of water from bran
• Saving energy during drying of the gluten

REFERENCES

Boni, L., "Enzymes in starch processing," unpublished manuscript.

Dorr-Oliver, B. V., technical information, Amsterdam.

Geyer, H.-U., "Zucker und Enzyme," Kali-Chemie, Hannover, Germany *Nachrichten* **69** (1984).

Geyer, H.-U., "Gleichzeitige Verwendung von Amyloglucosidase und Glucoseisomerase in einem Reaktor zur Herstellung eines Isomerase Sirups," German Discl. DE 39 26 609 (Feb. 14, 1991).

Hebeda, R. E., in Grayson, M. (ed.); *Kirk-Othmer Encyclopedia of Chemical Technology,* 3rd ed., Vol. 22, 1983, p. 499.

Hebeda, R. E., in Hui, Y. E. (ed.); *Encyclopedia of Food and Science Technology,* Vol. 4, Wiley, New York, 1992, p. 2490.

Hebeda, R. E., in Nagodawithana, T., and Reed, G. (eds.); *Enzymes in Food Processing,* 3rd ed., Academic Press, New York, 1993, p. 321.

Konieczny-Janda, G., "Beiträge zur Optimierung der industriellen enzymatischen Verzuckerung von Weizenstärke," paper presented at Starch Conference, Detmold, 1985.

Labout, J. J. M., *Die Stärke* **37**, 157 (1985).

Licht, F. O., *Internatl. Sugar Report* **118**, 591 (1986).

Radley, J. A. (ed.), *Starch Production Technology,* Applied Science Publishers, London, 1976.

Richter, G., and Grosser, D., *Die Stärke* **35**, 113 (1983).

Richter, G., *Lebensmitteltechnik* **6**, 330 (1986).

Richter, G., and Tegge, G., in *Proc. Internatl. Symp. Biotechnology and Food Industry,* Budapest, 1988, p. 46.

Röhm GmbH, product information.

Schenck, F. W., and Hebeda, R. E. (eds.), *Starch Hydrolysis Products. Worldwide Technology, Production, and Applications,* VCH, New York, 1992.

Shetty, J. K., and Marshall, J. J., Method for Determination of Transglucosidase," U.S. Patent 4,575,487 (1986).

Silver, S. C., "Wet-milling of starch containing grains—by steeping cleaned grain to soften it, then milling it with cellulase," World Patent 8,808,855 (1988).

SOE (formerly MS-USA), product information: Tenase, Taka-Therm, Diazyme L (1992).

SOEG (formerly MKC-Germany), product information: Optiamyl, Optitherm, Optidex, 1992.

Takamine, E. J., "Preparing and making taka-koji," U.S. Patent 525,820 (1894).

Tegge, G., *Stärke und Stärkederivate.* Behr, Hamburg, 1984.

Tegge, G., "Verzuckerung von Stärke—Vergangenheit und Zukunft," paper presented at *Biotechnica,* Hannover, 1985.

Vuilleumier, S., "World outlook for high fructose syrups," *Sugar y Azúcar* **8**, 27 (1983).

Weidenbach, G., Bonse, D., and Richter, G., *Die Stärke* **36**, 412 (1984).

Woods, L. F. J., and Swinton, S. J., in Tucker, G. A., and Woods, L. F. J. (eds.), *Enzymes in Food Processing,* 2nd ed., Blackie Academic & Professional (imprint of Chapman & Hall), Glasgow, London, New York, 1995, p. 250.

5.2 ENZYMES IN THE BREWING INDUSTRY

Reference: Power (1993)

5.2.1 Introduction

Brewing beer is almost as old as baking bread. The Sumerians produced an alcoholic beverage from fermenting bread grain. The Assyrians are known to have brewed beer about 4000 B.C.

The middle ages experienced a rapid expansion of brewing; beer was already brewed in the cloisters around 1000 A.D. This period may have also given rise to its name for the common latin word *biber,* which means beverage or drink. At that time, hops as a constituent of beer was established as well. Brewing in the fourteen and fifteenth centuries spread to the German Hanseatic cities, and in the sixteenth century to southern Germany. The "Reinheitsgebot" (purity law) of 1516, decreed by Duke Wilhelm of Bavaria, states that beer can be produced only from barley, hops, water, and yeast. This decree helped hold Bavarian beer in high esteem. This law was later modified to include malted barley. According to the beer tax law of November 29, 1939, beer is classified as "bottom-fermented beer" (lager beer) when produced from barley malt, hops, and water; and "top-fermented beer" (e.g., wheat beer), in which other malted grains, sugar, and colorants can be used.

Beer brewing is also widespread throughout Africa, where beer from millet has long been produced. In East Asia beer is brewed from rice; in South America, from corn.

The processes involved in brewing beer include malting in the malt house, wort preparation, and fermentation of the wort in the brewery.

5.2.2 The Malt House

The objective of the malting process is the enzymatic digestion of barley or wheat grains during germination. First the barley is cleaned and then steeped. This increases the kernels' water content of 44–48%. The steeped barley kernels are spread approximately 30 cm deep over the malt-house floor and kept at a temperature of about 15°C. Adequate aeration is mandatory for germination so that the low temperatures can be maintained. Germination is rapid and is completed after about 6–7 days. The "green malt" generated is kiln-dried as carefully as possible to retain high enzymatic activity. The temperatures used in kiln drying determine the malt quality, which, in turn, imparts the individual characteristic to a beer. Approximately 70–80 kg of dry malt is obtained from 100 kg of barley. The use of enzymes in the malting process has been described many times, but they are rarely employed, although soaking in a solution of *Trichoderma* cellulases is believed to help solubilize the malt during the mashing process. Gibberellic acid does, however, significantly increase the formation of amylolytic activity during malting. In the United States, its use has been approved since 1960.

5.2.3 The Brewery

Enzymes play a central role in brewing. In the traditional brewing process, the enzymes are formed during the malting operation. They are necessary to convert the high-molecular-weight polymer constituents of grains into low-molecular-weight compounds. These endogenous enzymes are the malt α- and β-amylases, which convert starch into fermentable sugars and nonfermentable dextrins, as well as proteases and peptidases that digest grain proteins into low-molecular-weight peptides and free amino acids. The latter are needed as nutrients for the yeast during fermentation. The protein digests are also involved in the beer's aroma and stability. The β-glucanases formed during the malting process are of special significance in successful brewing.

Further, there are a number of other important endogenous enzymes, namely, various phosphatases and the pentosanases that hydrolyze gums during malting.

Modern brewing technology employs, in addition to the malted barley described above, raw barley, and other raw grains such as rice, corn, or sorghum. The necessity of partially or completely replacing malt as a brewing addendum becomes important when malt is unavailable or excessively expensive. In such cases, the native malt enzymes are replaced with technical enzyme preparations (Fig. 5.16).

Ground malt or malt together with a given amount of raw grain and water are mashed at 37°C. The mash is then held at different temperatures for various time

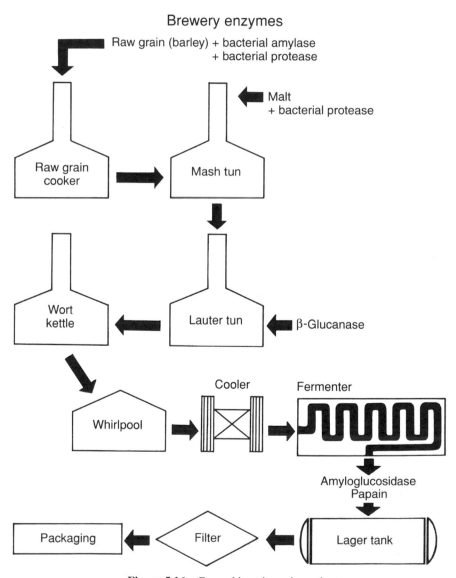

Figure 5.16   General brewing schematic.

periods (protease rests) to extract and solubilize the contents of malt and raw grain. At the end of the mashing process, on clarification, the spent grain is removed from the wort. The wort is delivered to the wort kettle and boiled together with hops or hop extract. During this process, the endogenous and exogenous enzymes are inactivated and certain protein fractions are coagulated and precipitated.

At this stage, solubilization and conversion of the hop constituents occur. Subsequently, the wort is filtered and cooled to the fermentation temperature. On the addition of yeast, fermentation is initiated; when completed, the yeast is removed and the green beer is transferred to the lager tank. The beer must then again be filtered prior to barreling or bottling.

5.2.4 The Mashing Process

Various mashing methods are used for solubilizing the constituents of malted or raw grains; in part, they have a traditional origin; examples are the Bavarian three-step mashing process (decoction or boiler mashing) and the British infusion mashing. Alternating thermal and enzymatic steps are employed in all the procedures in order to solubilize the high-molecular-weight polymers of the grains.

5.2.4.1 Decoction Mashing Decoction mashing includes a physical as well as an enzymatic malt extraction. The three-step mashing procedure (Fig. 5.17) is the best known and oldest method; the current decoction methods have been derived from it. In decoction procedures, depending on the mashing process employed,

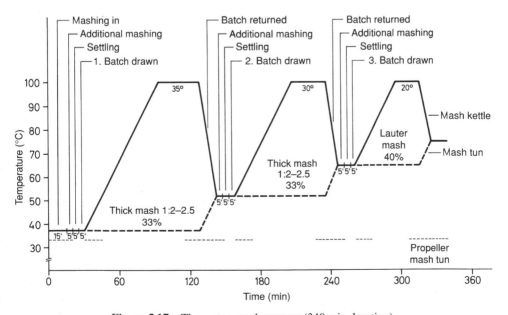

Figure 5.17 Three-step mash process (340 min duration).

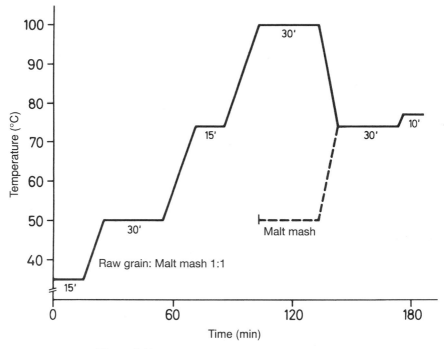

Figure 5.18 Method for processing 20% raw grain.

batches of mash are withdrawn and heated to 100°C; these boiled batches are then recombined with the main mash lot. This accomplishes pasting of a portion of the starch that then can be converted while the enzymes in the nonheated mash remain active.

This process is still in use today in brewing dark beers. In contrast to malt, the raw grains must first undergo pretreatment (raw-grain milling) to increase the availability of their constituents (Figs. 5.18 and 5.19).

Depending on the type and amount of raw grain, various mashing processes can be implemented. First, the raw grain is mashed with malt at a ratio of 1:1 and temperature of 35°C; the mash is then successively heated to temperatures of 50, 62, and 72°C, and held at each temperature for a determined period of time (rests) and then ultimately boiled at 100°C. The remaining malt is subsequently mashed and held at 50°C in the first rest. The boiled mash is cooled to 65–70°C and mixed with the pure malt mash.

Thus, an additional saccharification step can be conducted at 70–75°C. Subsequently, another rest can follow at 78°C, after which the lautering process is initiated.

5.2.4.2 Infusion Mashing The raw-grain or malt constituents are converted and solubilized only by enzymatic reactions. The boiler step is not used in this process (Fig. 5.20).

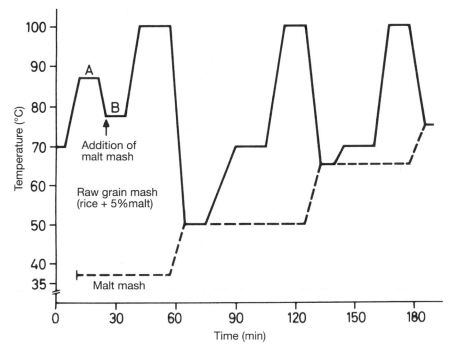

Figure 5.19 Method for processing raw grain with prior pasting (~20% rice): pasting; (b) liquefaction of raw grain.

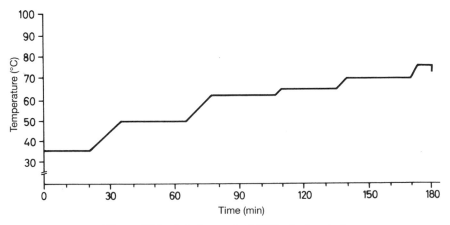

Figure 5.20 Infusion method (180 min duration).

The infusion method includes the following steps: first, a protein rest at 50°C for 30 min. Holding this step for a longer period is not required because the proteolytic activity and the activity of the endo-β-glucanases decrease. The maltose rest can be carried out in several steps in order to fully exploit the β-amylase activity. The rest of 70–72°C is usually held only until the iodine reaction is negative. However, because the boiling used in the decoction method is not used in this process, the rest time is extended to permit better saccharification. The mash is subsequently heated to 78°C and pumped into the lauter tun.

5.2.5   Brewing with Enzymes

In traditional brewing (Fig. 5.21), malt quality depends on several factors. These include the properties of the particular barley cultivar and where it is grown, climatic conditions, and the malting process itself. All these factors strongly influence the enzymatic activities of the malt. High-grade malts endowed with an optimal complement of enzymes are very expensive.

Variations in malt quality can give rise to difficulties during processing, such as during lautering of the mash, filtration of the wort, or in the green beer. However,

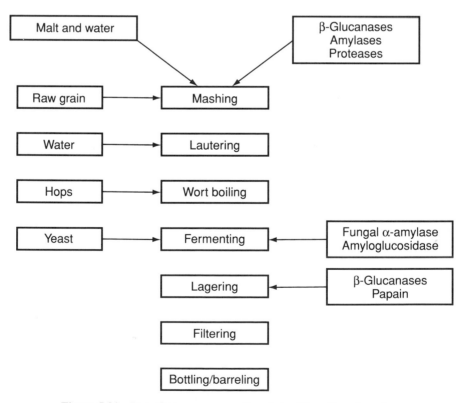

Figure 5.21   Use of enzyme preparations in traditional beer brewing.

TABLE 5.10   Purpose and Site of Enzyme Use

Enzyme	Application Purpose	Site
Bacterial proteases	Protein hydrolysis for increasing FAN[a]	Mash tun
Papain	Cold stabilization	Lager tank
Bacterial α-amylase	Starch hydrolysis	Raw-grain cooker
	Starch hydrolysis	Mash tun
Fungal α-amylase	Improving fermentation	Mash tun
		Fermenter
Amyloglucosidase	Improving fermentation	Mash tun
	Light beer	Fermenter
Bacterial β-glucanases	Improving extract	Mash tun
	Shorter purification times	Mash tun
	Improving wort filtration	Mash tun
	Improving beer filtration	Lager tank
Fungal β-glucanases	Improving extract	Mash tun
	Improving filtration	Mash tun

[a]Free amino nitrogen.

endogenous malt enzyme deficiencies can be compensated for by the targeted use of enzyme preparations with specific activities.

If barley or other cereals are brewed, the addition of exogenous enzyme preparations is absolutely essential for a reliable and economical process.

Table 5.10 lists the types and functions of enzyme preparations and these preparations are employed in brewing.

5.2.6   Brewing with Malt

If only poor-quality malt is available for brewing, the malt constituents can be solubilized with a modified mashing method or by the addition of enzymes.

5.2.6.1   β-Glucan Digestion   High-molecular-weight β-glucans are particularly significant in that they increase mash viscosity, thereby hampering mash lautering. Furthermore, the filterability of wort and beer is poor. Also, hazes, even gels formed from poorly solubilized malts, have been described for finished beer. Steiner (1968) has reported the amounts of β-glucans and pentosans in barley, malt, wort, and beer (Table 5.11).

Usually, some β-glucans are digested during malting; however, a large amount is solubilized at increased temperatures during mashing (McCleary and Nurthen, 1986). β-Glucan digestion is limited because the β-glucanases in malt are heat-labile and the poorly solubilized malt itself contains an inadequate amount of these enzymes.

However, the addition of β-glucanase preparations reported by Enari and Markkanen (1974) can compensate for this deficiency (e.g., 150 g Rohamalt MG

TABLE 5.11   Gum Contents in Barley, Malt, Hops, and Beer

	Contents (g/100 g)		
	Total Gums	β-Glucan	Pentosan
Barley (var.)	6.4	4.8	1.6
Malt	2.3	1.1	1.2
Wort	4.7	4.1	0.6
Beer	4.9	4.0	0.9

Plus [RM] per ton of malt) (Table 5.12). A review of microbial β-glucanases for brewing has been published by Borriss (1994).

5.2.6.2   Proteolysis   Proteins are inadequately digested in poor-quality malt. This results in a deficiency of the amino acids that must serve as nutrients for the yeast. In addition, undigested proteins will cause hazing in beer; consequently, problems of stability will arise. Sorensen and Fullbrook (1972) have studied the effects of various microbial proteases on proteolysis in mash. They found that alkaline bacterial protease is strongly inhibited by barley extracts. However, a neutral bacterial protease proved to be considerably more effective in this system.

A dose of 0.2–0.5 kg of enzyme preparation with an activity of 1.0 Anson unit per metric ton of malt is sufficient to generate adequate proteolysis during the protein conversion step. The soluble nitrogen content increases significantly and beer stability is improved.

5.2.6.3   Fermentability   The addition of amylolytic enzymes to wort improves its fermentability. Heat-stable bacterial α-amylase can be used for the maltose rest during the mashing process (e.g., 0.3–0.5 kg Rohalase HT [RM]/t malt). Continuing the starch conversion in the wort, 100–200 g of fungal α-amylase with an amylolytic activity of 45,000 SKB/g (e.g., Rohalase M3 [RM], or Fungamyl 1600 S [NO]) is added to each metric ton of wort.

TABLE 5.12   Analysis of a Tank of Beer Brewed with Malt

	Tank Beer	
Characteristics	Without Rohamalt MG Plus	With Rohamalt MG Plus
Turbidity (FUa)	16	13
Viscosity (mPA • s)	1.68	1.63
β-Glucan (mg/liter)	373	45
Filterability (g)	81	104
Filtering efficiency (ml/h)	5.3	6.2

[a]Formazan units (turbidity standard).

5.2.7 Use of Raw Grain in Brewing

To reduce brewing costs, an increasing number of brewers employ raw grain, ostensibly obtaining the same results (extracts, fermentable substances, brew stability, etc.) as with malt. Raw grains employed are sorghum, rice, wheat, maize, wheat-B starch, and sugar syrup. Only limited amounts of raw grain are permitted as brewing addenda in individual countries; in Europe, it varies from country to country, allowing 20–30% raw grain; in the United States, a maximum of 40% is permitted.

Raw grain must be milled finer than malt but not so fine as to cause clumping. When using both malt and raw grain, each must be milled separately. If the raw-grain portion is greater than 15%, then a coarser malt grist must be used to avoid problems in lautering. Mash filters are recommended when the raw-grain portion reaches 30–40%. The type of raw-grain processing depends on the raw grain selected. Thus, maize and rice grains must be mashed with lots of water because both imbibe considerable amounts of water. Using enzymes causes a rapid decrease in viscosity of the gelatinized starch. The iodine-reduction test measures the extent of starch conversion. Iodine color indication occurs only when the solubilized starch is pasted (gelatinized). The temperature at which the raw-grain conversion must be conducted depends on the pasting temperature. Iodine color indication for sorghum or rice is observed only at temperatures above 80°C (Table 5.13).

If a beer is brewed from malt and raw grain, parallel mashing of raw grain and malt is required. Subsequently, both are combined and saccharified at 68–72°C.

This temperature must be held until the iodine test standard is reached. However, a 45-min rest is recommended. An α-amylase at 0.5 kg per metric ton of grain with an activity of 6000 SKB/g (Rohalase AF [RM]) or, depending on the pasting temperature, a heat-stable bacterial α-amylase (e.g., Rohalase AT [RM]), at .05 kg/t grain with an activity of 4000 SKB/g, should be used for raw-grain conversion. The raw-grain conversion is conducted either under 1–2 bar pressure or at normal atmospheric pressure. If the malt has been extracted under pressure, problems in beer flavor can arise when the malt : grain ratio is greater than 1 : 1. Other variations in flavor arise from the use of oil-rich semolinas that, on lipid hydrolysis, generate fatty acids. These can also impair formation of the beer's head. Reduced amylolytic activity results in inadequate starch

TABLE 5.13 Pasting Temperatures

Raw Grain	Pasting Temperature (°C)
Barley	62–74
Barley malt	63–64
Malt	64
Maize	68–74
Rice	68–75
Wheat	56–66
Potatoes	52–58
Sorghum	69–75

hydrolysis. The availability of assimilable nitrogen for the yeast must be assured because of the very low solubility of raw-grain protein.

5.2.8   Brewing with Unmalted Barley

To reduce the cost of energy associated with malting, raw barley is used. However, using more than 15% unmalted barley causes problems during the brewing process because of limited enzymatic activity, primarily because the β-glucanase and proteases are lacking. A two-step mash process employing both malt and barley produces poorer results than when mashing with malt alone.

However, when enzymes are utilized, brewing with 30–40% unmalted barley is entirely possible. When compared to a standard mash (100% malt), the β-glucan is digested just as well with added β-glucanases. This way, a smooth filtration of wort and green beer is guaranteed. A bacterial α-amylase with an activity of 6000 SKB/g (e.g., Rohalase AF [RM]) should be added to the raw-grain cooker at 0.5 kg/t grain. To assure that an adequate amount of fermentable sugar remains available in the total mash, an amyloglucosidase, such as Rohalase HT [RM] or Diase [GB], should be added.

5.2.9   Beer Production with 30% Barley and 70% Malt

Barley is mashed in the presence of a bacterial α-amylase, such as Rohalase AF [RM] or BAN [NO] at 50°C; malt is mashed in parallel. After boiling the raw grain, both mashes are combined and the standard one-step mash process in the mash tun begins. To improve protein hydrolysis, a bacterial protease with an activity of 200 PU/mg (e.g., Rohalase P [RM]) can be added at 0.3–0.5 kg/t to the proteolytic step, which will yield about a 20% increase of formalin nitrogen (Fig. 5.22; see also Fig. 5.23).

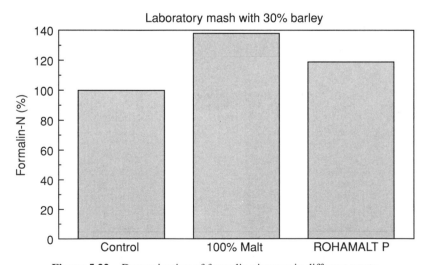

Figure 5.22   Determination of formalin nitrogen in different worts.

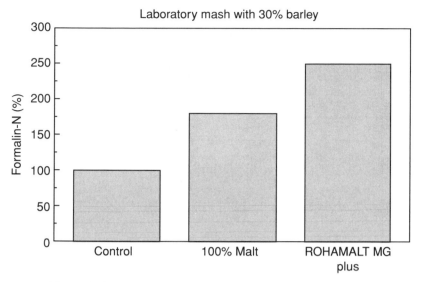

Figure 5.23   Determination of filtration volume of different worts.

5.2.10 Brewing with Corn
The use of technical enzymes in brewing beer from malt and corn has been success-ful and is outlined in the flowchart in Figure 5.24. The problems that normally de-velop with the addition of corn can be avoided by dosing the given enzymes appro-priately. Difficulties again include poor fermentability because of a very limited starch hydrolysis, the lack of free amino nitrogen due to inadequate proteolysis, and poor filterability because of excessive amounts of highly viscous β-glucans.

5.2.11 Brewing with 100% Sorghum

Brewing beer exclusively with sorghum (Fig. 5.25) has only recently become im-portant because importation of malt has been forbidden in some non-European countries. The key problem in this case is the poor fermentability of the wort be-cause the amino nitrogen for adequate nutritional support of the yeast is limited. Addition of a bacterial protease at 1 kg/t grain with an activity of 250 PU/mg (e.g., Rohamalt PSO [RM]) helps to liberate sufficient amino nitrogen. Because of the high pasting temperature of sorghum starch, a heat-stable α-amylase at 1 kg/t grain with an activity of 3000 SKB/g should be used for raw-grain extraction. For a better saccharification of mash, a combination of amyloglucosidase and fungal α-amylase is advantageous.

5.2.12 Analytic Monitoring of the Mashing Process

5.2.12.1 Proteolysis   Proteolysis is monitored by determining free amino nitro-gen (FAN). With a FAN content of 150–180 kg per ton based on a 12% wort, an ad-

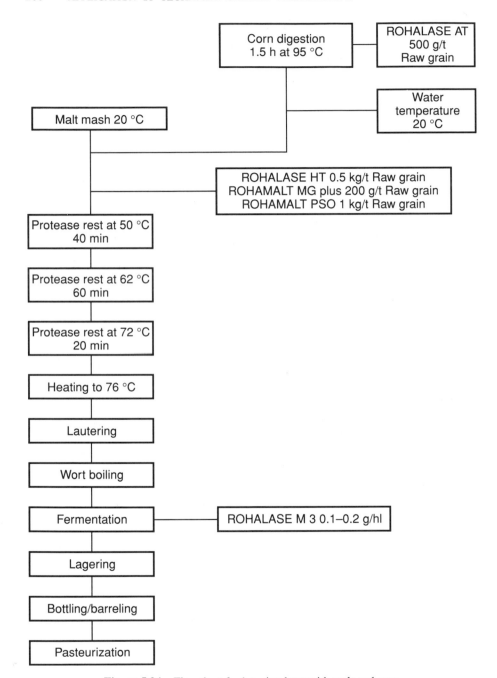

Figure 5.24   Flowchart for brewing beer with malt and corn.

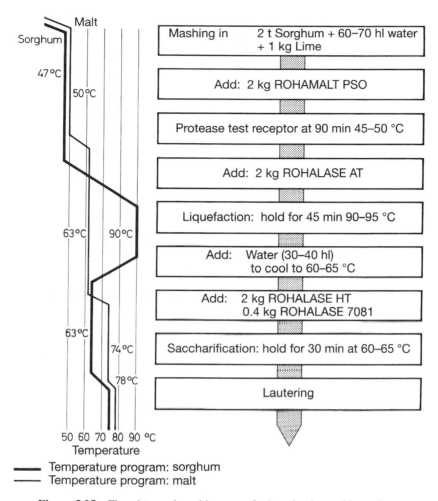

Figure 5.25   Flowchart and mashing steps for brewing beer with sorghum.

equate nutrition of yeast is assured. In addition, sufficient high-molecular-weight protein is still present to permit good stability of the head and body of the beer.

5.2.12.2   Starch Hydrolysis   The simplest way to determine the degree of starch hydrolysis is the iodine test. Inadequate hydrolysis is indicated by the development of a gray film formed by the various amounts of dextrins, which react with iodine to form a reddish to reddish-purple color. A poorly saccharified wort shows slow settling and poor clarification. Such conditions cause a drop in pH during fermentation, an increase in alcohol content, and increased haze. This brings secondary fermentation to a halt and damages the beer's flavor by producing, for instance, a yeast diacetyl flavor.

5.2.12.3 β-Glucan Digestion   Because direct determination of the β-glucan content is complicated, filtration measurements are usually conducted with wort and green beer. These are then compared with the filtration data obtained with standard brews. Varying malt and raw-grain qualities will typically give rise to long lautering times, caused by the high β-glucan content. This, however, is not attributable to β-glucan alone but also to the coagulates formed from proteins, polyphenols, and β-glucan.

5.2.13 Difficulties in Green-Beer Filtration

After completion of storage, the beer is again filtered. This can occasionally give rise to unexpected delays in the brewery's production run. Gumlike substances and hemicelluloses such as pentosans and β-glucans are responsible for poor filtration performance. Filterability is often unsatisfactory when poorly solubilized malt with little hemicellulase activity or when mixtures of raw grain are used. Improved filtration is achieved by adding hemicellulase preparations such as Rohamalt F [RM] (Fig. 5.26), Ceremix [NO], or Filtrase [GB] to the lager tank shortly before terminating lagering. The filtration test has been described by Goerth and Radola (1981).

5.2.14 Cold Stabilization of Beer

Beer stored at very low temperatures will develop a haze. The haze will disappear when the beer is warmed. Aged beer and especially pasteurized beers tend to develop a haze when chilled. This turbidity is the consequence of reactions between oppositely charged lipophilic colloids that precipitate in the same way as would an excessively dehydrated saturated solution. During warming, the precipitates again

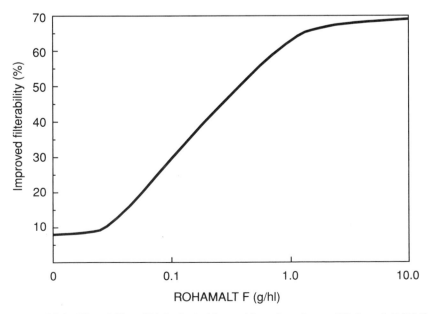

Figure 5.26   Filterability of high-alcohol beer with various doses of Rohamalt F [RM].

dissolve as a result of rehydration. The most important reactants are the polyphenols, which can form complex associations with carbohydrates and polypeptides.

Pepsin, ficin, bromelain, fungal, and bacterial proteases have been used, in addition to papain, to prevent cold haze. However, papain by itself or in combination with other proteases is the most frequently employed enzyme in cold stabilization. Papain activity is determined by the BAPA method (BAPA, Na-benzoyl-L-arginine-p-nitranilide), in which the activity of one BAPA unit corresponds to the amount of enzyme that converts 1 μmol of substrate per minute under defined reaction conditions (see Section 2.3.2.1).

Papain is an "intelligent" enzyme in that it acts only when haze is present, or it may act to prevent the development of the cold turbidity that previously hydrolyzed the protein. At a dose of 200–300 BAPA units of papain per hectoliter (26.4 gals) beer, e.g., Rohamalt Stab [RM] or Collupulin [GB], the amount of foam and its stability is not reduced. The Forcier test [*MEBAK Analysenmethoden* (1984); see also Drawert (1984)] is used for determining the haze stability of a beer. The beer is subjected to a cold–heat treatment, and the haze that develops is measured.

Other future options for cold stabilization may be the use of immobilized proteases or the use of tannase and polyphenol oxidase.

5.2.15   Light-Beer Production

Light beer, as distinguished from standard beer, is characterized by less wort and lower alcohol and residual sugar content and thus fewer calories (Fig. 5.27).

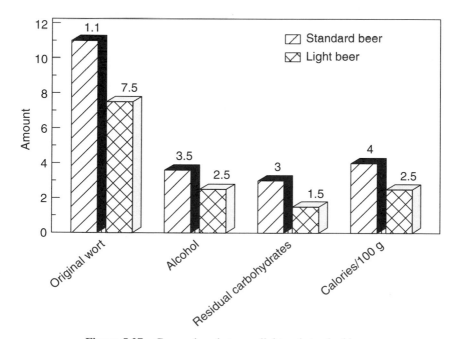

Figure 5.27   Comparison between light and standard beers.

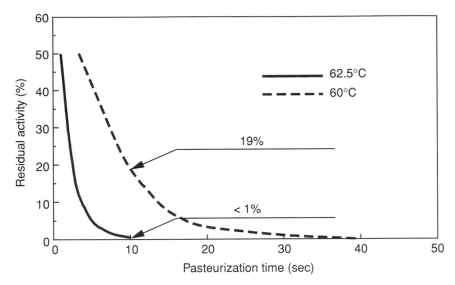

Figure 5.28   Inactivation of Rohamalt as [RM] in beer.

Light beer is brewed with malt or malt with a maximum of 50% raw barley grain. The production process is the same as that for standard beer; however, after boiling the wort and at the start of fermentation, a heat-stable amyloglucosidase is added, such as Rohalase AS [RM] (see Figs. 5.28 and 5.29), which converts dextrins during fermentation into glucose that is subsequently fermented by yeast.

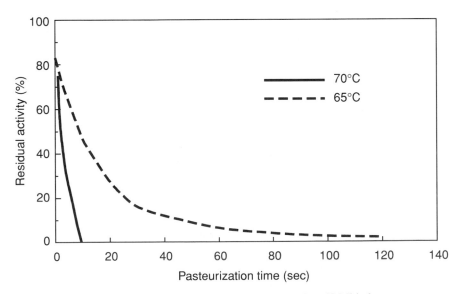

Figure 5.29   Temperature stability of Rohamalt as [RM] in beer.

To avoid secondary saccharification in beer, the use of a heat-labile amyloglucosidase (e.g., amyloglucosidase from *Rhizopus*) is recommended; this enzyme is then completely inactivated on pasteurization.

5.2.16 Diacetyl in Beer

Diacetyl formation can be prevented by addition of diacetyl reductase (Maturex [NO]) at the start of primary fermentation, when most of the α-acetolactate is being formed. The reaction is strongly pH-dependent. The activity at pH 5.0 is about three times higher than that at pH 4.0. In primary fermentation, the pH drops after the first day from 5.2 to 4.5 ([NO], technical information).

REFERENCES

Borriss, R., in Ruttloff, H. (ed.), *Industrielle Enzyme,* p. 728. Behr, Hamburg, 1994.

Drawert, E. (ed.), *MEBAK, Brautechnische Analysenmethoden,* Section 7.14.22, "Forcier-methode," in *Methodensammlung der Mitteleuropäischen Analysen Kommission,* 2nd ed. MEBAK, Freising-Weihenstephan, 1984.

Enari, T. M., and Markkanen, P. H., *Proc. Ann. Meeting Am. Soc. Brew. Chem.* 13 (1974).

Goerth, K., and Radola, R. J., *J. Inst. Brew.* **87,** 160 (1981).

McCleary, B. V., and Nurthen, E., *J. Inst. Brew.* **92,** 168 (1986).

MEBAK Analysenmethoden, 2nd ed., Section 7.14.22, "Forciermethod," 1984 (see Drawert reference, above).

Power, J., in Nagodawithana, T., and Reed, G. (eds.), *Enzymes in Food Processing,* 3rd ed., Academic Press, New York, 1993, p. 439.

Röhm GmbH, technical information.

Sorensen, S. A., and Fullbrook, P., "Application of enzymes in the brewing industry," *Tech. Q. Master Brew. Assoc. Am.* **9,** 166 (1972).

Steiner, K., *Schweiz. Brauerei-Rdsch.* **79,** 153 (1968).

5.3 ENZYMATIC ALCOHOL PRODUCTION

5.3.1 Introduction

Reference: Kreipe (1981)

Alcoholic beverages as well as alcohol for medicine and industry are produced in many countries from different starchy raw materials. Historically, the development started with producing brandy from sugar and starch. For many centuries, juices containing sugar or honey were fermented, and the alcohol produced was subsequently concentrated by distillation. The first reports about alcohol production date from the eleventh century. In Germany, taxing brandy as a source of income was first implemented in 1507.

Fortified spirits are directly distilled from wine without further processing. As such, depending on their origin, they enter the market as brandy or cognac.

Sugarcane and sugarbeet molasses often serve as the raw materials for alcoholic beverages and technical alcohol. Other raw materials include fruits such as cherries, pears, apples (for Calvados), plums (for Slivovic), and agave juice. The latter is the source for Mexican pulque.

Many types of alcoholic beverages are produced from starchy materials, including whiskey from malted barley and bourbon whiskey from corn and rye. Aquavit is obtained from potatoes or barley and vodka from rye, wheat, or potatoes. In East Asia sake is produced from rice.

Alcohol for industrial use is synthesized from ethylene by hydrogenation (OXO process). It is used as a solvent for paints and cosmetics and as a raw material for chemical compounds. It is employed in trade and domestically as fuel.

However, renewable natural energy sources such as sugar, starch, cellulose, and hemicellulose can serve as raw materials for producing alcohol.

In the 1970s, Brazil was particularly hard hit by the increased cost of crude oil. Oil costs exceeded $4 billion (U.S.) per annum. In 1975, Brazil, the country with the largest sugarcane production in the world, established a National Alcohol Commission that was empowered to coordinate the production of alcohol from molasses as a fuel additive: gasohol (Cheremisinoff, 1979). The target for the 1980s was the production of about 12 million metric tons of ethanol. Since 1977, 13–20% alcohol has been added to gasoline.

At that time, many companies in the United States, including large corporations, started producing alcohol from corn. New technologies were developed. There was an enormous increase in the demand, production, and consumption of amylases. Within a short time, the price of bacterial amylase and amyloglucosidase was cut in half.

In addition, whey, a waste product from cheesemaking and an environmental burden, was converted to ethanol by fermentation. In 1980, the potential production in the United States was estimated at about 89.8 million gallons of ethanol (Lyons, 1983). But when the price of crude oil dropped, these efforts were greatly reduced.

Naturally, cellulose, like starch, was also targeted as a raw material. Many research projects dealt with the enzymatic degradation of cellulose to fermentable sugars. However, the enzymatic conversion of cellulose still depends on a costly pretreatment of the fiber, and thus the cellulose-based technology is considerably more expensive than that of starch.

Foody (1992) discusses the technical and economic opportunities of generating alcohol from cellulose and related materials by means of enzymatic hydrolysis and the costs associated with such processes. According to this report, the cost of deriving alcohol from low-cost biomass could be as low as $0.77/gal. Further technical advances could reduce the cost by as much as $0.23/gal.

5.3.2   Starch Degradation

It was already mentioned in the preceding section that starch cannot be fermented directly. Starch must first be liquefied before it can be saccharified. This is accomplished with green malt, roasted barley malt, fungal malt, and microbial amylases. Prior to 1945 distilleries in Germany used only malt for saccharification; since then fungal malt has also been used. This process was developed in the United States by Underkofler in the late 1930s (Underkofler, 1972). Even earlier, mold cultures served as a source of saccharifying enzymes in the Orient. In the amylo process, steeped rice kernels are inoculated with spores of *Rhizopus delemar* and the mash, saccharified during an incubation of 24–30 h at about 37°C, is then fermented with yeast.

Fungal malt is the end product of a top-fermenting *Aspergillus oryzae* on a wheat bran nutrient medium. This procedure was later replaced by fungal amylase preparations obtained from bottom-fermenting species.

In addition to malt and the β-amylase preparations from barley or sweet potatoes, the following microbially derived enzymes are employed in distilleries.

1. For liquefaction:
   a. Bacterial α-amylase
   b. Heat-stable bacterial α-amylase
   c. Fungal α-amylase
2. For saccharification:
   a. Amyloglucosidase from *A. niger*, or from *Rhizopus* sp.
   b. Fungal α-amylase by itself, or in combination with amyloglucosidase
   c. Pullulanase for complete conversion in combination with other amylases
3. For reducing viscosity of the mash: β-glucanases

5.3.3 Alcohol Production

The following basic steps can be distinguished:

1. Processing the raw material by milling and heating or treating with acid or alkali.
2. Liquefying starch with α-amylases, or with acid to dextrins.
3. Saccharifying dextrins with β-amylases or amyloglucosidase to fermentable sugars.
4. Fermenting saccharified mash.
5. Distilling fermented mash. This yields unrefined alcohol and the wash as a by-product.
6. Rectifying crude spirits (first run, distilled alcohol, rectified alcohol and tails).

5.3.4 Starch Conversion

During the past 30 years, green malt and dry malt have been replaced in the distilleries by microbial enzymes for liquefying and saccharifying starch.

Short liquefaction periods for starchy raw materials are desired to ensure that saccharification can proceed during fermentation in 2–3 days. Until recently, the

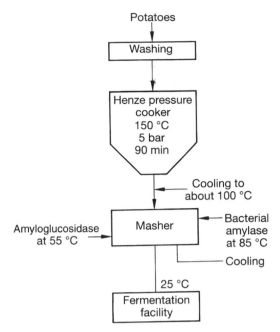

Figure 5.30 Flowchart of the Henze process.

Henze steam process (Fig. 5.30) which involved pressure-cooking potatoes at 140–180°C, was the process of choice.

Now, potato processing with a "cold-mash process" has gained importance. After chopping of the raw material in a blender, liquefaction is accomplished on addition of bacterial α-amylase from *B. subtilis,* at 65–75°C with a temperature maximum of 85°C (e.g., Optimash-L 840 [SO]). The energy requirement is thereby considerably reduced, particularly when energy recovery is possible. Milling raw grain into a fine meal is a prerequisite for the "mashing at saccharification temperatures" method employed in grain distilleries.

Ultrafine meals can be obtained using carborundum milling disks. In processing of rye and wheat, the grist is slurried in cold water (about 15°C) and the pH, if too low, is adjusted to 5.6–5.8. When malt is used, 4–5% malt is required (at 57°C), or a corresponding amount of technical enzyme preparation may be added. This process requires the addition of disinfectants.

5.3.5   Boiling Process

After reduction of the raw material into a finely ground meal, the liquefying enzymes are added and the starch mass is steam-heated to 80°C. Frequently, a protein rest at 50°C is staged to ensure that this temperature range is passed through slowly.

Experts in the field have different opinions about the yields and energy savings in milling or processing the raw materials. A few examples without evaluation are described below.

5.3.6 High-Pressure Steam Process

Pressure cooking serves to totally free the starch grains from their cellular associations. This occurs only on heating the mash in a pressurized vessel at temperatures >100°C. It also allows the introduction of thermostable bacterial α-amylases that can tolerate temperatures in excess of 100°C.

In the potato distillery—when exhausting the heated mash at constant temperature—water and lime are added to adjust the pH in the mash tub. After starting the discharge process and adjusting the temperature to 70°C, one-third of the allotted enzyme is added. After an additional 10 min of exhausting, the remaining two-thirds of the enzyme are added.

Alternatively, the rapid exhaust procedure is efficiently conducted at 90–100°C, and on cooling to 80–85°C, one-third of the enzyme is added. The remaining enzyme is added after the mash has cooled to 70°C.

In the grain distillery, white lime and water are added to the mash tub. After exhausting and cooling to 70–75°C, the entire enzyme lot (e.g., Optimash L 840 [SO] at 0.275 liter/t starch) is added. Given a high starch content, an aliquot of the enzyme is added at 80°C to rapidly lower the viscosity. The rest of the enzyme is then added at 70°C.

In the American batch process, soaked ground corn (about 25% dry substance) is treated with thermostable α-amylase (e.g., Termamyl 60 L [NO]) at 0.2–0.3 kg per ton starch), the mash is then heated with steam to about 150°C. Although the enzyme will be inactivated within a short time, it helps maintain a lower viscosity in the boiler. After this extraction, additional enzyme at 0.35–0.7 kg/t starch is added on cooling. When the liquefaction is completed, amyloglucosidase is added at 60°C to initiate saccharification.

In a process yielding ethanol for fuel, 100 kg of ground corn in 400 liters of water plus a thermostable bacterial α-amylase are initially heated to 65–71°C; then the temperature is raised to 160°C. On cooling to about 65°C, a fungal amylase from *A. niger* is added and the mash is saccharified with amyloglucosidase from *R. niveus* at 32°C.

5.3.7 Continuous Processes

A variety of these processes exist and were developed with the aid of technical enzymes. The advantages of continuous operation include better process control and better utilization of facility capacity. Figure 5.31 illustrates the continuous alcohol production process according to Misselhorn (1981).

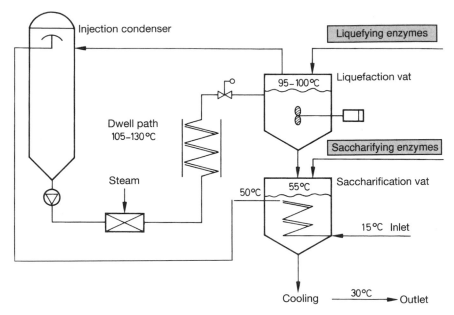

Figure 5.31   Continuous alcohol production process.

REFERENCES

Cheremisinoff, N. P., *Gasohol for Energy Production,* Ann Arbor Science Publishers, Ann Arbor, 1979.

Foody, B. E., in Heinemann, R., and Wolnak, B. (eds.), *Opportunities with Industrial Enzymes,* Wolnak, Chicago, 1992, p. 116.

Kreipe, H., in Kreipe, H. (ed.), *Handbuch der Getränketechnologie,* Ulmer, Stuttgart, 1981, p. 49.

Lyons, T. P., in Godfrey, T., and Reichelt, J. (eds.), *Industrial Enzymology,* The Nature Press, New York, 1983, p. 179.

Misselhorn, K., *Chem. Ing. Tech.* **53,** 47 (1981).

Underkofler, L. A., in Furia, T. O. (ed.), *Enzymes. Handbook of Food Additives,* 2nd ed., CRC Press, Cleveland, 1972.

5.4   ENZYMES IN FLOUR PROCESSING AND BAKING

References: Uhlig and Sprössler (1972), Barrett (1975), Dubois (1980), Hamer (1991, 1995), Kamel and Stauffer (1993), Sprössler (1993), Ter Haseborg (1988)

5.4.1   Introduction

Bread has always had a special meaning in human nutrition. Europeans obtain approximately half of their required carbohydrates and about one-third of their protein from bread; in addition, it contains other vital components such as vitamins, minerals, and trace elements.

The kernel of grain is by nature designed to sprout. It contains the necessary raw materials for growth: carbohydrates, proteins, and a multitude of enzymes. As mentioned in Section 5.2, cereal grains contain hydrolyzing enzymes whose activities are significantly increased on germination. These include amylases, proteases, hemicellulases, and lipases—enzymes that also play important roles in all processes involved in the manufacture of baked goods.

The amount, or rather the activity of native cereal enzymes, depends on grain variety, soil conditions, climate, and state of maturity of the grain at the time of harvest. Thus, the enzymatic makeup is due entirely to natural conditions and is not designed for the demands of the different processes involved in producing a wide variety of baked goods. The need for adding enzyme preparations is due to the fact that flour, especially high-quality flour, contains insufficient enzymes for particular applications. There are several causes for the lack of endogenases (enzymes), including, (1) mechanical harvesting, introduced in the 1920s, and (2) breeding of improved cereal varieties, which have led to high-quality yet enzyme-poor grain. Enzymes are also able to substitute for chemicals such as sodium meta-bisulfite that have been used for weakening the gluten in cookie production (Fig. 5.32).

Enzyme preparations that have traditionally been used include malt flour amylases, and more recently fungal α-amylases and the bacterial amylases employed to keep bread fresh. Various proteases modify or remove the gluten.

Pentosanase preparations, especially xylanases, recently have been used in baking. It was found that xylanases from *Aspergillus* (Rouau and Moreau, 1993) and *Trichoderma* (Wong and Saddler, 1992) increased the baking volume and optimized

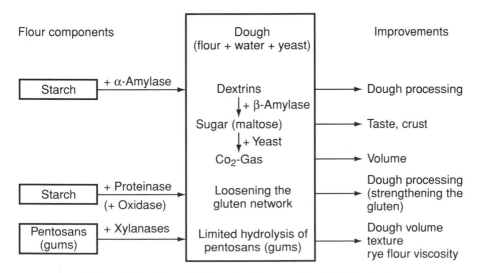

Figure 5.32 General schematic of enzymatic activity in dough.

the crumb structure. Such enzymes are used in Europe in many baking additive preparations, not only for the production of bread but also for wafers and crackers.

The action of special hemicellulases was described by Kormelink (1992). Other enzymes that yield variable results are glucose oxidase, which strengthens the gluten; and peroxidases and lipases, which intensify the aroma of bread. Lipoxygenase found in soy or bean flour is added in some countries to bleach the crump color or to strengthen the gluten.

5.4.2   Flour Components and Enzymes

5.4.2.1   Starch and Amylases   Some properties of starch, particularly its vulnerability to enzymatic attack, were discussed previously (Section 3.1.2.5). The intrinsic properties of starch have a great impact on baking technology and the characteristics of baked products. Before α-amylase is used, the native amylase activity of the flour must be determined precisely. The method of Sandstedt et al. (1939) is used to determine α-amylase activity (see Section 2.3.2.2). The α-amylase activity in grain is about 0.04 SKB/g; that of malt, about 50 SKB/g; and that of a fungal α-amylase preparation, about 5×10^2 to 1×10^5 SKB/g.

Analytic Methods   The activity of native amylases and that of added enzyme can be determined with various methods (see Fig. 5.33 for examples). In the United States, the determination of the maltose value (after Ramsey, 1922) is frequently used. The maltose value of wheat flour lies between 0.1 and 0.3 per 10 g of flour. This value should be between 0.25 and 0.3 in flour used for making white bread. Another method is the falling number (Perten, 1968), which determined the drop in

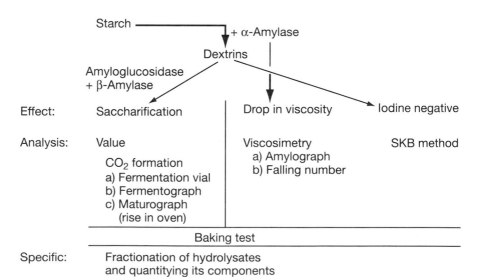

Figure 5.33   Methods for determining amylase activity.

TABLE 5.14 Maltose Value and CO_2 Formation in Flour on Addition of Malt and β-Amylase

Malt (g/100 kg flour)	β-Amylase[a] (g/100 kg flour)	SKB[b] 100 kg flour	Maltose Value with 2% NaCl	CO_2 Formation after 90 min (μl)
0	0	0	1.3	610
0	10	0	1.3	620
100	0	8,000	1.7	840
100	10	8,000	1.7	855
100	15	8,000	1.8	860

[a]Purified β-amylase (Serva, Heidelberg); flour grade 550.
[b]See Section 2.3.2.2.

viscosity on the influence of α-amylases. Because this method permits the determination of activity of only those amylases that are thermally stable, Sprössler (1982) developed a modified falling-number method by which the heat-labile fungal α-amylase activity can also be measured. The method and the instruments are marketed today under the designation FN 30 by the Falling Number Company.* (See also Tables 5.14–5.16 and Fig. 5.34.)

Another method of measuring amylase activity is by generating CO_2, which is produced by yeast fermenting maltose or glucose originating from the amylase-induced degradation of starch. The CO_2 formed under standardized conditions is a measure of enzymatic activity. Yet another method of assessing amylase activity in flour employs viscosity measurements with an amylograph. In wheat, this method permits a better determination of the thermostable bacterial amylases than the fungal α-amylases.

TABLE 5.15 Maltose Value and CO_2 Formation in Flour on Addition of Fungal α-Amylase and β-Amylase

Veron AV[a] g/100 kg flour	β-Amylase[b] g/100 kg flour	SKB[c] /100 kg flour	Maltose Value with 2% NaCl	CO_2 Formation after 90 min (μl)
0	0	0	1.3	630
0	10	0	1.3	615
10	0	15,000	1.9	920
10	10	15,000	1.9	905
10	15	15,000	2.0	930

[a]Fungal α-amylase prepation, Veron V (Röhm GmbH, Darmstadt).
[b]Purified β-amylase (Serva, Heidelberg); flour grade 550.
[c]See Section 2.3.2.2.

*Falling Number A/B, Svandanenvägen 34, S-12611 Stockholm 32, Sweden.

TABLE 5.16 Effect of Amyloglucosidase on Dough Properties and Falling Number of Flour with and without Fungal α-Amylase

Addition	AA[a] (SKB[c]/ 100 g)	AMG[b] (ml/ 100 kg)	Dough Characteristics								Falling Number (s)	
			Pasting		Viscosity Maximum		Temperature Maximum					
			WG357	WL1562	WG357	WL1562	WG357	WL1562			WG357	WL1562
Control	0.0	0.0	60.7	65.5	1040	655	88.0	90.3			488	659
	0.0	5.0	61.0	69.3	1005	660	88.0	91.0			480	667
Malted	7.5	0.0	60.7	65.5	140	65	73.0	79.0			331	401
wheat flour	7.5	5.0	60.7	65.5	140	55	73.0	79.0			310	406
Fungal	20.0	0.0	61.8	70.8	910	615	88.0	89.8			470	645
α-Amylase	20.0	5.0	61.0	72.3	780	600	88.0	91.0			470	619
Bacterial	0.4	0.0	61.0	65.5	660	405	86.5	89.0			450	640
α-Amylase	0.4	5.0	61.5	68.1	625	395	87.5	89.8			446	680

[a]α-Amylase prepation.
[b]β-Amyloglucosidase.

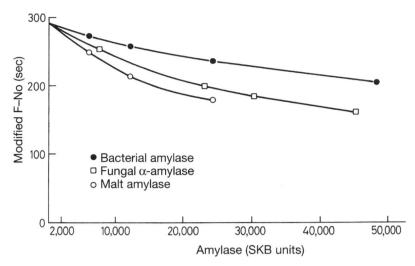

Figure 5.34   Modified falling number (FN): falling number as a function of amylase dose (Sprössler, 1982).

Amylase Activity in Dough and Its Impact on the Quality of Baked Goods   The following amylases are employed for processing flour used in making baked goods: fungal α-amylases, malt amylases, bacterial amylases, and amyloglucosidases.

The α-amylases hydrolyze broken starch grains in flour producing dextrins, which are further hydrolyzed by exoamylases to fermentable sugars (maltose and glucose). Flour contains about 6–10%, maximally 20%, damaged starch grains. With increased damage, the maltose value as well as the CO_2 generation of a given flour is increased, demonstrating that the enzyme activity in flour is substrate-limited. Because modern breadmaking requires a very short fermentation time, a high α-amylase activity is needed. In breadmaking, this activity must be heat-labile; otherwise dextrinization would continue in the baked goods after completion of the baking process.

FUNGAL α-AMYLASE   As discussed earlier regarding the specificity of fungal α-amylase, this enzyme preferentially hydrolyzes pasted starch and forms short-chain dextrins and predominantly maltose. Among the amylases, fungal α-amylase seems to be best suited for making bread (Table 5.17) or cookies.

Although the fermentation process can be potentiated with the addition of sugar, only the enzyme system itself produces the constant amount of sugars needed throughout the duration of the process, which is critical for a good stable fermentation. Uniform sugar formation ultimately produces an increased baked volume that, depending on the dose of fungal α-amylase, can be 5–15% above the control value. In addition to saccharification, α-amylase has a marked effect on the viscosity and pliability of the dough. Besides improving rheologic properties, a better crumb texture and taste of the final baked product can also be expected. High doses of fungal

TABLE 5.17   Mono- and Disaccharide Formation by Fungal Amylases in Wheat Bread[a]

Amylase	Enzyme (g/100 kg bread)	SKB/ 100 kg bread	Mono-, Disaccharides (g/kg bread)
Malt	100	7,000	2.5–3.0
	200	14,000	4.0–4.5
Fungal amylase preparation	10	15,000	4.2–5.2
(Veron AV)	15	22,500	6.0–8.0
	20	30,000	7.5–8.8
Fungal amylase preparation	10	22,000	7.8–9.5
+ AMG (Veron AG)	15	33,000	11.5–12.5

[a][RM] data; AMG, amyloglucosidase; SKB, see Section 2.3.2.2.

α-amylase preparations (about 10^5 SKB in 100 kg flour) affect the keeping quality of baked goods because of side activities present in the preparations. Fungal α-amylase also hydrolyzes rye starch. The rye flour amylogram indicates that the viscosity is reduced (see Fig. 5.47a).

In the oven, various chemical and physical reactions take place in the dough. Initially dough viscosity decreases, allowing higher enzymatic activity. Temperature optima and thermostability of the enzymes used are therefore very important. Differences can be observed depending on where such reactions unfold, that is, on the surface of the dough or in the dough center. The proteins coagulate and release their water to the starch, which now continually gelatinizes and is, especially during the initial phase, further modified by the fungal α-amylase (a shift in water-binding pattern also occurs; see Fig. 5.35). If malt or bacterial α-amylase is used, enzymatic

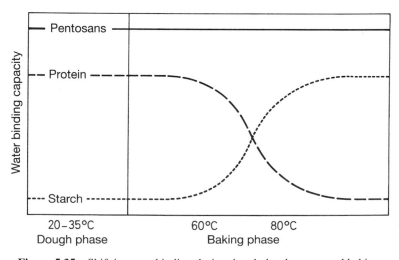

Figure 5.35   Shift in water binding during dough development and baking.

activity is retained even at temperatures of 85–100°C. The dough dehydrates at the surface, followed by the formation of a solid crust, which develops an especially strong color with amyloglucosidases and other amylases.

Dosage of Fungal α-Amylase   In grist mills, fungal amylases are added to standardize the diastatic power or "lift" of the flour (Uhlig and Sprössler, 1972). The quantity to be added depends on the amylolytic activity of the natural untreated flour. In Europe, 20,000–40,000 SKB per 100 kg of flour is required; this amount can fluctuate considerably from year to year as a result of variable harvest conditions.

In most countries, addition of fungal α-amylase to flour in the mill is permitted.

Baking ingredients often contain high fungal α-amylase activity so that as much as 150,000 SKB in 100 kg of flour may be present. An identical amount of malt would cause considerable weakening of the dough and adversely affect crumb texture.

BACTERIAL α-AMYLASE   Bacterial α-amylase acts like fungal α-amylase or malt when added to wheat starch. Because the activity optimum is at pH 6–7, this α-amylase would perhaps be more suitable for flour processing than fungal amylase if it were less stable at high temperatures. Only very unstable bacterial amylases can be used in very specific cases when processing wheat flour (Schulz and Uhlig, 1972). One such application is to maintain bread freshness by delaying retrograding of amylose in the crumb. When employing such enzymes, the range between a dose with no effect and an overdose is very narrow. An excess amount of enzyme will produce a weakened, if not pasty, crumb. One preparation employed in various countries over the past 20 years is Veron F 25 [RM]. Its effects will be described later. In order to keep rye, light rye, and whole rye bread fresh, heat-labile bacterial amylase preparations can be dosed at significantly higher levels than those used in wheat-based baked goods (Maleki et al., 1972).

5.4.2.2   *Protease Cleavage within Gluten*   During dough preparation, gluten protein binds a portion of the water and expands forming a latticelike structure. Proteases cleave bonds within the gluten proteins and thereby cause a softening of the network (Fig. 5.36). The dough's resistance to stretching decreases and its extensibility increases. Adequate extensibility of the dough is essential for gas retention as the dough expands from the carbon dioxide gas formed by the fermenting yeast. In the oven, the dough should rise uniformly.

The activity of endogenous proteases and peptidases in cereal grains increases with kernel maturation and reaches high levels on sprouting. The endogenous proteases, which resemble papain (see Section 3.2.2.2), are activated by reductants and inhibited by oxidants. Reed and Thorn (1957) observed that small amounts of cysteine will activate proteases, but this effect may also be due in part to the fact that cysteine can cleave the intramolecular disulfide bridges in the gluten. This renders the substrate more accessible to proteolytic attack.

Gluten is a complex between two types of proteins: gliadin and glutenin. The cleavages catalyzed by different proteases have completely different effects on the modification of these proteins (Hamer, 1991).

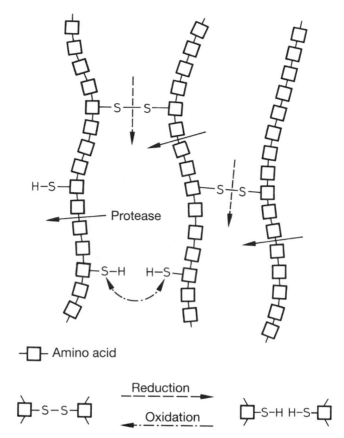

Figure 5.36   Gluten protein structure and protein hydrolysis [simplified model, Sprössler (1993)].

For specific applications, care must be taken in selecting from the many technical protease preparations available. Certain fungal proteases with an activity optimum in the neutral or weakly acid pH range hydrolyze gluten with considerable
specificity.

In the initial phase of hydrolysis only selected linkages are cleaved. Examples of
this are provided by the neutral protease from *Aspergillus oryzae* (Veron PS [RM]
and the fungal Protease 31,000 [SOEG]) discussed earlier (see Section 3.2.4.3).

In contrast, neutral bacterial protease cleaves the gluten in a nonspecific manner
that can subsequently lead to complete gluten hydrolysis.

Analytic Methods for Proteases   Activity determinations with analytic chemical
methods must be distinguished from the practical methods used in flour technology.
The analytic chemical methods are based on hemoglobin proteolysis. These are basically modifications of the Anson method; the American Association of Cereal

Chemists (AACC) (1962) method with the activity expressed in hemoglobin units (HbU) per gram. The results obtained with the various assays for different enzymes can be compared only by orders of magnitude. For papain, one Anson unit corresponds to about 75,000 HbU/g.

Different methods exist for determining protease activity in flour or dough. The faringraph is employed in the United States, where proteases have long been used in treating protein-rich wheat varieties. Bowlby et al. (1953) have compared the Brabender units (BUs)* of the farinograms with the data from other analytic methods. They found that gluten proteolysis and the milk clotting test with fungal or bacterial proteases, papain, and trypsin provide values comparable to the BU obtained with the farinograph (amylograph or viscograph).

Other methods that determine protease activity as a function of gluten characteristics are the extensograph and the alveograph, which is utilized primarily in France and Belgium.

Both methods measure the proteolytic effect on the viscoelastic properties of the gluten. Figures 5.37–5.41 illustrate the effect of various proteases on these measurements. The activity of the enzyme preparations is recorded in Hb units. For papain, 2000 units Hb/g correspond to about 1 Anson unit per gram.

The diagrams clearly illustrate the aggressive action of bacterial protease and that of papain on the physical structure of gluten. Bacterial proteases dramatically increase extensibility, but also reduce the elastic quality of the gluten. Fungal proteases increase extensibility only gradually, but improve gluten elasticity. Fungal

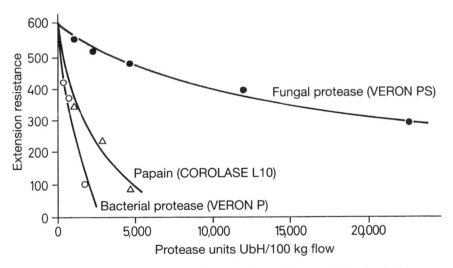

Figure 5.37 Effect of proteases on the extension resistance of doughs [extensogram, Sprössler (1993)].

*Brabender unit; amylograph or viscograph unit. (Brabender oHG, Duisburg, Germany, 1985)

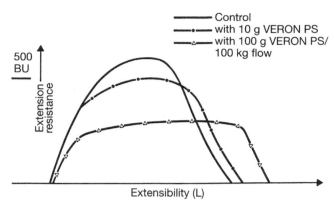

Figure 5.38   Effects of fungal protease (Veron PS [RM]) on dough extensibility [extensogram, Sprössler (1993)].

proteases are more dose-tolerant and do not decrease the dough's performance in the alveograph.

Kiefer et al. (1982) described an interesting method for determining the characteristics of the gluten protein itself, free from influences found when characterizing gluten in dough. This capillary viscosimetric determination is very sensitive in detecting changes in dough elasticity as a function of protease action.

The glutograph measurement on isolated gluten, a test method described by Seitz and Dörfner (1987), can also record enzymatic gluten degradation. The isolated gluten, after a 3-min relaxation period, is subjected to continuous shearing.

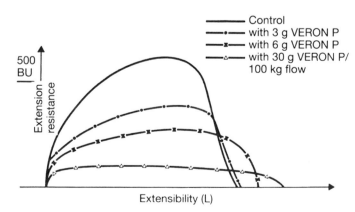

Figure 5.39   Effect of bacterial protease (Veron P [RM]) on dough extensibility (extensogram).

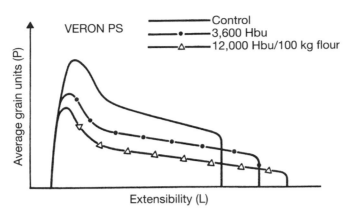

Figure 5.40   Effect of fungal protease (Veron PS [RM]) on dough extensibility (alveogram; dough resting time 135 min).

The change in extensibility of the various gluten samples is recorded in seconds (see Fig. 5.42).

The enzymatic effect on gluten can be expressed more graphically using the glutograph, which measures the effect over longer time periods ([RM] data). Introducing additional incremental measurements, in this case an added 3×20 min, clearly illustrates the effects of different proteases on shearing times. Samples of gluten-containing doughs were fermented for 20, 40, and 60 min and then read with the extensograph. The studies were conducted with an ascorbic acid–treated German wheat flour (Fig. 5.43) and an untreated American wheat flour (Fig. 5.44).

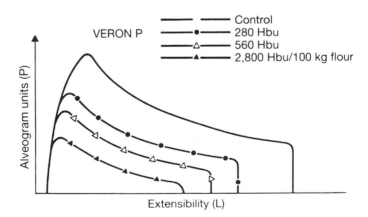

Figure 5.41   Effect of bacterial protease (Veron P [RM] on dough extensibility (alveogram; dough resting time 135 min).

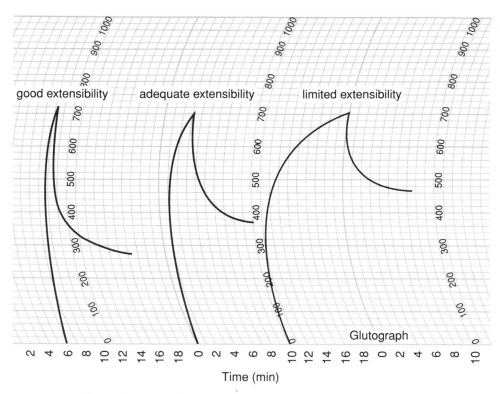

Figure 5.42 Extensibility diagrams for various glutens (glutograph).

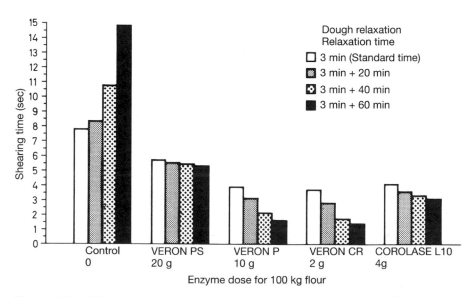

Figure 5.43 Effect of various enzyme preparation on gluten relaxation in German wheat flour (glutograph).

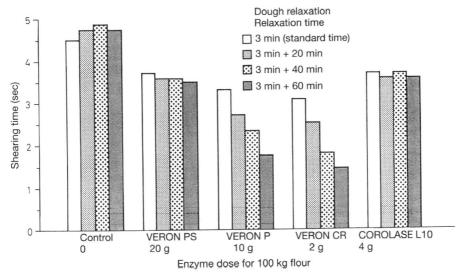

Figure 5.44 Effect of various enzyme preparations on gluten relaxation in American wheat flour (glutograph).

Commercially, a drastic reduction in gluten strength requires treatment with bacterial proteases, papain, or bromelain. Specific fungal proteases are preferentially employed in modifying gluten for making bread and rolls.

5.4.2.3 Pentosans (Gums): Pentosanases

The significance and function of pentosans have been addressed in the sections on the plant cell wall (see Section 3.1.4.1) and xylanases (see Section 3.1.7.2). As mentioned earlier, rye contains about 8% pentosans or gums and wheat contains about 3% of these oligosaccharides. Both water-soluble and water-insoluble pentosans can be isolated from wheat. This fraction, free of starch and proteins, is called "tailings" (Bechtel et al., 1964). Pentosans can absorb up to 10 times their water-content weight and form gels (McCleary et al., 1986). In rye, they assume the role of wheat protein with regard to water uptake. According to Hoseney and Faubion (1981), gel formation is the result of cross-linkages formed between the pentosans and proteins (Fig. 5.45).

Pentosans affect the freshness of bread. Kulp and Bechtel (1963) found that by adding increasing amounts of tailings to a wheat flour dough, the bread volume is reduced and the crumb becomes coarser. According to Rugbjerg (1987), modification of dough by pentosans improves dough fermentation tolerance, increases dough expansion, and improves dough freshness. Cauvain (1985) reports that only a few cellulases affect bread volume but their effectiveness does not exceed that of fungal amylases.

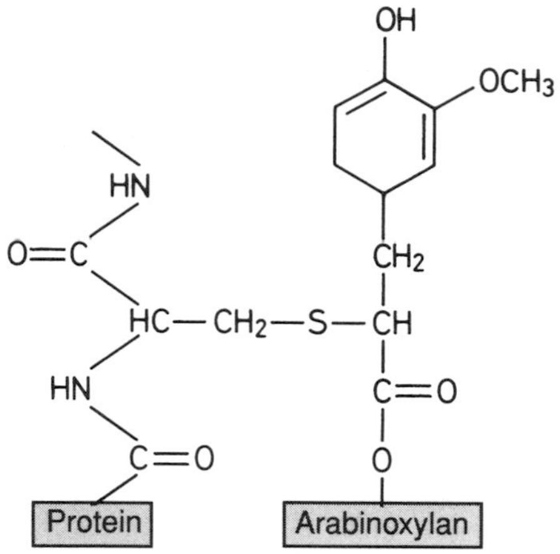

Figure 5.45   Cross-linking of protein by arabinoxylan (Hoseney and Faubion, 1981).

McCleary (1986) described the effect of a purified, 1,4-β-xylanase from *Tricho-derma viride* on isolated pentosans and the resultant dough characteristics. He found that the doughs become weak and sticky even when only small amounts of pure xylanase were added. Treating a suspension of flour with xylanase reduced flour viscosity and permitted a better separation of the starch and gluten components of the flour.

Maat et al. (1992) observed that the addition of xylanases increase the loaf volume, with no or negligible detrimental effects to other loaf characteristics, such as loaf height or crumb quality. Kormelink (1992) produced xylanases by cultivation of *Aspergillus awamori* and by separation he isolated different types of xylanases. Gruppen (1992) studied the action on water-unextractable non-starch cell wall material (above-mentioned tailings) and found that one of the xylanase fractions (26,000 daltons) caused a rapid swelling of the tailings, followed by a decrease in swelling at longer incubation time. This xylanase fraction also caused an increase in loaf volume between 10 and 20% depending on the quality of the flour.

Kosmina (1977) reports that pentosans treated with cellulase prior to their use can counteract the otherwise inhibitory effect of insoluble pentosans on dough expansion.

The baking performance of rye flour depends on the amount and the ability of the native rye pentosans to swell. Dry harvests yield a rye flour with a high amylogram that will produce a tough and dry dough (Sprössler, 1977). During dough preparation, the rye pentosans (gels) swell and the dough becomes tough, the volume decreases, and the crumb cracks. The addition of pentosanases will alleviate these difficulties by breaking down the pentosans into swellable, yet still viscous, fragments.

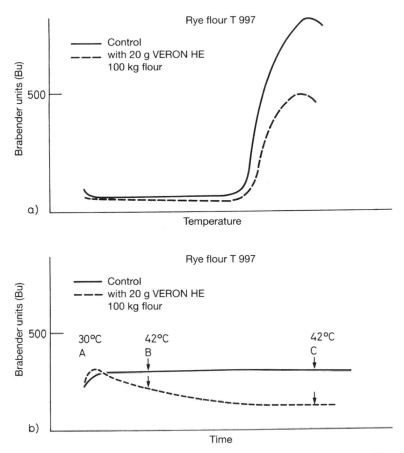

Figure 5.46 Effect of Veron HE [RM] pentosanase on (*a*) amylogram and (*b*) swelling curve of rye flour T 997.

Figures 5.46*a* and 5.46*b* illustrate the effect of a pentosanase preparation (Veron HE [RM]) on the amylogram and swelling curve of a rye flour.

Weipert (1988) compared the effects of fungal amylases, bacterial amylases, and proteases, as well as pentosanases on the rye flour amylogram and swelling curve. Surprisingly, the proteases, as did the pentosanases, significantly reduced the viscosity (Figs. 5.47 and 5.48).

Pentosanase Analysis   Currently, there is no analytic enzymatic method that provides a clear correlation between activity and the effect of pentosanases in commercial applications; therefore, the methods of choice are empirically developed assays that determine pentosanase activity by such means as the decrease in the previously described swelling curve and/or the viscosity of a flour suspension. The change in

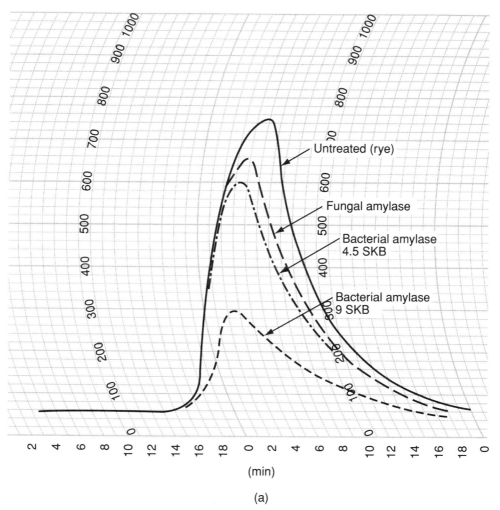

(min)

(a)

Figure 5.47 Effect of various enzymes on rye flour viscosity as shown in amylograms (a), (b), and (c) [Brabender oHG., Duisburg, Germany] (Weipert, 1988).

viscosity of a rye gel extract by the action of hemicellulases can also be employed as an analytic procedure. Various enzyme producers use xylan (the pentosan from wood) as the analytic substrate for standardizing their preparations, although the results obtained in this manner do not allow conclusions regarding the enzymes' efficacy in practical applications.

5.4.2.4 Lipoxygenase (EC 1.13.11.12) The enzyme is present in almost all legumes. Soybeans contain about 200 times more lipoxidase than wheat. So far no microorganisms have been found that produce this enzyme in commercially useful amounts.

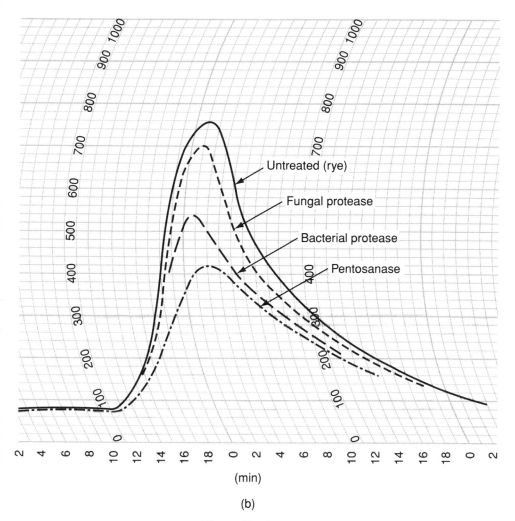

(min)

(b)

Figure 47 *Continued*

Lipoxidase catalyzes the oxidation of unsaturated fatty acids that contain a *cis,cis*-1,4-pentadiene group:

$$R - CH \overset{cis}{=\!\!=} CH - CH_2 - CH \overset{cis}{=\!\!=} CH - R^1$$

linoleic acid

$$\downarrow O_2$$

$$R - CH \overset{cis}{=\!\!=} CH - CH \overset{trans}{=\!\!=} CH - \overset{\displaystyle OOH}{\underset{|}{CH}} - R^1$$

fatty acid hydroperoxide

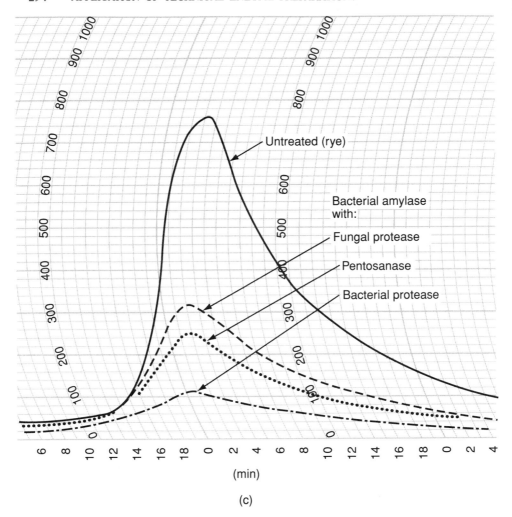

Untreated (rye)

Bacterial amylase
with:

Fungal protease

Pentosanase

Bacterial protease

(min)

(c)

Figure 47 *Continued*

Polyunsaturated fatty acids are thus converted to fatty acid hydroperoxides and aldehydes that can react in the dough in many ways. Also, sulfhydryl groups can be oxidized to disulfide bridges.

The use of soybean flour for bleaching (the undesirable carotinoids) and improving the baking quality of wheat were patented in the United States in 1934. Presently, 1% soybean flour is used primarily in the United States. In France, bean flour is used for the same purpose. (See also Section 5.4.1.)

5.4.2.5 Sulfhydryl Oxidase (EC 1.8.3.2)   This enzyme has recently been isolated from *Aspergillus niger* [FR]. Scott (1989) showed that sulfhydryl oxidase

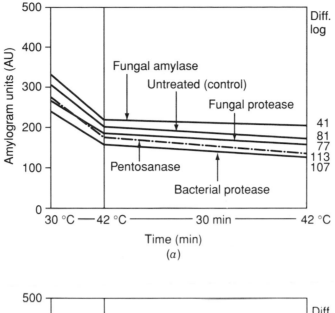

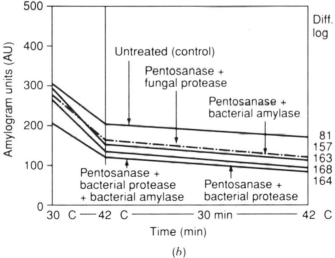

Figure 5.48 Effect of various (a) hydrolases, and (b) combinations of hydrolases on the swelling curves of rye flour.

(SO) stabilizes weak dough during breadmaking. It catalyzes the oxidation of sulfhydryl to disulfide groups as shown in the following reaction scheme:

$$2R-SH + O_2 \xrightarrow{\text{SO}} R-S-S-R + H_2O_2$$

Strengthening the gluten can be achieved by sulfhydryl oxidase alone or in combination with glucose oxidase (GO). Doughs stabilized in this manner produce an improved baked volume and an even porous structure of the crumb.

In the search for a suitable replacer for bromate, Haarasilta et al. (1989) (Cultor Ltd., Patent 1989) reported the effect of ascorbic acid together with glucose oxidase and sulfhydryl oxidase on the sulfhydryl groups of flour. The SH groups are oxidized, but to obtain a good effect in baking, a combination with hemicellulases is necessary. This may offer an opportunity to replace the chemical oxidants still permitted in a few countries.

5.4.3   Enzyme Preparations and Their Effects on Various Baking Processes

Enzyme preparations find widespread application in today's baking ingredients and baking industry. The enzymes, in addition to affecting the viscosimetric and rheologic properties of flour dough (see Table 5.18), also provide the analytical data used to standardize flour quality in flour mills and in the baking industries' quality assurance programs. Besides malt flour, fungal amylases and fungal proteases, enzyme mixtures are also employed to achieve specific processing and baking objectives. Such mixtures possess at least one primary activity (usually fungal amylase) but frequently also have a number of secondary activities such as amyloglucosidase and pentosanase, as well as hemicellulase. The enzyme preparations used in the present baking and baking ingredients industries can, therefore, have very targeted missions in the processing and quality of baked goods.

Different countries may have different objectives when employing enzymes. Determining factors are the available natural resources, traditional processing methods, and the means of marketing and distribution.

For example, the American baking industry requires an increased production tolerance of large dough lots and an improved freshness and stability of rolls and

TABLE 5.18   Effect of Enzymes on Baking Properties

Enzyme	Effect on Dough	Effect on Baked Product
Fungal α-amylase	Suppleness Reduced viscosity Increased leavening	Increased volume Improved texture
Bacterial α-amylase	—	Improved storage life
Amyloglucosidase	—	Increased volume Improved texture Increased browning
Fungal proteases	Dough relaxation Lowered extensogram Increased extensibility	Increased volume Improved texture
Hemicellulases	Improved fermentability	Increased volume Improved texture and storage life
Lipoxygenase	—	Crumb bleaching

breads. Increasing the baked volume (expansion) is no longer required because of the good baking characteristics of American wheat flours, which inherently produce adequate dough expansion. Enzymes are also employed to reduce kneading time and to shorten or eliminate sponge preparation.

By comparison, the individual tradesperson-oriented baking in central Europe primarily requires the optimization of baked volume coupled to a great processing tolerance of the variable flour qualities. Freshness (i.e., softness) plays a minor role because of short distribution routes and different consumer tastes. Crispness of the crust serves as a measure of freshness for many European rolls and bread varieties (e.g., French baguettes and many other varieties of rolls).

The applications discussed here are currently used by the baking and baking ingredient industries who employ suitable and thoroughly tested enzyme preparations with oxidizing and reducing agents together with a large variety of emulsifiers.

5.4.3.1   *Retarding Staling*   Staling of bakery products is characterized by a loss in product freshness due to a decrease in moisture level, resulting in crumb firmness.

The staling mechanism has been studied by various investigators, and it is now generally accepted that staling is due to a gradual transition of starch from an amorphous structure to a partially crystalline state (Hebeda, et al. 1990). The degree of firming is quantified by various instrumental methods (penetrometer, panimeter, compressimeter).

The baking ingredients and baking industries employ emulsifiers to improve freshness. Emulsifiers such as mono- and diglycerides of fatty acids and sodium- and calcium-stearyl lactylates are quite effective and yield a significantly improved softness of the baked goods' crumb. The shelf life of bread can be extended by adding monoglycerides.

Because of their strong dextrinizing action, bacterial amylases have a distinct effect on crumb softness (Sprössler, 1985), but crumb elasticity can decrease. Because of the high thermostabilities of these enzymes, a precise control of the baking temperatures is required. For specific baked products, high baking temperatures must also be reached in the crumb.

Cole (1981) describes the use of fungal amylases, slightly stabilized in a sugar solution, to maintain freshness.

A patent granted to Novo Industri (Nielsen et al., 1982) defines the combination of bacterial amylases with pullulanases as an effective means to extend stability and freshness. According to Qi Si and Simonson (1994), the antistaling effect of a thermostable maltogenic exo-acting amylase "Novamyl" (NO) is due to a high amount of small starch fragments, mainly maltose, produced in situ, which prevents the interaction between starch and gluten. The enzyme is significantly more effective at retarding crumb hardening than fungal α-amylase.

Hebeda et al. (1991) described an amylase derived from *Bacillus megaterium* that is more heat-resistant than the fungal amylase but, unlike other bacterial amylase, is inactivated in the oven. This antistaling enzyme, with the tradename Multifresh [EB], prolongs baked good freshness by as much as 75%.

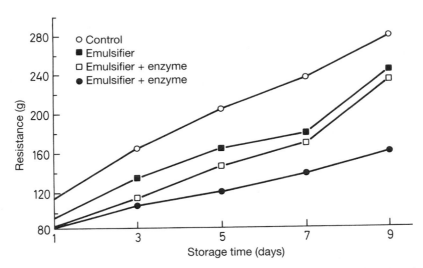

Figure 5.49 Improving bread freshness with complex enzyme preparations (amylases, hemicellulases).

Himmelstein (1989) reported the effect of an enzyme complex (thermostable fungal enzymes) capable of producing a baked crumb with distinctly greater softness without a concomitant loss in crumb elasticity. Furthermore, improved fermentation stability and increased baked volume were obtained as well (Fig. 5.49).

5.4.3.2 Interrupting Fermentation: Frozen Doughs Frozen doughs, especially in the U.S. market, are held for extended periods of time. In most cases, the dough's fermentation time increases with age and the baked volume decreases. It is assumed that the leavening ability of the yeast weakens with time and the gluten network loses its gas-retaining ability. Amylase preparations and specially formulated enzyme complexes can counteract these developments ([RM] data; Fig. 5.50).

5.4.3.3 Mixed-Grain and High-Fiber Enriched Breads Since 1972 a consumer-driven growth in specialty breads with an addition of high-fiber products (whole wheat, wheat bran, multigrain concentrates, soybean fibers, oat husks, microcrystalline cellulose, etc.) was observed. With the addition of 15–30% of this fiber-rich material, the swelling and the water-absorption properties of the dough change drastically. The result is slack or sticky doughs or in other cases, excessively stiff doughs. Slower or faster water absorption can also influence the kneading properties, and the rising, fermentation, and the poststiffening and stickiness of the dough, leading to bread of inferior volume and quality. Both dough defects can be regulated by the ad-

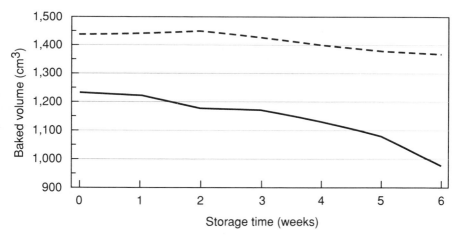

Figure 5.50 Baked volume of frozen doughs (Kaiser rolls) with and without enzyme treatment with 12g/100kg Veron FD Super.

dition of two different xylanase products derived from *Aspergillus* or *Trichoderma reesei*.

The *Aspergillus* xylanases hydrolyze only a few bounds within the pentosans, causing a swelling of insoluble pentosans and leading to a stabilizing effect on slack and sticky doughs. The dosage of such preparations depends on various factors, such as the type, particle size, and the added quantity of the higher-fiber material. In baking tests with 30% whole wheat flour, a low dosage of *Aspergillus xylanase* leads to optimal stabilization and baking volume. High rates lead to a faster fermentation and reduction in baking volume. The *Trichoderma* enzymes hydrolyze soluble and insoluble pentosans and β-glucans to smaller, rapidly swelling fragments and make it possible to control stiffening and postswelling effects in the dough. Also in this case the dosage depends on the parameters of the high-fiber material mentioned above. A high overdosage leads to an undesirably flat and slack dough (Himmelstein, 1989).

5.4.3.4 Enzymes for Baked Goods with Long Shelf Life

The use of enzymes to prolong shelf life of baked goods (cookies, crackers, and wafers) (Table 5.19) is quite dependent on the flour supply and the processes used to make the various baked goods. Primarily low-protein flours are required. In many countries, however, low-protein flours are not available in adequate quantities. Various papers describe the effect of proteolytic enzymes on gluten-rich flours (Rotsch, 1966).

The dough's pH, or the various pH levels developed during processing the sponge for a variety of baked products, plays a major role in determining the appropriate enzymes to prolong shelf life. Then, if the pH optima of the enzyme prepara-

TABLE 5.19   Enzymes for Baked Goods with Extended Shelf Life

Enzyme	Improvements in Baked Goods
Fungal α-amylase	Texture
Bacterial α-amylase	Volume, dough cohesiveness
Bacterial proteases	Product quality
Papain	Dough tolerance
	Crumbliness
	Browning uniformity
Hemicellulases	Texture

tions are also taken into account, a suitable enzyme for a given application can readily be selected (Table 5.20).

Therefore, cracker doughs with a pH range of 6.0–3.5 can be treated with acid fungal proteases. The fungal protease improves the extensibility and consequently also the laminarization of folding of the doughs. In addition, amylases and proteases can positively influence the leavening time of the sponges as well as the product's taste and aroma.

Proteases and amylases with a neutral pH optimum can be used in single-step cracker and cookie doughs where the pH progresses from the weakly acidic to the alkaline range. Amylases have an added advantage in that they impart a better cohesiveness to cookie and cracker doughs. In addition, amylases can enhance dough expansion during the primary baking and the subsequent browning.

Gaines and Finney (1989) described the effects of cellulases and hemicellulases in cookie baking. These enzymes can prevent moisture uptake, particularly in doughs of natural fiber-rich baked goods such as Graham crackers and whole grain biscuits or cookies.

TABLE 5.20   pH Optima of Proteases Used in Baking

Source	Enzyme	pH Optimum
Cereal proteases	Serine enzyme	3.8 or 4.4
Papain	Sulfhydryl enzyme	6–7
Bacterial protease		
Neutral	Metalloenzyme	~7
Alkaline	Serine enzyme	10–11
Fungal protease (Aspergillus)		
Acid	—	4–6
Neutral	Metalloenzyme	7.5
Alkaline	Serine enzyme	9–10

In summary, the use of enzymes in cookie and cracker production provide the following advantages:

1. In production
   a. Permitting the use of protein-rich flours
   b. Reducing energy requirements for kneading
   c. Shortening dough resting time
   d. Facilitating the rolling and cutting of dough
   e. Maintaining shape of cut dough
2. In improving baking properties
   a. Improving crumbliness
   b. Smoothing edges and surfaces
   c. Reducing hairline cracks
   d. Uniform browning
   e. Providing uniform size and weight for packaging

Bacterial proteases are very useful in the production of wafers in applications requiring the use of protein-rich flours. The protease hydrolyzes the gluten, thereby lowering batter viscosity and preventing gluten precipitation. The escape of moisture is also facilitated. The wafers acquire an even texture and color (Drechsel and Ruttloff, 1975).

It has also been observed that addition of bacterial protease can initially significantly increase batter viscosity. However, complex protease preparations that contain carbohydrase activity can be used to prevent such viscosity increases. On such enzyme addition (Veron W), the viscosity decrease sets in immediately ([RM] data; Fig. 5.51).

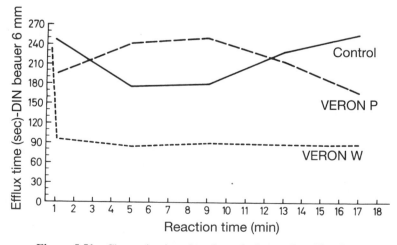

Figure 5.51   Change in viscosity of a wafer batter (Brazilian flour)

REFERENCES

Am. Assoc. Cereal Chemists (AACC), 22–60 (1962).

Barrett, F. F., in Reed, G. (ed.), *Enzymes in Food Processing,* 2nd ed., Academic Press, New York, 1975, p. 301.

Bechtel, W. G., Geddes, W. F., and Jilles, K. A., in Hlynka, I. (ed.), *Carbohydrates in Wheat Chemistry and Technology,* 1st ed., Am. Assoc. Cereal Chemists, St. Paul, MN, 1964, p. 277.

Bowlby, C., Tucker, H., Miller, B. S., and Johnson, J. A., *Cereal Chem.* **30,** 480 (1953).

Cauvain, S. P., "Effects of some enzymes on loaf volume in the CBP (Chorleywood bread process)," *FMDRA Bull.* **1,** 11 (1985).

Cole, M. S., *Getreide, Mehl und Brot* **35,** 60 (1981).

Cole, M. S., "Antistaling baking composition," U.S. Patent 4,320,151 (1982).

Drechsel, W., and Ruttloff, H., *Bäcker und Konditor* **7,** 212 (1975).

Dubois, D. K., "Enzymes in baking," *Am. Inst. Bakery* **II**(10–12) (1980).

Gaines, C. S., and Finney P. L., *Cereal Chem.* **66,** 73 (1989).

Gruppen, H., *Structural Characteristics of Wheat Flour Arabinoxylans,* doctoral dissertation, Univ. Wageningen, CIP-DATA Koninklijke Bibliotheek, The Hague, 1992.

Haarasilta, S., Pullinen, T., and Tammersalo-Karsten, I., "A method of improving the properties of dough and quality of bread," Eur. Patent Appl. 0338 452 A1 (1989).

Hamer, R. J., in Tucker, G. A., and Woods, L. F. J. (eds.), *Enzymes in Food Processing,* AVI, New York, 1991, pp. 168, 181.

Hamer, R. J., in Tucker, G. A., and Woods, L. F. J. (eds.), *Enzymes in Food Processing,* Blackie Academic & Professional (imprint of Chapman & Hall), New York, 1995, p. 190.

Hebeda, R. E., Bowles, L. K., and Teague, W. M., *Cereal Foods World* **35,** 453 (1990).

Hebeda, R. E., Bowles, L. K., and Teague, W. M., *Cereal Foods World* **36,** 619 (1991).

Himmelstein, A., paper presented AACC Meeting, Washington, DC, 1989.

Hoseney, R. C., and Faubion, J. M., *Cereal Chem.* **58,** 421 (1981).

Kamel, B. S., and Stauffer, C. E. (eds.), *Advances in Baking Technology,* Blackie Academic & Professional (imprint of Chapman & Hall), New York, 1993.

Kiefer, R., Jinja, K., Kempf, M., Belitz, H.-D., Lehmann, J., Sprössler, B., and Best, E., *Zeitschr. Lebensmittel Untersuch. Forschg.* **174,** 216 (1982).

Kormelink, F. J. M., *Characterization and Mode of Action of Xylanases and Some Accessory Enzymes,* doctoral dissertation, Univ. Wageningen, CIP-DATA Koninklijke Bibliotheek, The Hague, 1992.

Kosmina, A., *Biochemie der Brotherstellung,* VEB, Leipzig, 1977, p. 325.

Kulp, K., and Bechtel, W. G., *Cereal Chem.* **40,** 493, 665 (1963).

Maat, J., Roza, M., Verbakel, J., Santos da Silva, J. M., Bosse, M., Egmond, M. R., Hagemans, M. L. D., v. Gorcom, R. F. M., Hessing, J. G. M., v.d. Hondel, C. A. M. J. J., and Rotterdam, C., in Visser, J., Belman, M. A., Kusters-Van Someren, M. A., and Voragen, A. G. J. (eds.), *Xylans and Xylanases,* Progress in Biotechnology Series, Vol. 7, Elsevier, Amsterdam, 1992, p. 349.

Maleki, M., Schulz, A., and Brümmer, J.-M., *Getreide, Mehl und Brot* **26,** 216 (1972).

McCleary, B. V., *Internatl. J. Biol. Macromol.* **8,** 349 (1986).

McCleary, B. V., Gibson, T. S., Allen, H., and Gams, T. C., *Die Stärke* **12,** 433 (1986).

Nielsen, G., Diers, I., Outturup, H., and Norman, B., "Debranching enzyme product and its use," Eur. Patent Appl. EP 63 909 (1982).

Perten, H., *Cereal Sci. Today* **18,** 682 (1968).

Qi Si, J., and Simonson, R., *Novo Nordisk Report EF 9414394, Proc. Internatl. Symp. AACC/ICC/CCOA,* Beijing, Nov. 1994.

Ramsey, L. A., *Am. Inst. Baking Bull.* 8 (1922).

Reed, G., and Thorn, J. A., *Cereal Sci. Today* **2,** 280 (1957).

Robinson, D. S., *Food Biochemistry and Nutritional Value,* Wiley, New York, 1987, p. 554.

Röhm GmbH, technical information.

Rotsch, A., *Brot und Gebäck* **11,** 213 (1966).

Rouau, X., and Moreau, D., *Cereal Chem.* **70,** 626 (1993).

Rugbjerg, U., "Application of enzymes in baking," paper presented at 2nd Meeting on Industrial Application of Enzymes, Barcelona, 1987.

Sandstedt, R. M., Kneen, E., and Blish, M. J., *Cereal Chem.* **16,** 712 (1939).

Schultz, A., and Uhlig, H., *Getreide, Mehl und Brot* **26,** 218 (1972).

Scott, H. D., in Whitaker, J. R., and Sonnet, P. E. (eds.), *Biocatalysts in Agricultural Biotechnology,* p. 176; ACS Symp. Series, No. 389 (3rd Chem. Congr. North America, 195th Am. Chem. Soc. Meeting, Toronto, 1988) ACS, Washington, DC, 1989.

Seitz, W., and Dörfner, H.-H., *Getreide, Mehl und Brot* **1,** 16 (1987).

Sprössler, B. G., and Uhlig, H., *Getreide, Mehl und Brot* **26,** 210 (1972).

Sprössler, B. G., *Die Mühle und Mischfuttertechnik* **122,** 406 (1985).

Sprössler, B. G., *Die Mühle und Mischfuttertechnik* **114,** 235 (1977).

Sprössler, B. G., *Die Mühle und Mischfuttertechnik* **119,** 425 (1982).

Sprössler, B. G., in Nagodawithana, T., and Reed, G. (eds.), *Enzymes in Food Processing.* 3rd ed., Academic Press, San Diego, New York, 1993, p. 293.

Ter Haseborg, E., *Alimenta* **27,** 1 (1988).

Ter Haseborg, E., and Himmelstein, A., *Cereal Foods World,* 419,421,423 (1988).

Uhlig, H., and Sprössler, B. G., *Die Mühle und Mischfuttertechnik* **109,** 15 (1972).

Weipert, D., *Die Mühle und Mischfuttertechnik* **125,** 314 (1988).

Wong, K. K. Y., and Saddler, J. N., *Progr. Biotechnol.* **7,** 171 (1992).

5.5 ENZYMES IN JUICE- AND WINEMAKING

5.5.1 Introduction

References: Pilnik and Voragen (1991, 1993), Lea (1995), Villettaz (1993)

The production of fruit and vegetable juices is economically important in that it makes valuable components of fresh fruits and vegetables available to the consumer even when the fruits and vegetables are out of season.

This requires methods for extracting the liquid component of fruits, the juice, which is then clarified either by filtration or, if kept as extracted, stabilized pasteurization.

The fruit juice industry began to develop in the early 1930s, at which time juice yields were generally unsatisfactory and it was often difficult to filter the juice to acceptable levels of clarity. Advances in industrially suitable fungal pectinases, along with increased knowledge about fruit components, especially pectins, helped resolve these difficulties. The combined know-how initiated developments that led to improved taste and quality of fruit juices as well as production economics (Mehlitz, 1934).

The pectinases used were essentially mixtures of various glycosidases, primarily capable of pectin hydrolysis, and were simply called *filtration enzymes*. Mehlitz (1930) described the increase in filtration efficiency of various juices after enzyme treatment (Table 5.21).

The value of pectinases and their effectiveness at fruit pH and temperatures of up to 55°C was demonstrated not only for the clarification of fruit juices but also for an improved pressing of chopped fruits, called *fruit mash* (Koch, 1956; Krebs, 1955).

Although individual processing techniques will vary significantly with the kind of fruit and the desired product, the most important process steps can be summarized in the following flowchart:

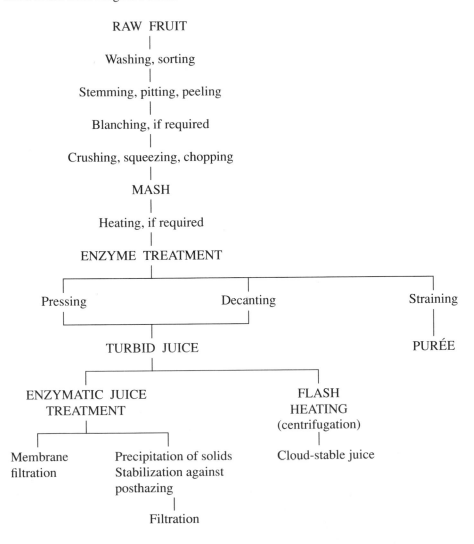

TABLE 5.21 Increase in Filtration Efficiency with Enzymes

Type of Fruit	Acceleration Factor[a]
Strawberries	5.3
Sweet cherries	1.8
Sour cherries	1.5
Raspberries	1.4
Red currants	1.6
Gooseberries	17.2
Apples	2.3

[a]Untreated equals 1.0.

Enzymes used in mash and juice processing must accomplish different tasks. Previously, when only one all-purpose enzyme was used, processing often was not completed satisfactorily.

Current enzyme preparations used in fruit juice (see summary of purposes and activities in Table 5.22) and winemaking possess specific capabilities for degrading hydrocolloids, depending on the raw material and the product desired.

Pectins, as distinguished from other polymer substances present in fruits, have the greatest influence on processing. Therefore, pectinases are, even at the current state of technology, the most important group of enzymes employed in this industry. If maximum degradation of polymers is desired, cellulases and hemicellulases are combined with pectinases. Amylases and arabanases hydrolyze starch and araban in stone-fruit juices, thereby preventing potential turbidity in packaged juices due to precipitation of such polymers.

5.5.2 Individual Processes: Current State of Technology

5.5.2.1 Mash Treatments An optimal mash composition is sought by the customary methods of extracting juice by pressing, that is, improving the ability of juice to separate from the fruit pulp. Juice viscosity is decreased by hydrolyzing the soluble pectin (Pilnik, 1970).

Protopectin, important for the firmness of mash solids, however, must stay intact so that a sufficient number of drainage channels for juice runoff remain. A mash is considered to have good pressing quality if most of the juice runs freely off the press without force. This has the further advantage that a press can be loaded with larger quantities of mash, thereby increasing press capacity. The drained pulp can be squeezed dry under pressure generating high yields of additional juice (Lüthi and Glenk, 1968; Fig 5.52). The extracted residue or pomace is relatively dry and does not stick to the press cloth.

The juice obtained in this manner contains only small amounts of solids and can later be clarified without difficulty.

In the event that a homogenous purée is desired from a fruit or vegetable mash, a strainer or decanting centrifuge is used to remove coarse fruit parts and fibers. In this case, the mash solids are enzymatically digested by partial hydrolysis of the

TABLE 5.22 Use of Technical Enzyme Preparations in the Fruit Juice Industry

Objective	Activity	Enzymes[a]
Improving pressing and extraction of apple extraction of apple mash	Primarily hydrolysis of soluble pectin while preserving protopectin; decreasing juice viscosity; maintaining texture; partial hydrolysis of cell wall components	PG (with limited protopectinase activity) + PE (with little PTE); cellulase–hemicellulase complex
Improving pressing of berry and stone fruit mashes; high color yields	Fast drop in juice viscosity; breakdown of fruit tissue	Acid- and thermostable pectinase (with PG/PE and PTE components.
Producing purées of high viscosity with strainers, thus replacing boiling	Partial hydrolysis of protopectin (cell wall cement); this tissue digest results in cell suspensions; no further hydrolysis of soluble pectin	Nearly pure PG with a good protopectinase activity; limited cellulase activity
Improving concentration of purées by maintaining cloud stability; production of fruit powders	Additional partial hydrolysis of cellulose; partial digestion of cell walls	Additional C_1-cellulase activity
Separating cellulose fibers from mash; producing cloud-stable vegetable juices of low viscosity	Partial hydrolysis of protopectin; additional hydrolysis of soluble pectin to medium-sized fragments with limited formation and precipitation of acid moieties; removal of hydrocolloids from cellulose fibers	PG/PTE with limited PE activity; hemicellulase complex
Producing clear juices by continuous extraction	Complete hydrolysis of all pectins and mucous substances	PG/PTE + PTE with hemicellulase complex; for some fruit varieties, C_1-cellulase is added

Application	Function	Enzyme characteristics
Clarifying fruit juices by destabilizing cloudy matter; membrane filtration	Complete hydrolysis of all pectins and extensive degradation of branched polysaccharide side chains	PG/PE with some PTE; with high hemicellulase activities
Preventing posthazing from retrograded starch in filtered juices or in juice concentrates from pip fruits	Saccharification of solubilized starch	Amyloglucosidase at temperatures of ~50°C; fungal α-amylase at ambient temperature
Preventing posthazing from reassociated araban in juice concentrates from pip fruits	Degradation of araban to nonprecipitable fragments	Pectinase with high arabanase activity
Producing concentrates from citrus and tropical fruit juices, maintaining cloud stability	Partial degradation of soluble pectin with only limited pectic acid formation	PTE/PG with only small amounts of PE
Clarifying citrus juices with high acidity	Total pectin hydrolysis	PG/PE with good activity at pH 2 and acid stability
Clarification of ciders by flotation	Deesterification and gelling of pectin	Purified PE, without noticeable depolymerizing activities
Increasing color yields in red winemaking; inhibiting foam during fermentation of flash-heated musts	Pectin degradation supported by partial proteolytic hydrolysis of the tissue; modification of thermally denatured protein	PG/PE and PTE with additional acid protease activity
Eliminating filtration problems in young wines made from partially fouled berry harvests; clarification of press wines from thermovinification	Digesting residual pectin; hydrolysis of β-glucan and other hemicelluloses	PG/PE and PTE high in β-glucanase activity and other hemicellulases

[a]PE, pectin esterase; PG, pectin glucosidase; PTE, pectin transeliminase.

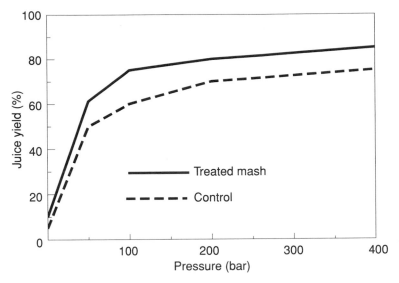

Figure 5.52   Juice yield from pressing apple mash.

protopectin. This requires that sufficient quantities of uniformly dispersed fruit pulp still remain in the final product (Ruttloff, 1973; Bielig and Wolff, 1973).

Apple Mash Treatment   Apples should be processed with a grinder or apple mill avoiding crushing with dull blades. Mash obtained in this manner has a good pressable texture, but a single pressing step yields only 75–80% juice. Stored apples produce less juice and have a pomace that sticks to the press cloth. This is the result of protopectin hydrolysis that occurs during ripening, which separates cells and softens the fruit tissue. The enzymatic treatment of apple mash was therefore first introduced for stored fruit (Eid and Holfelder, 1974). Concurrently, an extraction technology was developed (Voordouw et al., 1974) that operates in the manner shown in Figure 5.53.

  The apples are washed and ground; a suitable pectinase is added at the mill, especially if the apples have a soft texture. In the mill, the enzyme preparation is mixed thoroughly with the mash, an important step, because the mash in the vat is not stirred. Stirring and heating the mash to temperatures above 35°C causes the mash to disintegrate and yield a pulp with poor pressibility. The tank to which the enzyme is added is filled at the top and continuously emptied at the bottom. A dwell time of approximately 30–60 min should be maintained. The treated mash is transferred to a continuous-belt press for an initial pressing yielding 50–60% of the juice. The pulp residue is subsequently completely expressed, usually with a horizontal press. To assure satisfactory expression, water is added prior to this extraction. If there was no enzyme treatment prior to the first pressing, an appropriate pectinase is added to the water used for the final extraction in order to enhance extraction efficacy. There is only an insignificant increase in total pressed juice volume; however, the sugar content rises considerably so that the final sugar content exceeds the normal concentration by 90% (Bielig and Rouwen, 1976).

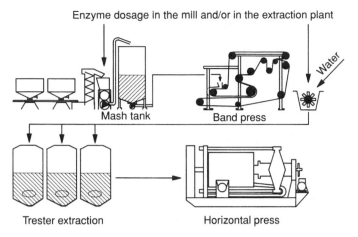

Figure 5.53 Diagram of juice pressing technology.

The pectinases developed for apple mash contain a high percentage of pectin esterase (PE) and pectin glycosidase (PG), the latter has only a limited hydrolytic effect on protopectin. Pectinase treatment dramatically lowers juice viscosity. Deesterified pectin fragments are generated that show little hydrating ability and thus reduce pomace stickiness. After an adequate first expression of juice, the pomace acts as an aid in subsequent pressings. A combination of cellulase and hemicellulase perforates the cell walls through which the retained juice can easily be exchanged by the extraction water (Pilnik and Voragen, 1981). Industrial experience has shown that mash treatment in this manner increases press capacity by 30–50% and enhances juice yields by 5–10% (Fig. 5.54).

Treatment of Berry and Stone-Fruit Mashes While improvements in the current technology of apple mash production have been made, the enzymatic treatment of soft fruit mashes was already a classic and indispensable procedure for obtaining worthwhile quantities of pressed juice (Wucherepfennig et al., 1966). Berries differ significantly from pip fruits in a number of characteristics that are important for enzyme treatment; for instance, they are loaded with pectin and require a correspondingly greater enzyme dose (Fig. 5.55).

Further, the acidity is high with a pH between 2.8 and 3.2, which reduces pectinase activity.

With increasing reaction time, the enzyme becomes inactivated in the acid pH range, especially at operating temperatures of 50–55°C.

The berries' pigments and tannins also precipitate proteins, which can limit or inhibit enzymatic activity (Wucherpfennig et al., 1970).

Because of these problems, a pectinase is used that is relatively acid-stable and uninhibited by tannins. A sufficiently high dose of pectin transeliminase (PTE) assures a rapid drop in viscosity, disintegration of the mash, and release of the desired pigments.

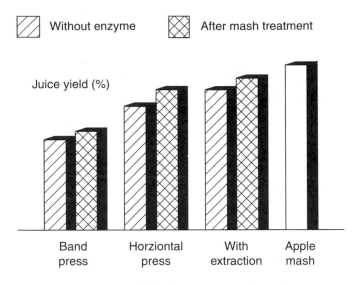

Figure 5.54 Increase of juice yield after enzyme treatment.

Plant Tissue Maceration   In contrast to mash treatment that is designed to improve pressibility and results in firm mash solids, maceration (softening) involves the complete disintegration of the fruit tissue; the soluble but high-molecular-weight pectin, however, is still largely retained. Alternatively, boiling, despite drawbacks produces the same result. Less damaging to the end product is the enzymatic mac-

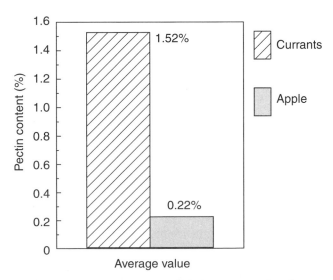

Figure 5.55 Pectin content of black currants and apples.

eration with a very pure pectin glycosidase. This enzyme primarily attacks the protopectin that cements the cells, cleaving only a few nonesterified groups to retain a soluble, high-molecular-weight pectin. The intact plant cells begin to separate from one another, forming a cell suspension. The result is a fine and homogenous fruit purée that can be easily strained and thereby freed of rough fibers, peels, and other matter. This procedure is applied in those situations where, along with the liquid fruit components, a finely dispersed fruit pulp is desired in the end product (Sulc and Ciric, 1968; Sulc and Vujicic, 1973).

Vegetables of solid texture, such as carrots and celery, or stone fruits such as peaches and apricots, are routinely used as raw materials. The purées obtained are processed for fruit and vegetable baby food formulas, as basic ingredients in yogurt and ice cream, or for turbid multifruit juices. An advantage of enzymatic maceration is the good stability of the turbid juices against pulp sedimentation and clarification.

Processing of carrots is illustrated in Figure 5.56.

The raw material is cleaned, peeled, sliced, and blanched. The dry mash, after cooling to 45–50°C, is adjusted to pH 4.5–4.8 with an edible acid and treated with a macerating pectin glycosidase (PG). The mash is stirred slowly during the reaction. The mash disintegrates into a homogenous, creamy purée. After ca. 30 minutes, the remaining solids can be separated with a strainer or with a decanter from the pulpy macerate.

If, instead of a viscous purée, a thin watery fruit macerate is sought, a C-1 cellulase is used in combination with PG. Vigorous stirring with strong shearing forces generates an intimate contact between the cellulose substrate and cellulase enzyme. The cell walls are hydrolyzed and a significant decrease in viscosity ensues. After separation of the fluid pulp from residual fibers, the pulp can be highly concentrated. Thus, less water needs to be removed in the subsequent drying process (Schmitt, 1983). A technically simple method to produce a drinkable, fluid and

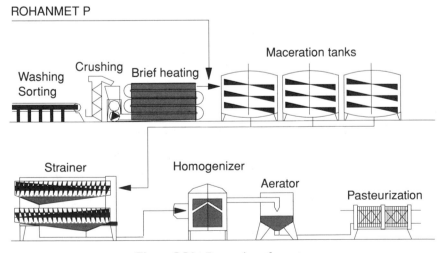

Figure 5.56 Processing of carrots.

cloud-stable juice that resembles the process described earlier (Schmitt, 1988) is available. The enzyme preparation employed performs the following special tasks:

1. The fruit tissue is disintegrated with a macerating PC. In order to retain some pressible pulp in the mash, the process is not fully completed.
2. The soluble pectin is largely depolymerized by PTE. Because PE is absent, there is no formation of galacturonic acid groups, which tend to coagulate as a consequence of their negative charges; this means that the juice remains cloud-stable.
3. Various hemicellulases can solubilize gums that stick to cellulose fibrils. The enzymatic action facilitates the removal of microfibers from the liquid by simple straining or pressing. The high viscosity of whole-fruit juices is due primarily to the presence of such fibrous material and to a lesser extent to the presence of high-molecular-weight pectin.

Figure 5.57 illustrates the change in viscosity of a carrot mash treated with the special enzymes described.

Liquefaction of Fruit Mashes   The mash must be liquefied for optimal yields of soluble dry matter for extraction without water, or in preparation for continuous de-juicing with decanting centrifuges, or for membrane filtration procedures. Liquefaction involves an extensive degradation of protopectin, soluble pectin, and other hydrocolloids. Combinations of special pectolytic enzyme preparations containing hemicellulases that have a great capacity for digesting cell wall components are used (Pilnik et al., 1975; Beilig et al., 1971).

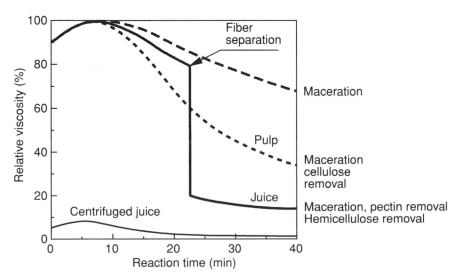

Figure 5.57   Time course of viscosity change in carrot mash.

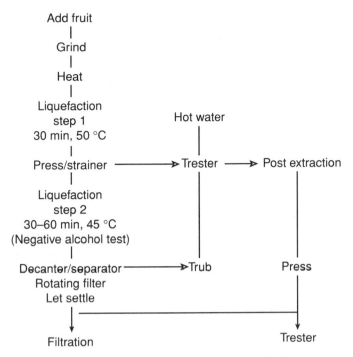

Figure 5.58   Liquefying berries and stone fruits.

Arabanase, galactanase, mannanase, glucanase, laminarinase, and even proteases are of special importance in this process (Feldmann, 1987).

Liquefaction takes place in mash tanks under vigorous stirring at sufficiently high temperatures. To avoid offtastes, stones and skins are removed by straining the mash after the first liquefaction step. After complete digestion of colloids, the solids are eliminated by decanting or microfiltration (Bott et al., 1986).

Figure 5.58 outlines a method for producing berry juice by continuous dejuicing.

5.5.2.2 Production of Clear Juices and Concentrates

To save storage and transportation volume, the majority of fruit juices produced today is concentrated to 68–72% dry solids, which are semifinished products. The volatile essences obtained by evaporation and rectification are stored separately as essence or aroma concentrates.

Two steps are required to prepare a cloudy press juice for filtration and concentration. Each step is conducted separately: first the enzymatic treatment and then the clarification with fining agents. Figure 5.59 illustrates the individual steps for making apple juice concentrate.

The press juice still contains high-molecular-weight pectin and, depending on the maturity of the apples used, also some starch granules; when extracting pomace araban from the cell walls, tannins and other insoluble particulates, responsible for

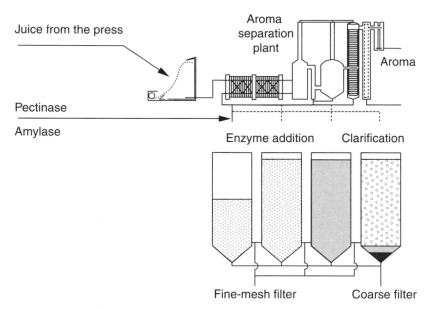

Figure 5.59   Procution of apple juice concentrate.

hazes, are also present. The juice is heated to 95–100°C to evaporate the volatile essences. This process solubilizes the starch. When eliminating the essences with an evaporator, the juice is also concentrated to 80–50% of its original volume and cooled to 50°C (Grampp, 1976).

The enzyme preparations are added when filling the enzyme treatment and clarification tanks. When the tanks are fully charged (1–2 h), the enzymes are left to react for another 10–15 min. The fining agents are added only at this time, in order to precipitate particulates and to absorb tannins that could cause haze in the filtered juice. After settling of the fining flocks, the supernatant is filtered to clarity; the sediment is then coarsely filtered with rotating vacuum filters and again filtered to clarity. The clarified filtrate is concentrated in vacuo to 72° Brix. The processes of enzyme treatment and clarification are described below.

Pectin Degradation   Because complete pectin hydrolysis is a prerequisite for successful clarification (Krug, 1968), the pectinases used in clarification contain especially high proportions of PG and PE, along with a lesser amount of PTE (Endo, 1965).

Figure 5.60 illustrates the drop in apple juice viscosity during pectin hydrolysis; the PTE initially decreases viscosity, but a combination of PG/PE advances complete hydrolysis more rapidly.

Customarily, pectin degradation is monitored by the alcohol test. This is done by mixing one part of turbid juice from the treatment tank with two parts of acidified ethanol in a test tube without shaking. When pectin is present, a gel plug develops that contains haze particles and air bubbles. The gels form at the surface of the mixture and can become rather solid. The test samples removed from the juice while in-

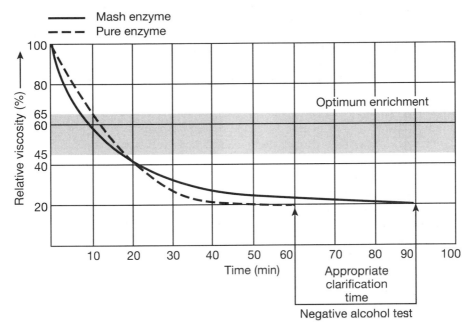

Figure 5.60 Change of viscosity in apple juice during pectin degradation.

creasing pectin hydrolysis show progressively less gelling until none occurs when the endpoint is reached. The particles left will then merely coagulate and settle to the bottom. On further treatment beyond this point, the desired side effect of the pectinase preparations, namely, hemicellulase activity, degrades the colloids bound to the solid particles. The polysaccharide side chains once freed from the pectins will now also be depolymerized. The alcohol test shows that the particles causing turbidity no longer coagulate, but will instead form a homogenous suspension of colloid-free solids in the alcoholic mixture.

Juices that pass this critical alcohol test react thoroughly and rapidly with fining agents. Total degradation of colloids is especially important when juices are directly subjected to membrane filtration without prior treatment with flocculating agents (Dietrich et al., 1985).

Starch Degradation Starch solubilized by heating is not affected by clarification and filtration and thus presents a risk for subsequent juice hazing. During storage of the concentrate or after bottling, starch retrogrades and appears as a white, foggy haze. To avoid this, the starch is saccharified by adding a thermostable amyloglucosidase. Starch hydrolysis and pectin hydrolysis take place simultaneously; that is, amyloglucosidase and pectinase are added together provided starch is present. The presence of starch, or its absence on successful hydrolysis, is checked with the iodine color test. This test is performed by layering a 0.1 N iodine solution over a juice sample and observing the interface. The presence of starch is indicated by a dark-blue color.

Fungal α-amylase, an endoenzyme, appears to be more suitable than amyloglucosidase in lowering the molecular weight of starch but can tolerate an optimal temperature of only 30°C at a juice pH of about 3.5 (Krebs, 1971).

This fact, and the knowledge that clarifying agents are as effective at 50°C as at 18–20°C, resulted in the development of hot clarification (Grampp, 1976).

Araban Degradation   Araban can also cause turbidity. Araban consists exclusively of arabinose and is a component of the cell wall of pip fruits.

On additional extraction of apple pomace, araban occasionally enters the press juice. Araban cannot be removed from the juice with fining agents and will precipitate during cold storage concentrates as microscopically minute twin spheres. In stored or packaged juices araban hazes have not yet been observed. Pectinases used in juice processing today contain sufficient active arabanase that additional dosing is unnecessary. It is important, however, that the arabanase be an *endo*-arabanase that rapidly generates low-molecular-weight degradation products that do not precipitate (Schmitt, 1985).

Clarifying and Stabilizing Agents   These are nonenzymatic substances that bind undesirable particles in the juice by either physicochemical reactions or adsorption. They cause coagulation into coarse flocks that are easily removed by filtration.

Dissolved gelatin serves as the most important flocculating agent; because of its positive charge, it strongly attracts and binds the negatively charged juice tannins and those bound to haze particles, thereby producing sizable conglomerates. Because an insufficient amount of gelatin will not perform adequately, a preliminary test is required to determine the precise amount required for a given juice. Because gelatin is a protein, it may not be added in excess because proteins precipitate, causing haze in packaged juice.

Bentonite, a clay with a high swelling capacity, is introduced as an aqueous colloidal suspension. Because of its planar crystalline structure with embedded sodium and calcium ions, it is characterized by a separation of charges, enabling the bivalent bentonite to bind electropositive as well as electronegative compounds.

Silica sol is colloidal hydrated polysilicic acid, which, because of the high electron density on its surface, serves primarily as a link between external positively charged particles. Silica sol adsorbs those gelatin particles that did not participate in prior flocculation reactions.

The following is a typical clarification procedure, which begins with the addition of bentonite and thorough mixing. Turbid conglomerates promptly form. Then gelatin is added, producing floccules of haze particulates, bentonite, tannins, and protein. Subsequent addition of silica sol adsorbs excess gelatin and solidifies the flocculate, which can settle into compact sediments (Dickmann and Görtges, 1982; Görtges, 1982; Possmann and Wucherpfennig, 1972)

5.5.2.3 Special Production Processes   Processing tropical fruits and especially citrus fruits can be improved with the application of special pectinases. This involves the realization of objectives that at first appear contradictory. For most, the juices obtained should have a stable and homogeneous turbidity. Yet, the viscosity should be sufficiently low to allow high concentration. Juice output should increase with enzymatic treatment of mashes and pomaces without requiring clarification of the additional juice or extract produced.

Citrus Processing   The production of orange juice concentrates often coincides with undesirable gel formation due to the presence of pectin; this is especially true for early ripening varieties such as Hamlin or with the use of certain high-yield juice extractors. With specific partial degradation of pectin while maintaining residual viscosity, the concentratability can be improved with no loss in cloud stability. Pectinases with limited pectin esterase (PE) activity are used for this purpose. Figure 5.61 illustrates the decrease in viscosity in an orange peel extract.

Pulp-wash is the aqueous extract of strainer residues and contains the finely dispersed fruit pulp. The extract should be concentrated as much as possible. The following advantages are obtained when a low-PE pectinase is present on addition of the extraction water: higher yields in soluble dry solids by about 15–20%, decreasing the viscosity of the extract and therefore increasing the ability for volume juice concentration.

Citrus oil is obtained by centrifuging washwater from abraded citrus peels. The water recycled following removal of the oil becomes enriched with pectin. A standard clarifying pectinase is used to continuously degrade the pectin during the operation, thereby improving oil separation on centrifugation.

Clarified citrus, lime, and grapefruit juices are also produced and concentrated. The pectin has to be degraded at an extremely low pH (1.5–2.2). After centrifugation to reduce the juice solids, the pectin is hydrolyzed with a PG/PE pectinase until

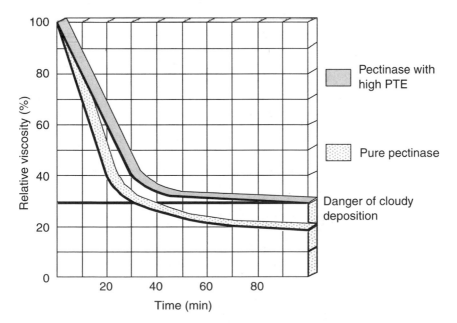

Figure 5.61   Time course of viscosity change in orange peel extract.

the alcohol test is negative. The pectinase preparation should have an exceptionally good acid stability. The reaction temperature should not be above 25–30°C, to avoid damaging the enzyme. The turbidity from the pectin degradation can be precipitated with silica sol.

Processing of Tropical Fruits   In contrast to citrus processing, the emphasis here is on treating the mash, which is often difficult to extract by straining. This applies particularly to mango, guava, papaya, and maracuja, in which 5–20% yield increases can be attained with mash treatment. Macerating pectinases containing PG alone or with PTE and some PE are used. Figures 5.62 and 5.63 outline the process, yields, and concentratability of the pulp juices.

5.5.3   Winemaking

Enzymes have now become an integral part of enologic methods along with the ancient knowledge of winemaking. They are employed to improve the quality of wines as well as reduce winemaking costs. Generally, the standard pectinases from fruit juice technology are used for treating mash, prefermentation must, and young wines, and for producing unfermented clear grape juice.

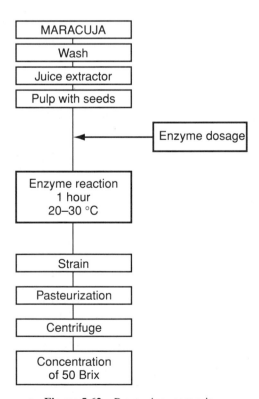

Figure 5.62   Processing maracuja.

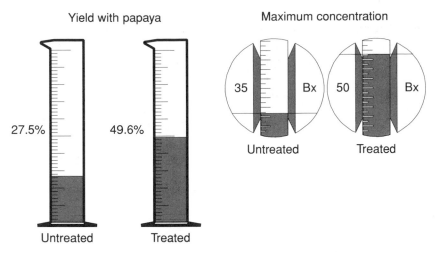

Figure 5.63   Yield and concentration of papaya juice.

Special enzyme preparations contain, in addition to the main pectolytic activity, a protease or β-glucanase and offer a number of advantages in special applications.

5.5.3.1 *Processing of Grape Mash*   Pectinases are used for those varieties of white grapes that are difficult to press; they can increase the yield by an additional 20% of the free-run juice. To extract the aroma of the so-called bouquet varieties, a pectolytic mash treatment is used as well.

Several methods exist for treating red wine mash:

The traditional skin fermentation process includes the addition of pectinase, which reduces fermentation time because of good initial liquefaction. Color extraction is increased.

Rosé wine is obtained by letting the mash stand prior to pressing in order to extract some of the pigments. A pectinase treatment reduces the holding time.

Thermovinification includes methods in which the mash is heated to extract the color; enzymatic activity is optimal at temperatures of 45–50°C, thereby improving color extraction and free-run juice and press juice yields.

Pectinase alone may frequently produce unsatisfactory results with color-sensitive red wines. The initially increased color extraction is lost again during production and storage. Added protease activity results in a better color stability.

5.5.3.2 *Must and Young Wine Treatment*   The partial removal of turbidity from the must, that is, the removal of gums or preclarification, is an important step in eliminating oxidation products and assuring a clean fermentation. This process is accelerated with pectinase. This also applies to must clarification and filtration when preparing the sweet reserve that is filter-sterilized for storage. This clarified, unfermented grape juice (sweet reserve) is added to fermented wine, giving it added sweetness prior to bottling.

White grape musts are often flash-heated prior to fermentation to kill wild yeast strains. Cultured yeast is added to initiate fermentation. This frequently gives rise to intensive foaming during fermentation, requiring ample tank space. Pectinases and proteases reduce foaming and improve clarification and filtration of young wines. Because the denatured, native proteins are preferentially degraded, the amount of bentonite required to stabilize the proteins is reduced.

Young wine from *Botrytis*-infected harvests, often involving premium varieties used for "selected" or "berry-selected" wines—especially young wines from standard varieties with heavy *Botrytis* contamination, resist the final polishing filtration prior to bottling. Microscopic amounts, as little as 50 ppm of β-glucan, a metabolite of *Botrytis* fungi, are responsible for this problem.

In such situations, treatment of the young wine with a β-glucanase preparation is indicated.

Figures 5.64a–5.64c illustrate the use of enzymes in the production of white and red wine.

(a) ROHAPECT Red wine making:
    skin fermentation process

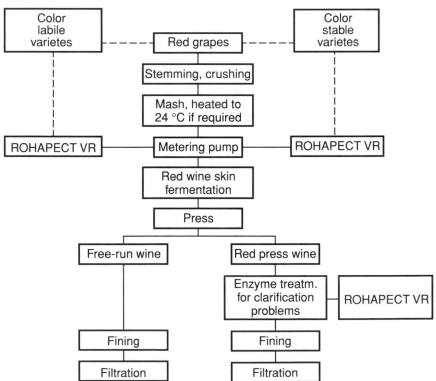

Figure 5.64 Winemaking: (*a*) red wine, skin fermentation; (*b*) red wine, thermovinification; (*c*) white wine. ROHAPECT is a trade name for pectic enzyme producted by Röhm.

(b) ROHAPECT Red wine making:
    Thermo vinification

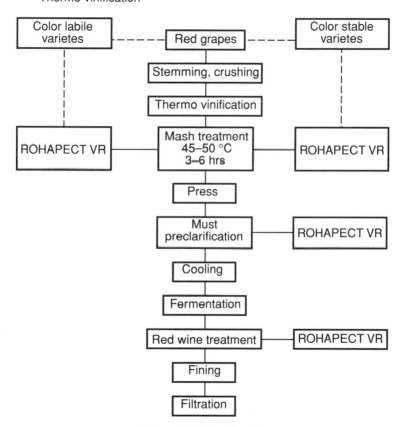

Figure 64 *Continued*

(c) ROHAPECT White wine making:

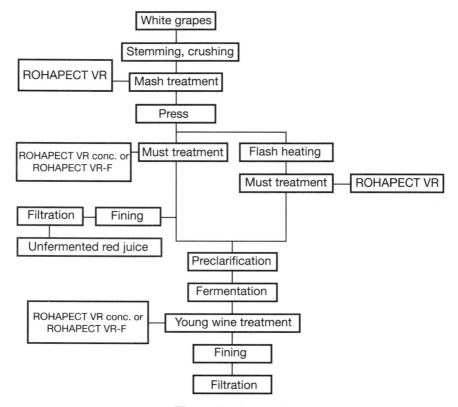

Figure 64   *Continued*

REFERENCES

Bielig, H.-J., and Rouwen, F. M., *Fl. Obst* **43,** 426 (1976).

Bielig, H.-J., Wolff, J., and Balcke, E., *Fl. Obst* **38,** 408 (1971).

Bielig, H.-J., and Wolff, J., *Fl. Obst* **40,** 413 (1973).

Bott, E., Helmfort, H., and Lehmann, H., *Fl. Obst* **53,** 504 (1986).

Dickmann, H., and Görtges, S., *Fl. Obst* **49,** 582 (1982).

Dietrich, H., Wucherpfennig, K., and Scholz, R., *Fl. Obst* **52,** 324 (1985).

Eid, K., and Holfelder, E., *Fl. Obst* **41,** 88 (1974).

Endo, A., *Agric. Biol. Chem.* **29,** 129, 137, 222, 229, 234 (1965).

Feldmann, G., *Confructa Studien* **31** (1987).

Görtges, S., *Fl. Obst* **49,** 93 (1982).

Grampp, E., *Fl. Obst* **43,** 382 (1976).

Koch, J., *Die Fruchsaftindustrie* **1,** 66 (1956).

Krebs, J., *Fl. Obst* **22,** 6 (1975).

Krebs, J., *Fl. Obst* **38,** 142 (1971).

Krug, K., *Fl. Obst* **35,** 322 (1968).

Lea, A. G. H., in Tucker, G. A., and Woods, L. F. J. (eds.), *Enzymes in Food Processing,* 2nd ed., Blackie Academic & Professional (imprint of Chapman & Hall), New York, 1995, p. 223.

Lüthi, H. R., and Glenk, U., in Schorrmüller, J. (ed.), *Handbuch der Lebensmittelchemie,* 2 ed., Vol. 5, 1968, 122.

Mehlitz, A., *Biochem. Ztschr.* **221,** 217 (1930).

Mehlitz, A., *Ztschr. Lebensm. Untersuchg. Forschg.* **68,** 91 (1934).

Pilnik, W., *Fl. Obst* **37,** 430 (1970).

Pilnik, W., and Voragen, A. G. J., *Fl. Obst* **48,** 261 (1981).

Pilnik, W., and Voragen, A. G. J., in Fox, P. F., ed., *Food Enzymology,* Vol. 1, Elsevier, New York, 1991, p. 303.

Pilnik, W., Voragen, A. G. J., and De Vos, L., *Fl. Obst* **42,** 448 (1975).

Pilnik, W., Voragen, A. G. J., in Nagodawithna, T., and Reed, G. (eds.), *Enzymes in Food Processing,* 3rd ed., Academic Press, New York 1993, p. 363.

Possmann, P., and Wucherpfennig, K., *Fl. Obst* **39,** 46 (1972).

Ruttloff, H., *Industrielle Enzyme,* Steinkopf, Darmstadt, 1973.

Schmitt, R., *Fl. Obst* **50,** 23 (1983).

Schmitt, R., *Confructa Studien* **29,** 22 (1985).

Schmitt, R., *Fl. Obst* **55,** 321 (1988).

Sulc, D., and Ciric, D., *Fl. Obst* **35,** 232 (1968).

Sulc, D., and Vujicic, B., *Fl. Obst* **40,** 130 (1973).

Villettaz, J.-C., in Nagodawithana, T., and Reed, G. (eds.), *Enzymes in Food Processing,* 3rd ed., Academic Press, New York, 1993, p. 423.

Voordouw, G., Voragen, A. G., and Pilnik, W., *Fl. Obst* **41,** 282 (1974).

Wucherpfennig, K., Brethauer, G., and Ratzka, D., *Fl. Obst* **33,** 458 (1966).

Wucherpfennig, K., Millies, K. D., and Landgraf, H., *Fl. Obst* **37,** 87 (1970).

5.6. ENZYMES IN THE DAIRY INDUSTRY

References: Fox (1982), Richardson (1975), Cheeseman (1981) Desmazeaud (1985), Brown (1993), Law and Goodenough (1991, 1995)

5.6.1 Milk Proteins and Casein Micelle

The main components of milk are proteins, lipids, lactose, minerals, vitamins, and a large number of enzymes such as oxidases, phosphatases, peroxidases, catalases, amylases, and lipases. (See also Table 5.23.)

Casein makes up about 80% of the total milk protein. It is a phosphoprotein consisting of monomer chains that have a molecular weight between 20,000 and 30,000 daltons. Caseine monomers form aggregates of about 1 million daltons.

TABLE 5.23   Approximate Composition of Milk and Whey

Components	Milk (%)	Whey (%)
Water	87.6	93.7
Lipids	3.3–3.8	0.5
Protein	3.3	0.8–1.0
Casein	2.6	—
Whey protein	0.5–0.7	0.8
Lactose	—	4.85
Calcium	0.12–0.15	0.05–0.09
Phosphorus	0.1	—

In the presence of calcium, large aggregates, the so-called submicelles, are formed. These structures combine to form casein micelles in the presence of calcium phosphate. Micelles can reach sizes of 50–250 nm. Casein can be differentiated into four proteins: (1) the $\alpha\text{-}s_1$ casein, makes up 40% of the total casein and has a molecular weight of 23,600 daltons and is nearly insoluble in 0.03 M $CaCl_2$ at 37°C; (2) the $\alpha\text{-}s_2$ casein, which makes up 10% of the total casein; (3) the β-casein, which constitutes 32% of the total casein; and (4) the κ-casein, which represents 15% of the total casein. The β-casein and the κ-casein are both soluble at 4°C. Under the conditions described above, the κ-casein remains soluble at 37°C. The soluble κ-casein presumably forms an insoluble complex with α-casein that stabilizes the micelles. All caseins are highly phosphorylated and precipitate with Ca^{2+} (>6 mM).

These micelles are destroyed when the milk is acidified. The casein is precipitated along with other milk components. A complete precipitation occurs at pH 4.6, which is the isoelectric point for casein. The acid precipitate is granular and brittle, in contrast to the precipitate obtained with chymosin.

5.6.2   Cheesemaking: 7000 Years of Biotechnology

Reference: Fox (1988).

More than 1000 cheese varieties are produced worldwide. Cheesemaking (Fig. 5.65) is the most traditional method of preserving a food as valuable as milk. On the action of rennet, the casein in milk clots and the milk components are separated into a solid phase and a liquid phase.

5.6.2.1   Rennet Activity   Milk clotting on treatment with rennet (chymosin) is accomplished in several stages. The first stage involves the hydrolysis of the peptide bond between phenylalanine (105) and methionine (106) in κ-casein. Porcine pepsin A, pepsin C, and the protease from *Mucor miehei* also cleave this bond. The protease from *Endothia parasitica* cleaves the bond between serine (104) and

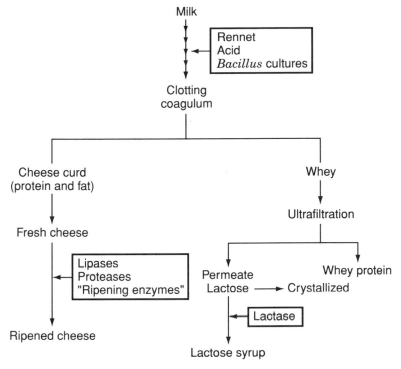

Figure 5.65   Cheesemaking and enzymes used in this process.

phenylalanine (105) (Droehse and Foltmann, 1989), thus liberating the C-terminal third segment of κ-casein. This reaction denatures the κ-casein, which usually stabilizes the complex (which is made up of the four types of casein) against precipitation by calcium ions. The cleaved glucopeptide is quite hydrophilic because of the large complement of sugars and polar amino acids and therefore remains soluble.

$$\kappa\text{-casein} \xrightarrow{\text{chymosin}} para\text{-}\kappa\text{-casein} + \kappa\text{-casein macropeptide}$$

In the second stage, the casein micelles aggregate in the presence of calcium after ~95% of the κ-casein has been cleaved. The extent of κ-casein proteolysis can be determined by the amount of protein soluble in 12% trichloroacetic acid (TCA) or by the quantity of *para*-κ-casein precipitated (Cheeseman, 1981). The milk's viscosity increases during aggregation, and a gelatinous structure develops. Rennet retention in the cheese curd depends on the pH. Between 3 and 10% rennet is retained in the pressed curd [GB]. The proteases that remain in the curd participate in the cheese ripening processes.

The mechanism of gelling and its dependence on temperature and pH were described by Dalgleish (1979). The ability of the gels to release water is due to the ability of the rennet–milk gels to contract (syneresis) when sliced. The third step involves a gentle hydrolysis of casein in the curd, the degree of which depends on the enzyme used.

The following factors influence rennet activity:

- The amount of rennet required depends on the cheese variety being produced. For a standard rennet with 15,000 units per milliliter, quantities of 5–50 ml per 100 liters of milk have been specified (Burgess and Shaw, 1983). Higher doses do not improve curdling, but rather cause an increased proteolysis during the third step.
- Gelling is promoted with increased temperature. The firmness of the curd increases up to 40°C and decreases thereafter.
- Calcium addition (see Fig. 5.66) produces harder gels. The optimal amount of calcium chloride is 100 g per 100 liters of milk.
- Low pH promotes gelling; gels with a coarse granular texture are formed. The dependence of rennet activity on pH is more pronounced with microbial rennet than with pure calf rennet.
- Many proteases from animal gasters, plants, and microorganisms have rennet activity. In addition, they possess a nonspecific activity for other peptide bonds in casein. This nonspecific proteolysis influences the degree of total casein precipitation and curd stability. The protein loss is usually greater with microbial rennet than with calf rennet.
- Nonspecific proteolytic activity can adversely influence cheese ripening or impart a bitter taste to cheese.

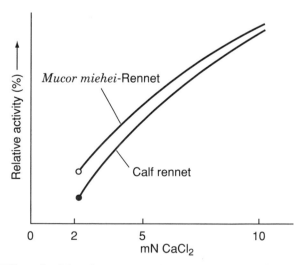

Figure 5.66   Effect of calcium ion concentration on activity of calf rennet and renin from *Mucor miehei*.

- It has long been known that heating milk above 65°C inhibits rennet-mediated curdling. This inhibition can be prevented if milk is acidified or treated with calcium chloride prior to heating.

5.6.2.2 Rennet Preparations According to Martin et al. (1981), various rennet preparations are used in cheesemaking. Several mixtures of pure chymosin with beef pepsin and microbial rennet enzymes are employed. Because the latter are more thermostable than mammalian chymosin, thermally labile microbial enzymes have been developed for processes in which pasteurized whey should not have rennet activity. Destabilization is achieved by oxidizing the enzyme. Also, more labile rennet preparations are preferred in making cheeses with extended shelf life.

Figure 5.67 compares the temperature stability of microbial rennet preparations [NO] with that of calf rennet.

With calf stomach as the traditional source for chymosin on the decline and the use of microbial rennet substitutes restricted because these substances often produce adverse flavors, genetic engineering technology has provided preparations that are identical to those for natural calf chymosin. Several companies have marketed recombinant chymosin under various tradenames, such as Maxiren [GB], Chymogen [CH], and CHY-Max [PF]. Gist-Brocades [GB], Genencor [GR], Novo Nordisk [NO] and Pfizer [PF] are leaders in this field. For supplemental reading on this technology, the articles by Teuber (1993) and Wegstein and Heinsohn (1993) are recommended.

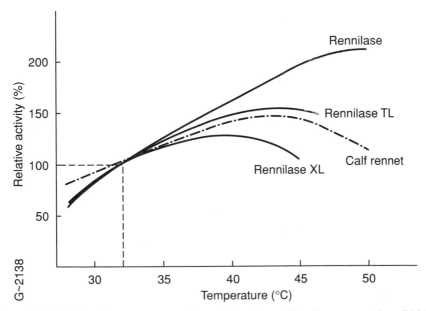

Figure 5.67 Effect of temperature on the activity of some Rennilase preparations [NO] as compared to calf rennet.

5.6.2.3 Rennet Dosing The amount of rennet to be added depends on the type of cheese to be made and the curdling time desired. For instance, 10–30 ml of rennet with 1:10,000 Soxhlet units is sufficient to curdle 100 liters of milk.

In a process described for making cheddar, a standard rennet at a dose of 150 ml per 1000 liters of milk at 31°C is employed.

When comparing the activity of two rennet samples, the preparations are diluted with tap water to approximately 1000 Soxhlet units. Milk is brought to rennet temperature, the pH is adjusted, and calcium chloride is added.

After 10 ml of the diluted enzyme preparation is added to 1000 ml of milk, the mix is stirred lightly and the time to first curdling is determined. This time period should be 120–360 s. Comparing the curdling time with a standard determines the relative activity of the sample. Martin et al. (1981) compared other rennet enzymes with rennin (chymosin) as a standard.

5.6.3 Cheese Ripening: Supporting the Natural Processes

Taste and texture of cheese are altered by the enzymatic conversions taking place during the ripening process. These include the native milk enzymes discussed earlier and the exogenous enzymes from rennet that are added to cheese and influence its ripening process. This process is further affected by the native microbiological flora of milk and the added starter cultures. Proteolytic (Grappin et al., 1985; Brown, 1993) and lipolytic degradative reactions are significantly involved; these cause the formation of even-numbered fatty acids (from C_4 to C_{10}), odd-numbered methyl ketones (from C_5 to C_9) and α-ketoacids. All these degradative products play a role in the ripening process (Ney, 1987). In most cases, regulations specify minimum time periods for the ripening process. Longer ripening times, however, produce the desired intensity of flavor. Cheddar, for example, can ripen between 2 and 12 months, thus developing a mild, sharp, or very sharp flavor. Longer ripening periods require more energy and storage costs and higher investment expenditure. According to Arbige et al. (1986), approximately 2 billion pounds of cheddar cheese are produced in the United States annually. The bulk of this cheese requires 3–9 months of ripening to achieve the desired flavor. The ripening costs are reported to be $2.50 per 100 lb for a ripening period of 6 months, and $10 per 100 lb for the development of a "12-month flavor" (Katokocin, 1984). Therefore, it is clear that systems are being sought that will allow shortening of the ripening period. Several enzymatic processes are in use:

1. Addition of extracts from *Lactobacilli* sp. such as *L. helveticus, L. bulgaricus,* or *L. lactis* (Desmazeaud, 1982). N. F. Olson and associates at the University of Wisconsin in Madison (USA) have studied combinations of proteases, peptidases, lactase, and starter cultures of *L. helveticus* in cheddar ripening.

2. Addition of microencapsulated enzymes together with rennet, or subsequently incorporating them into cheese curd (El Soda and Desmazeaud, 1981). Acceleration of cheese ripening with technical enzyme preparations is influenced by two factors. When adding the enzymes to milk before its subsequent curdling with

rennet, 90% of the enzymes remain in the whey. On the other hand, adding the enzymes directly to the curd does not result in an ideal distribution. Lecithin-encapsulated enzymes (Law and Kirby, 1987), however, can solve these problems. Enzymes can be packaged with milk fat, sorbitan monostearate, and lecithin (Magee and Olson, 1981). Diacetyl and acetoin flavors can be produced with extracts from *Streptococcus lactis* packaged in this manner (Magee et al., 1981). A protease isolated from *Micrococcus caseolyticus* (Rulactin) incorporated into liposomes was used by Alkhalaf et al. (1986) for the production of Saint Pauli cheese. Up to 60% of the liposomes were retained in the cheese curd. Neutral bacterial protease encapsulated with egg phosphatidyl choline or soybean lecithin can significantly accelerate cheese ripening (Kirby et al., 1987).

3. Addition of unpackaged technical proteases or lipases (Law and Wigmore, 1983), as described in many patents and publications (Jolly and Kosikowski, 1975). The specificity of the proteases and lipases play a decisive role in this process. According to Arbige et al. (1986), the ratio of tricaprylin to tributyrin activity (TCU/TBU) constitutes a good criterion for the usefulness of a lipase esterase in ripening cheese (Table 5.24). This means that lipase preparations with relatively high tributyrin activity are less suited than milk lipase. A preparation from *A. oryzae* is described in a patent by Arbige and Neubeck (1987) in which the TCU/TBU ratio is greater than 0.75.

According to this patent, the lipase esterase is added as virtually insoluble particles to the curdling process. During coagulation, the lipase preparation is incorporated and thoroughly dispersed within the curd and is not lost to the whey.

This lipase preparation from *A. oryzae,* found to produce a distinct cheddar aroma, was the product of a joint venture between Chr. Hansen, Inc. [CH] and Genencor; it was marketed by [CH] under the tradename Flavor-Age until 1993.

TABLE 5.24 Relative Specificities of Lipase Preparations (Arbige et al., 1986)

Preparation and Source[a]	TCU/TBU Ratio[b]
Calf rennet paste	0.00
Aspergillus niger [NO]	0.02
Calf lipase esterase	0.04
Pancreatic lipase [SOE]	0.09
Goat pregastric enzyme	0.17
Lamb pregastric enzyme	0.19
Candida cylindraceae [MSJ]	0.31
Aspergillus niger [AM]	0.73
Mucor sp. [AM]	0.67
Milk lipase	0.76
Aspergillus oryzae	1.59

[a]See Appendix A.
[b]TCU, tricaprylin (C-8) activity; TBU, tributyrin (C-4) activity.

5.6.3.1   Acceleration of Cheddar Ripening with Neutral Bacterial Protease [Reference: Ridha (1984)].

Neutral bacterial protease (0.001 to 0.01%) is mixed with freeze-dried starter cultures and milled cheese curd and incubated at 10°C. Progress of ripening is determined by the ratio of soluble versus insoluble nitrogen (ripening index). The ripening process can thus be shortened from 8 to 2 months; there is some danger, however, that flavor and texture may be adversely affected (Ney, 1987).

Lipases can also enhance the typical cheddar flavor. Free C_6 and C_{10} fatty acids are the most important for aroma (Arbige et al., 1986). The *Aspergillus* lipase esterases are very specific for liberating these acids. By contrast, cleaving C_4 fatty acids produces a rancid taste; and the presence of free C_{12} fatty acids imparts a soapy taste to the cheese. The use of enzymes in making soft cheddar cheese spread is described by Wargel et al. (1980). The starter is a milk protein concentrate containing 50% dry substance, lactose, and fat (20%). The protein is partially hydrolyzed with microbial acid protease, and the fat is partially lipolyzed with an *Aspergillus* derived lipase (Italase C and Capalase KL). The lipolysis limits should be from 0.005 to 0.18% of C_2–C_{12} fatty acids. Proteolysis should generate 60–70% peptides with molecular weights between 5000 and 25,000 daltons. Lipolact K1 [RM] is an enzyme preparation especially developed for these processes.

5.6.3.2   Production of Italian Cheese Varieties

The highly aromatic or piquant flavor of certain cheese varieties such as Romano, Provolone, Asiago, and Fontina is accomplished by increased lipolysis with animal-derived lipase esterases. In these traditional processes, special enzymes are prepared from the complete gasters of calves, lambs, or goats. Alternatively, commercial pregastric lipase esterases are added to the rennet (Puhan and Morgenthaler, 1982).

The typical aroma of Provolone or Romano cheeses is based on a content of 1.7–5.7 mg butyric acid per gram of extract (dry substance).

Microbial lipases from *A. niger* or *Mucor miehei* can also be used for this process; 2–10 g of such preparations are required for 100 liters of milk.

Microbial rennet enzymes and lipase esterases are, in addition to the typical Italian cheeses, used predominantly for making Feta, Domiatti, and Ras Egyptian cheeses (El Salam et al., 1978). For the latter, Italase K or Capalase KL together with rennet are added to a 50% mixture of water buffalo and cow milk.

Cabrini et al. (1983) report on the use of a neutral protease for the production of Crescenza, Caciotta, Italico, and Grano cheeses. They found that the protease digests the α-s_1 fraction of casein but does not attack β-casein. The best result was obtained with Caciotta cheese. The authors generally recommend the use of neutral bacterial proteases for the manufacture of soft cheeses with short ripening times.

Fedrick et al. (1986) studied accelerated ripening with a neutral fungal protease from *A. oryzae*. Rhozyme P 11 (50,000 HbU) [GR] in doses of 0.001–0.1% was used. The results were acceptable with less than 0.05%. Bitterness arose with higher doses, which also produced a faulty texture. At a dose of 0.05%, the TCA-soluble nitrogen (mg tyrosine/g cheese) rises from 3 to 5.5 mg; with a storage time of 6 months, the values are double those of the controls.

Omar and El Zayat (1986) recommended accelerated ripening of Gouda with a system that employs a yeast catalase (0.2 g Maxilact [GB] per 1 liter of milk), in addition to the animal lipase esterase.

5.6.4   Aroma and Flavor Production

References: Ruttloff (1982), Moskowitz and Noelck (1987)

This area includes the manufacture of enzyme-modified cheeses (EMC), the production of cheese aroma and flavor concentrates, and increased butter flavor. In the United Sates, EMCs are made primarily from natural cheeses (Nelson, 1970); the enzymes are added during either cheese production or ripening. Incubation under controlled conditions is required for constant flavor formation. The flavor-forming mechanism aided by these enzymes is not fully understood. However, it is known that amino acids and peptides have an important role in cheese flavor development (Law and Goodenough, 1995). EMCs are used in concentrations of 1–2% as flavor intensifiers and contain, in concentrated form, the most important flavor components of a fully ripened cheese. EMCs are available for cheddar, Swiss, Provolone, Romano, Parmesan, Mozzarella, and Roquefort cheeses to add various intensities of their individual aromas and flavors. They are used for making processed cheese, cheese spreads, and cheese powders and for flavoring baked goods and salad dressings.

Lipases with a hydrolytic index greater than 1 are preferentially used for making EMC and flavor concentrates:

$$\frac{EA}{LA} > 1$$

where EA = tributyrin activity
 LA = olive oil activity.

According to Feldman and Dooley (1977), such lipolytic systems are produced by *Mucor miehei* and can be applied as follows:

1. *Enzyme-Modified Cheeses:* First 1–100 g of lipase is added to 1000 liters of milk dosed with either the starter cultures or rennet. Then, to intensify the blue cheese flavor, 8–15 g of lipase is added to 1000 liters of milk along with the starter cultures.

2. *Modified Butterfat:* Lipase is added at 39°C to butter fat and subsequently incubated for 1–3 days at 22–37°C; the enzyme is then inactivated by briefly heating to 70°C at pH 5.

3. *Modified Butter:* Lipase is added with lactic acid to cream and incubated for approximately 4 h at 37°C. The modified butter is used in baked goods.

4. *Modified Milk Powder:* After 0.001 to 0.01% lipase is added to reconstituted milk having 9–14% dry substance, the mixture is incubated at 4–20°C and subsequently heated to stop the reaction. This modified milk powder is used to enhance the milk flavor of mixed drinks and ice cream.

5.6.4.1   Other Aroma and Flavors   According to Mick (1982), lipolysis of emulsified butterfat with an *Aspergillus* lipase at pH 6.0 produces a Camembert cheese flavor; use of a *Mucor* lipase esterase [RM] yields a sour milk flavor; another *Aspergillus* lipase [RM] generates a freshly baked bread flavor.

Modified Butterfat   For some foods, an increased butter flavor is desired. Among these are baked goods. This flavor is generated by the partial hydrolysis of butterfat with a suitable lipase:

- Add 1 kg of butterfat melted at 40°C; then add 100 g of lecithin + 100 ml of 0.2 M phosphate buffer, pH 6.5.
- Then add 6 g of Lipase RY [AM], dissolved in the above.
- Mix and homogenize for 5 min at 1000 rpm (revolutions per minute).
- Incubate for 1–5 h at 40°C.
- Heat to about 80°C for 10–20 min.
- Cool and package.

Product: Enzyme-modified butterfat.

Blue Cheese Flavor
Reference: Moskowitz (1979)

There are many blue cheese varieties. The best known is Roquefort, but also Gorgonzola, Stilton, Bavaria-Blue, and many others are well known for their characteristic aroma, thanks to the metabolites of a blue fungus, *Penicillium roqueforti*. The most important flavor components are various methyl ketones and the secondary alcohols of C_5–C_{11} compounds. The most significant ketones are 2-heptanone, 2-hexanone, 2-nonanone, and their corresponding alcohols.

   P. roqueforti produces a protease and several lipases. The latter have been obtained by fermentation, and have been isolated and characterized. The specific lipases can be used in cheesemaking and in lipid technology.

   A blue cheese flavor is currently produced by submersed fermentation of *P. roqueforti*. Substrates can be whey, milk, or lipase-modified butterfat (Jolly and Kosikowski, 1975; Ruttloff et al., 1981).

   After sterilizing the medium and cooling to 24–26°C, the culture medium is inoculated with spores or mycelia of *P. roqueforti*. After about 70 h, the medium is heated to 130°C to kill the organism and inactivate the enzymes. The resulting product may have a 20-fold flavor intensity over that of the natural cheese. It is used for making soft cheese, for flavoring salad dressings and baked goods, and for making powdered blue cheese.

5.6.5   Cold Sterilization of Milk with Lysozyme

Lysozyme has been isolated by Jolles and Jolles (1961), from human milk but was found only in small quantities in cow milk. As mentioned in Section 3.1.3.3, lysozyme can lyse the cell walls of certain bacteria. Therefore, it was suggested to add lysozyme as a bactericide to baby food that contains cow milk.

In cheesemaking, contamination by *Clostridium tyrobutyricum* produces gas in Dutch and Swiss cheese. The vegetative cells of this organism are lysed by lysozyme (Lodi et al., 1983). The enzyme can generally be used for cold sterilization of various foodstuffs and beverages (Nagano and Sato, 1974).

5.6.6   Cold Sterilization of Milk with Hydrogen Peroxide and Catalase

In some countries, hydrogen peroxide is used for the cold sterilization of milk. Treatment of milk with hydrogen peroxide has the advantage of destroying harmful organisms but does not affect milk enzymes and useful bacteria. Any excess of hydrogen peroxide is destroyed by catalase treatment.

Hydrogen peroxide is added to milk at a concentration of 0.02% with the temperature maintained at 30°C for 20 min. Subsequently, a diluted catalase solution, adjusted to 20 Keil units (KU), is added (i.e., ~10 g of Catalase L [SOE] per 1000 liters of milk).

5.6.7   Lactose Hydrolysis

Several procedures for hydrolyzing lactose in milk and whey with lactose preparations are available. The enzyme is selected according to the pH of the substrate. Yeast lactase has an approximate pH optimum at 6.5–7.0 for hydrolyzing lactose in milk and sweet whey. For splitting lactose in acid whey, the lactase from *A. oryzae*, is the enzyme of choice (see Section 3.1.3.4).

1. *Milk as Substrate:* Lactose hydrolysis permits the lactose-intolerant consumer to drink milk. Because lactose hydrolysis makes the milk sweeter, the addition of sugar to milk beverages can be reduced (Reimerdes and Gottschick, 1981a, 1981b). Powdered milk with hydrolyzed lactose is currently produced on a technical scale.

2. *Sweet Whey as Substrate:* In the manufacture of ice cream, about 25% of the skim milk powder can be replaced with powdered whey. An increase in the amount of whey may cause the lactose to crystallize in the ice cream, giving it a grainy texture. After cleaving the lactose in powdered whey, the proportion of whey in ice cream can be increased to more than 50% (Gregory, 1982). The increase in sweetening power is equivalent to 410 g of sucrose in 1 kg of powdered whey (750 g of lactose) with 80% of the lactose hydrolyzed.

3. *Fermented Milk Products:* Lactose hydrolysis in cheese production and ripening has already been discussed. For the production of yogurt, acid formation is accelerated by addition of lactose. The products become more creamy and milder in taste.

4. *Whey and Whey Concentrates for Fodder and Feed:* Lactose hydrolysis in whey permits greater whey concentration without the risk of lactose crystallization. Hydrolysis also makes whey more tolerable as a feed component; thus, it is possible to increase the proportion of whey in feed.

5. *Lactose Syrup Production:* A syrup with 65–75% dry substance can be produced from whey or deproteinized whey (permeate from ultrafiltration) after hydrolysis of 80–95% of the lactose. This syrup is extraordinarily well suited for manufacture of baked goods.

5.6.8 Process of Lactose Hydrolysis

Free enzymes, especially yeast lactase, are used primarily in the hydrolysis of milk. Free fungal lactase or, more economically, carrier-bound fungal lactase can be employed for the continuous hydrolysis of sour whey.

1. *Batch Processing of Milk:* Milk is charged with 0.5 g per liter of yeast lactase with 5000 ONPG [*o*-nitrophenyl-*p*-nitranilide (*o*-nitrophenyl glycoside)] units per gram and incubated for 4–5 h at 40°C depending upon the desired degree of hydrolysis. Alternatively, incubation may be conducted at 5–10°C for 18–24 h. The product is then pasteurized at 72°C for 15 s.
2. *Yogurt Production:* A well-known process involves the prehydrolysis of milk with lactase as described above. Günther and Bürger (1982) describe a continuous process in which both the lactases and the starter cultures are added to the milk. Because the pH decreases from 6.5 to 4.6 during yogurt ripening, a mixture of yeast and fungal lactase is added to assure optimal lactase activity over this pH range during the ripening process. Thereby, it is possible to obtain 95% lactose hydrolysis. A yogurt with hydrolyzed lactose has the following advantages:
   a. The incubation time can be reduced.
   b. The consistency is improved.
   c. The product is judged "to have a milder taste."
   d. Syneresis is reduced.
   e. This yogurt can be enjoyed by individuals with lactose intolerance.
3. *Continuous Hydrolysis:* The continuous hydrolysis of lactose with carrier-bound lactase was described in Section 4.4.3.

5.6.9 Potential Use of Other Enzymes for Product Modification in Milk Processing

Glucose oxidase has been proposed as a stabilizing antioxidant for milk products such as mayonnaise, butter, and powdered milk. It seems to be purely a question of cost-effectiveness whether enzymes will replace chemical antioxidants.

Glucose oxidase, which produces glucuronic acid from glucose, can acidify milk in situ during the production of cottage and Mozzarella cheeses (Rand and Hourigan, 1975).

5.6.9.1 *Superoxide Dismutase*   Fox (1982) discussed the use of superoxide dismutase, which reduces oxygen free-radicals. Thus, the lipoxidation of milk can be inhibited significantly. The hydrogen superoxide produced can be removed with catalase or peroxide (Hill, 1979):

$$2\,O_2^- + 2\,H^+ \longrightarrow H_2O_2 + O_2$$

5.6.9.2 Sulfhydryl Oxidase This enzyme is found in cow milk. It can also be produced from *A. niger* ([FR], 1987). It oxidizes sulfhydryl groups to disulfides such as cysteine and cystine, and oxidizes reduced ribonuclease and reduced chymotrypsin. It is different from other oxidases such as xanthine oxidase or glutathione oxidoreductase:

$$2\,RSH + O_2 \longrightarrow R - SS - R + H_2O_2$$

This enzyme has the capacity to remove the cooked taste from ultrahot milk. Swaisgood et al. (1982) have isolated the enzyme from milk and bound the purified enzyme to porous glass or ceramic. A continuous-flow reactor for milk was designed. See Section 5.4.2.5 for an application of this enzyme in breadmaking.

5.6.9.3 Transglutaminase Ikura et al. (1980) have employed the transglutaminase reaction for cross-linking proteins to produce protein gels or to covalently incorporate individual essential amino acids into proteins. In this way the methionine content of casein can be increased. Gel-forming proteins are of general interest to the food processing industry.

REFERENCES

Alkhalaf, W., El Soda, M., Desmazeaud, M. J., and Gripon, J. C., *Microbiol., Aliments, Nutr.* **4,** 111 (1986).

Arbige, M. V., Freund, P. R., Silver, C. S., Zelko, J. T., *Food Technol.* **40,** 91 (1986).

Arbige, M. V., and Neubeck, C. E.,, "Lipolytic enzyme derived from a *Aspergillus* microorganisms having an accelerating effect on cheese flavor development," U.S. Patent 4,636,468 (1987).

Arbige, M. V., and Neubeck, C. E., "Lipolytic enzyme derived from a *Aspergillus* microorganism having an accelerating effect on cheese flavor development," U.S. Patent 4,726,954 (1988).

Brown, R. J., in Nagodawithana, T., and Reed, G., *Enzymes in Food Processing,* 3rd. ed., Academic Press, New York, 1993, p. 347.

Burgess, K., and Shaw, M., in Godfrey, T., and Reichelt, J. (eds.), *Industrial Enzymology,* The Nature Press, Macmillan, London, 1983, p. 260.

Cabrini, A., DiCapua, E., Mucchatti, G., and Neviani, E., *Latte* **8,** 247 (1983).

Cheeseman, G. C., in Birch, G. G., Blakebrough, N., and Parker, K. J. (eds.), *Enzymes and Food Processing,* Applied Science Publishers, London, 1981, p. 195.

Dalgleish, D. G., *J. Dairy Res.* **46,** 653 (1979).

Desmazeaud, M. J., *Ind. Aliment. Agric.* **3,** 195 (1982).

Desmazeaud, M. J., *Laits et Laitiers* **2,** 583 (1985).

Droehse, H. B., and Foltmann, B., *Biochim. Biophys. Acta* **995,** 221 (1989).

El Salam, M. H. A., El Shibiny, S., El Bagoury, E., Ayad, E., and Fahny, N., *J. Dairy Sci.* **45,** 491 (1978).

El Soda, M., and Desmazeaud, M. J., *Milchwiss.* **36,** 140 (1981).

Fedrick, I. A., Aston, J. W., Nottingham, S. M., Dulley, J. R., *New Zeal. J. Dairy Sci. Technol.* **21,** 9 (1986).

Feldman, L. I., and Dooley, J. G., "Lipolytic enzyme flavoring system," U.S. Patent 4,065,580 (1977).

Finnsugar [FR], technical information, 1987.

Fox, P. F., in Dupuy, P. (ed.), *Utilisation des enzymes en technologie alimentaire,* p. 135. *Internatl. Symp.,* May 5–7, Techn. Doc. Lavoisier, Paris, 1982.

Fox, P. F., *Biotechnol. Appl. Biochem.* **10,** 522 (1988).

Grappin, R., Rank, T. C., and Olsen, N. F., *J. Dairy Sci.* **68,** 531 (1985).

GRAS, *Title 21 United States Code of Federal Regulations,* 170.30 (f) (2), 1985.

Gregory, K. W., in Dupuy, P. (ed.), *Utilisation des enzymes en technologie alimentaire, Internatl. Symp.,* May 5–7, Techn. Doc. Lavoisier, Paris, 1982, p. 249.

Günther, E., and Bürger, E., in Dupuy, P. (ed.), *Utilisation des enzymes en technologie alimentaire, Internatl. Symp.,* May 5–7, Techn. Doc. Lavoisier, Paris, 1982, p. 243.

Hill, R. D., *CSIRO Food Res. Quart.* **39,** 33 (1979).

Ikura, K., Kometani, T., Sasaki, R., Yoshikawa, M., and Chiba, H., *J. Agric. Biol. Chem.* **44,** 1567 (1980).

Jolles, P., and Jolles, J., *Nature* **192,** 1187 (1961).

Jolly, R. C., and Kosikowski, F. V., *J. Dairy Sci.* **58,** 846 (1975).

Katkocin, D. M., *Food Engineering,* 85 (May 1984).

Kirby, C. J., Brooker, B. E., and Law, B. A., *Internatl. J. Food Sci. Technol.* **22,** 355 (1987).

Law, B. A., and Goodenough, P. W., in Tucker, G. A., and Woods, L. F. J. (eds.), *Enzymes in Food Processing,* 1st ed., AVI, New York, 1991, p. 98.

Law, B. A., and Goodenough, P. W., in Tucker, G. A., and Woods, L. F. J., (eds.), *Enzymes in Food Processing,* 2nd ed., Blackie Academic & Professional (imprint of Chapman & Hall), New York, 1995, p. 114.

Law, B. A., and Kirby, C., *North Eur. Food Dairy J.* **53,** 104 (1987).

Law, B. A., and Wigmore, A. S., *J. Dairy Res.* **50,** 519 (1983).

Lodi, R., Oggioni, F., Vezzoni, A. M., and Carini, S., *Latte* **19,** 41 (1983).

Magee, E. L., and Olson, N. F., *J. Dairy Sci.* **64,** 600 (1981).

Magee, E. L., Olson, N. F., and Lindsay, R. C., *J. Dairy Sci.* **64,** 616 (1981).

Martin, P., Collin, J. C., Garnot, P., Dumas, P. R., and Mocquot, G., *J. Dairy Res.* **48,** 447 (1981).

Mick, S., *Über das Aroma von Sauerrahmbutter und dessen Modifizierung durch Anwendung technischer Lipasen,* Diplom thesis, Technical University München-Weihenstephan, 1982.

Moskowitz, G. J., in Peppler, H. J., and Perlman, D. (eds.), *Microbial Technology,* 2nd ed., Vol. II, Academic Press, New York, 1979, p. 201.

Moskowitz, G. J., and Noelck, S. S., *J. Dairy Sci.* **70,** 1761 (1987).

Nagano, Y., and Sato, I., "Process for sterilizing foodstuffs and beverages," U.S. Patent 3,852,476 (1974).

National Research Council/Food and Nutrition Board, *Food Chemicals Codex,* National Academy Press, Washington, DC, 1981.

Nelson, J. H., *J. Agric. Food Chem.* **18,** 567 (1970).

Ney, K. H., *Gordian* **1/2,** 16 (1987).

Olson, N. F., personal communication, 1987.

Omar, M. M., and El Zayat, A. I., *Dtsch. Lebensmittelrdsch.* **82,** 152 (1986).

Puhan, A., and Morgenthaler, M., *Dtsch. Molkerei-Zeitg.* **103,** 141 (1982).

Rand, A. G., and Hourigan, J. A., *J. Dairy Sci.* **58,** 1144 (1975).

Reimerdes, E. H., and Gottschick, W., *Lebensmitteltechnik* **9,** 402 (1981a).

Reimerdes, E. H., and Gottschick, W., *Lebensmitteltechnick* **10,** 466 (1981b).

Richardson, G. H., in Reed, G. (ed.), *Enzymes in Food Processing,* 2nd ed., Academic Press, New York, 1975, p. 362.

Ridha, C., *J. Dairy Sci.* **12,** 63 (1984).

Ruttloff, H., Quehl, A., Leuchtenberger, A., Rothe, M., and Engst, W., poster, 2nd Aroma Symp. Socialist Countries, Rydzyna, Poland, 1981.

Ruttloff, H., *Die Nahrung* **26,** 575 (1982).

Swaisgood, H. E., Sliwkowski, M. X., Skudder, P. J., and Janolino, V. G., in Dupuy, P. (ed.), *Utilisation des enzymes en technologie alimentaire, Internatl. Symp.,* May 5–7, Techn. Doc. Lavoisier, Paris, 1982, p. 229.

Teuber, M., *Food Reviews Internatl.* **9,** 389 (1993).

Wargel, R. J., Greiner, S. P., and Hettinga, D. H., "Verfahren zur Herstellung von Käse," German Discl. 29 42 411 (1980).

Wegstein, J., and Heinsohn, H., in Nagodawithana, T., and Reed, G. (eds.), *Enzymes in Food Processing,* 3rd ed., Academic Press, New York, 1993, p. 71.

5.7 ENZYMES IN THE MEAT INDUSTRY

References: Etherington (1991), Etherington and Bardsley (1995)

Sigmund Schwimmer (1981) titled his detailed discussion of the biochemical processes that follow the death of an animal, the termination of its blood circulation and cessation of oxygen supply, "muscle to meat." He describes the effect of enzymatic as well as nonenzymatic reactions on the texture, color, and taste of meat. Not all factors determining meat texture are currently known.

The most important properties that the meat consumer demands are juiciness, good chewability without loss of firm texture, color, and taste. It has been reported (Rhodes et al., 1955) that 75% of all American housewives are disappointed when shopping for steak quality, especially steak tenderness.

Native meat enzymes, the cathepsins, play a special role in tenderizing meat by controlled aging.

Many ethnic populations have long used meat tenderization as a common kitchen practice. Wrapping meat in papain leaves or dipping or immersing in papaya or pineapple juice is a standard practice in tropical countries.

Technical meat tenderizing enzymes have been used since about 1940. Papain in different preparations is the most frequently used enzyme. Health considerations are not an issue, because the papaya fruit itself is a food. Incomplete data are available about the tenderizing mechanism, but many of the factors that influence this process are known. In addition to pH, temperature, and the enzyme quantity used, the delivery of the proteolytic enzyme into the tissues and its distribution therein is important. For this, several methods exist.

5.7.1 Surface Treatment

Usually, meat slabs are sprayed with a liquid papain solution, or are dipped into enzyme solutions. Distribution within the meat is assured by needle piercing; in the home this is done by piercing the meat with a fork. Generally, 5–10 ppm of a commercial papain preparation is used.

5.7.2 Injection Treatment

Two methods are employed: antemortem and postmortem injections. In commercial meat tenderizing, the latter is performed with a multiinjection system in which the enzyme solution is injected under pressure into the meat slabs. This method is not used in Germany.

5.7.2.1 Antemortem Injection

1. A papain solution is injected into the animal before slaughter. The incubation period prior to slaughter should be equal to 60% of the time required for one blood circulation cycle (Hogan et al., 1964).
2. A reversibly inactivated papain solution is injected prior to slaughter. In this case, papain is inactivated with hydrogen peroxide–catalase, which causes oxidation of the free sulfhydryl groups. Reactivating the enzyme is accomplished during the initial phase of heating the meat (Hogan, 1966).

5.7.3 Meat Tenderizers

There are many patents and publications describing enzymes that are used in meat tenderization. In addition to papain, bromelain, ficin, pancreatic proteases, pancreatic elastase, collagenase, and microbial proteases are described. Papain preparations are the most widely used.

Various compounds are added to papain for spraying or dusting the preparations onto meat. Most preparations contain sodium chloride and sodium glutamate. In addition, polyphosphate, protein hydrolysate, glycerin, and oil emulsions may be added.

The appropriate enzyme quantity required when dusting the preparation is 0.1 g of a technical papain preparation (having an activity of 100-g tryosine units per gram of protein) for each kilogram of meat. For meat injection, approximately 2–3% (v/wt) of 0.005% papain solution is required.

REFERENCES

Etherington, D. J., "Enzymes in the meat industry," in Tucker, G. A., Woods, L. F. J. (eds.), *Enzymes in Food Processing,* 1st ed., AVI, New York, 1991, pp. 128–160.

Etherington, D. J., and Bardsley, R. G., in Tucker, G. A., and Woods, L. F. J. (eds.), *Enzymes in Food Processing,* 2nd ed., Blackie Academic & Professional (imprint of Chapman & Hall), New York, 1995, p. 144.

Hogan, J. M., "Meat tenderizing composition," U.S. Patent 3,235,468 (1966).

Hogan, J. M., Orono, M., and Bernholdt, H. F., "Method of tenderizing meat," U.S. Patent 3,163,540 (1964).

Rhodes, V. J., Kiehl, E. R., and Brady, D. E., *Res. Bull.* No. 583, Univ. Missouri, College of Agriculture, Columbia, MO, 1955.

Schwimmer, A., *Source Book of Food Enzymology,* AVI, Westport, CT, 1981, p. 481.

5.8 PROTEOLYSIS

References: Adler-Nissen (1986, 1993), Löffler (1986), Schwimmer (1981)

Numerous applications of proteolysis exist in different areas of food processing. In brewing beer, soluble nitrogen can be increased by hydrolyzing proteins with papain or neutral bacterial protease. The final beer product is also stabilized with proteases to avoid cold precipitation of proteinaceous substances. Wine clarification and wine stabilization also involve protein degradation by microbial proteases. In baking, especially in cookie production, the partial degradation of the gluten protein yields improved dough properties, which results in a higher quality product.

Other examples of proteolysis in commercial applications are given in the chapters on tanning (Section 5.9) and laundry detergents (Section 5.10).

The greatest quantities of protease used in food production are those employed to curdle milk during cheesemaking. The limited proteolysis yields a precipitate: the formation of an insoluble cheese curd. Such reactions are also known for other proteins and enzymes. Proteolysis of albumin by subtilisin leads to the formation of an insoluble peptide called *plaque albumin* (Güntelberg and Ottesen, 1954). Producing cheeselike products from soybean such as Tofu, Miso, Juba, Natto, and Temphe also involves similar reactions. In Japan, 350,000 t of soybeans are used annually for manufacturing these products. Treatment includes the partial hydrolysis and denaturation of the soybean protein. The protein is subsequently precipitated by heat treatment or by the addition of calcium. In some cases, additional fermentation is employed to develop the particular taste and characteristic texture of the product. Tofu should possess a soft, smooth yet firm texture. Its elastic gel con-

TABLE 5.25   Proteolytic Enzymes

Enzyme	pH Range	Temperature Maximum (°C)	Specificity[a]
Pepsin	2–4	40–45	5
Acid fungal protease (*A. saitoi*)	2.4–4	45	9
Papain	4–8	60–65	9
Neutral fungal protease (*A. niger*)	5–7.5	50	9
Neutral bacterial protease	6–8	55–60	6
Pancreatic protease	6–9	45–50	5–11
Alkaline bacterial protease	7.5–9.5	60	7
High-alkaline bacterial protease	8–12	55	9

[a]Specificity is defined as the number of bonds hydrolyzed during degradation of the B-chain of insulin. The data are based on average reaction times of 1–2 h. With increased reaction time, additional bonds are hydrolyzed. The number of peptide bonds hydrolyzed also strongly depends, of course, on the enzyme formulation of the technical-grade preparation. As mentioned earlier, it should be understood that the specificity "5–11" given for the pancreatic proteases indicates that they differ significantly in their composition. The data obtained on insulin B-chain hydrolysis do not automatically apply to other substrates in that they differ in chemical composition and susceptibility to proteolytic attack.

tains about 6% protein, 3.5% lipids, 1.9% carbohydrate, and 88% water (Pomeranz, 1985).

5.8.1   Enzymes for Protein Hydrolysis

For protein hydrolysis, a number of enzymes are presently available that cleave proteins with various degrees of specificity over a pH range of 2–12 (see Section 3.2). For quick reference and review, the pertinent proteases are listed in Table 5.25.

5.8.2   Substrates

The natural substrates discussed here are the plant and animal proteins that are used in nutrition. Protein content, as defined by food chemists, is based on the amount of nitrogen as the characteristic constituent of protein. The protein in a given proteinaceous food is determined by multiplying its Kjeldahl nitrogen content by a specific factor. The factors used are 6.38 for milk protein, 5.7 for wheat gluten, 5.7 for soybean protein, and 5.55 for gelatin. Generally, a factor of 6.25 is used in calculations of protein content.

The degree of hydrolysis (DH) is an important unit of measure in proteolysis. This unit is defined as the ratio between the bonds cleaved and the total peptide bonds present (Adler-Nissen, 1976). The number of peptide bonds that are cleaved by proteolysis is called the "hydrolysis equivalent" (h). It is determined by analyzing the number of amino groups by formaldehyde titration (Sörensen, 1908) or ninhydrin titration (Moore and Stein, 1948):

$$DH = \frac{h}{h_{tot}} \times 100\%$$

The value for h_{tot} is calculated from the amino acid composition of the protein. Adler-Nissen (1976) has provided a detailed discussion on the calculating the degree of hydrolysis (DH) for soluble and insoluble proteins.

Soybean hydrolysis can be calculated according to the following formula:

$$DH = \frac{NB}{MP} \times \frac{1}{h_{tot}} \times B$$

where NB = normality of base
  MP = total protein mass (kg)
   B = NaOH consumption (liters)
   $\frac{1}{h_{tot}}$ = total number of peptide bonds in a given protein per kilogram (for soy proteins, the factor is 7.75)

5.8.3 Objectives of Protein Hydrolysis

Two major problem areas may be solved by enzymatic hydrolysis: (1) insoluble proteins can be solubilized, and (2) the functional properties of proteins can be changed.

There are various reasons for solubilizing proteins. These include utilizing meat wastes in slaughterhouses or fish protein wastes from fish processing, processing skin remnants from tanneries into feed products, or removing the hemoglobin when processing blood solids into a protein concentrate.

On the other hand, the beverage industry has an interest in protein-enriched beverages. Because this involves mainly fruit juices, the proteins must be soluble in the acid pH range and possess a good organoleptic quality. The soybean hydrolysate ISSPH, soluble at its isoelectric point, has attained great importance in the United States.

In addition to total proteolysis, a new development arose early in the 1970s involving the enzymatic modification of proteins. A pepsin-modified soybean protein is described in an U.S. patent by Gunther (1974) that has improved whippability. Since then, there have been many examples of improved protein properties based on such partial proteolysis. A few examples include wettability, hydratability, foaming ability, emulsifying properties, and gelling. Soybean protein and milk or whey protein usually serve as substrates.

5.8.4 Bitterness

Protein hydrolysates that are used in the manufacture of food must have acceptable taste qualities. A meatlike or neutral taste is preferred. Unfortunately, increased hydrolysis of most proteins is associated with significant bitterness. This is particularly true for casein hydrolysates, but also occurs with soy, peanut, and potato protein digests. However, egg and gluten protein hydrolysates are less affected. Collagen hydrolysates are not bitter. According to Ney (1971), bitterness is associated with certain peptides, especially hydrophobic peptides that are generated during proteolysis.

Hydrophobicity is measured by the Q value, which is defined as the free energy required for transferring the amino acid side chains from ethanol to water. The Q value is expressed as calories per mole of amino acid. Ney found that all peptides with a Q value above 1400 cal/mol are bitter. Guigoz and Solms (1976) have compared the Q values of 206 peptides for their bitterness. They confirmed Ney's rule with some exceptions. On gelatin hydrolysis, for example, no bitter peptides arise, although the Q values for the cleaved peptides are greater than 1400 cal/mol. Sixty-one bitter peptides have been isolated from casein hydrolysates, all having chain lengths of 2–15 amino acids except for one 24–amino acid peptide. It was found that bitterness depended more on the substrates than on the specificity of the protease.

To date, no protease has exhibited the ability to solve this problem. The relationship between the formation of bitterness and the degree of hydrolysis (DH) (see Fig. 5.68) could only be slightly modified by selectively combining certain proteases. In other words, the bitter point, that is, the point at which bitterness first occurs, could be shifted to a somewhat greater degree of hydrolysis.

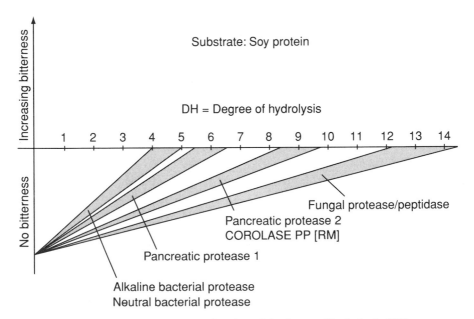

Figure 5.68   Bitterness as a function of the degree of hydrolysis (DH).

Because formation of bitterness depends largely on the reaction conditions during enzymatic hydrolysis, the pH, temperature, substrate, enzyme concentrations, and reaction time must be precisely monitored and kept constant.

5.8.5 Methods for Debittering Protein Hydrolysates

Many studies have been undertaken to avoid the formation of bitterness during hydrolysis or alternatively, to subsequently remove the bitterness formed. This is important because the use of proteases for hydrolysis or for protein modification depends on the fact that the taste of such products must be acceptable. The following methods were developed toward this end:

1. Using peptidases to avoid the formation of bitter peptides
2. Employing the plastein reaction to form larger nonbitter peptides
3. Selectively removing bitter peptides by chromatography
4. Treating the hydrolysate with activated charcoal as a hydrophobic absorbent for the hydrophobic peptides (Murray and Baker, 1952)
5. Masking the bitter taste

5.8.5.1 Use of Peptidases A noticeable improvement can be expected from combinations of proteases with exopeptidases such as carboxypeptidase or aminopeptidase.

Using carboxypeptidase from wheat (Umetsu et al., 1983) or pancreatin that contains both carboxypeptidases and proteases, bitterness can be removed from hydrolysates. Clegg and McMillan (1974) produced casein hydrolysates with papain and subsequently incubated the hydrolysate with a homogenate from porcine kidney for 24 h at 40°C to remove the bitterness.

It was observed that when fungal proteases were used, the tendency to produce bitterness was reduced (Drepper et al., 1981). This may be due to the fact that most fungal proteases also have some peptidase activity.

Plainer and Sprössler (1989) significantly reduced bitterness in enzymatic hydrolysates by treatment with technical peptidase preparations, obtaining hydrolysates with no bitterness having a DH of 4–5% with casein and a DH of 10–13% with whey protein. For gluten, a DH of 30% was obtained without bitterness. (See also Fig. 5.69.)

Debittering can be achieved with a partial (~50%) hydrolysis of sodium caseinate or soya with preparations containing endoproteases, exopeptidases, and specifically leucin aminopeptidase (Flavourzyme [NO]). An alternative for solving the bitterness problem was pursued in the enzyme laboratory of Röhm GmbH, Darmstadt, where hydrolysis tests with carrier-bound fungal protease (Corolase PN [RM]) were undertaken. The improvement in reducing bitterness was considerable; yet, a continuous hydrolysis process with casein failed because of unmanageable reactor contamination.

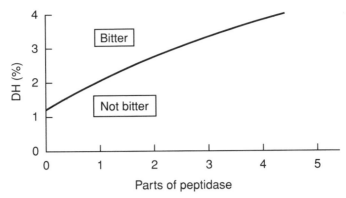

Figure 5.69   Shifting the onset of bitterness to a higher level of hydrolysis by increasing peptidase activity during casein hydrolysis (Plainer and Sprössler, 1989); 1 part Corolase 7092 RM, fungal protease from *A. niger* and 0–5 parts Corolase 7093 RM, technical exopeptidae preparation from *A. niger.*

5.8.5.2   *Debittering with the Plastein Reaction*

The plastein reaction uses reversible protease activity for the synthesis of new proteins from the protein hydrolysate pool. Satterlee (1981) discussed the various theories concerning the mechanism of this protein reaction. The protein is initially cleaved into a mixture of low-molecular-weight peptides. A reverse reaction, resynthesis, mediated by the proteases, sets in when a critically high peptide concentration is reached.

Fujimaki et al. (1971, 1974) first used the plastein reaction to remove bitterness in protein hydrolysates. Because of this characteristic, the reaction has received much attention. The mechanism of the plastein reaction was first believed to be a resynthesis of protein. However, subsequent theoretical and empirical research have disproved this hypothesis. The most likely mechanism is a one-step transpeptidation reaction that produces new hydrophobic peptides (Yamashita et al., 1975). It permits the insertion of essential amino acids into a protein and is, therefore, a means of protein modification. One can use it to generate nutritionally balanced proteins. Long-chain (e.g., C_{12}) alkyl esters can be incorporated into food proteins. The new proteins generated by this transpeptidation reaction were found to have exceptionally good emulsifying properties and are therefore of interest to the food industry (Watanabe et al., 1982, Arai et al., 1986).

5.8.5.3   *Masking Bitterness*

Another means of reducing bitterness is masking it by adding other protein hydrolysates, primarily gelatin hydrolysates, or by cohydrolysis with carbohydrates. Addition of hydrocolloids such as guar flour, pectin, or maltodextrin also reduces the bitter taste (Sprössler and Uhlig, 1981). The same result can be obtained by the addition of organic acids such as malic, lactic, or phosphoric acid. However, these procedures have currently been only moderately successful.

5.8.5.4 Removing Bitter Peptides by Adsorption

Removing bitter peptides by chromatographic methods is possible but is still not economically feasible. A somewhat more practical procedure involves the adsorption of hydrophobic bitter peptides to activated charcoal. The extraction of the bitter peptides with an azeotropic *n*-butanol/water mixture has been successful in a number of cases.

5.8.6 Hydrolysis of Soy Protein

5.8.6.1 Isoelectrically Soluble Protein Hydrolysate

The raw material for production can be a soy protein isolate or a concentrated extract from soy meal or soy flour (with or without fat) at pH 4 (Olson and Adler-Nissem, 1979). Cellulases can be added to improve such extracts:

- Mix water at 55°C, soy protein, or concentrate with 8% protein at pH 8 and 0.6 liter of 2% Alkalase [NO].
- Hydrolysis—stirring and keeping the pH constant with a pH-stat at 8.0 for 1.5–2 h until the DH reaches 10%.
- Enzyme activation—adjust pH to 4.0–4.2; keep at 50°C for 0.5 h.
- Centrifugation.
- Filtration.
- Activated-charcoal treatment; at 50°C for 30 min with ~0.1% activated charcoal.
- Filtration, concentration, and subsequent spray drying (Olson and Adler-Nissem, 1981).

The isoelectric solubility increases from 5% in the starting material to approximately 42% in the hydrolysate. The capacity to emulsify increases from 100 ml/g to 280 ml/g and the whippability from 20 to 200%. A similar process with soy protein and a pancreatic protease having a high peptidase activity (Corolase PP [RM]) yields a hydrolysate that has no bitter taste with a DH > 20. The hydrolysis reaction is performed with 0.05% Corolase PP showing 225,000 LVU/g at pH 8.5 and 50°C. Subsequently the hydrolysate is adjusted to pH 5.0 with malic acid. The reaction solubilizes about 75% of the protein at pH 5.

For similar applications, mixtures of proteases can be recommended or alternatively, a stepwise hydrolysis with selected enzymes (Feldman et al., 1974). For example, the alkaline bacterial protease can be used to initially solubilize the protein; polypeptides can then be split into peptides with a neutral bacterial protease. Similar processes operating at a pH range of 4.5–6 with papain, bromelain, or acid fungal protease have also been described.

5.8.6.2 Modifying the Functional Properties of Proteins

Attempts have been made since the early 1970s to modify the properties of soy proteins with enzymes. Initial studies with a fungal protease from *A. oryzae* that modifies isolated soy protein demonstrated an increase in emulsifying capacity with, however, a concomitant reduction in the stability of the emulsion (Puski, 1975). Similarly, the aerating or whippability was increased, but again with a simultaneous reduction in the foam stability.

Adler-Nissen and Olson (1979) produced a soy protein hydrolysate with a combination of alkaline and neutral bacterial proteases (pH range 7.0–8.0). The results showed that with a low DH, the emulsifying capacity and foam volume increase, but both decrease with increasing DH. After systematic studies of this hydrolysis, the optima for both the increase in emulsifying capacity and foam stability was found to lie at a DH of 5.

Functional protein hydrolysates with improved foam stability can be obtained when the soy protein isolate is purified by ultrafiltration. If, in addition, short-chain peptides are removed from the hydrolysate by ultrafiltration, foam stability can be improved.

Studies on continuous hydrolysis of proteins in membrane reactors have been summarized by Cheryan and Deeslie (1980). According to this review, it is possible to produce protein hydrolysates with a continuous process. However, with this system, considerable problems are encountered with substrate contamination in that the substrate presents an ideal nutrient medium for contaminating organisms.

5.8.7   Hydrolysis of Blood Protein

Most of the fresh blood collected during slaughter is processed by centrifugation, yielding blood plasma and precipitated blood solids. Plasma has superior emulsifying properties, and after processing, it is used in the food industry.

The cell fraction containing 70–75% of the blood protein finds limited use in the food industry because of its color and astringent taste. This fraction, often referred to as *blood solids* (see Fig. 5.70), is converted into a powder by spray drying that can be used in animal feed. An enzymatic process removing the heme, and thus the color, was developed by Adler-Nissen (1978).

After hydrolysis to 18% DH with alkaline bacterial protease at pH 8–9, the hydrolysate is acidified to pH 4.8–5.0. Heme is precipitated, and the enzyme is inactivated. Filtration follows, and bitterness is removed with activated charcoal. The spray-dried product is nearly colorless and can be used by the food industry for producing food concentrates.

Drepper et al. (1981) described a similar process. The enzyme used was an acid fungal protease (Corolase PS [RM]). The hydrolysis was performed at pH 4.5–5.0 and 50°C for ~6 h. According to the authors, little bitterness is encountered, even without activated-charcoal treatment.

5.8.8   Hydrolysis of Fish Protein

Fish protein hydrolysates have existed, in one form or another, for over 2000 years. The recipe used for a Roman fish sauce is quite simple: "Large and small fish are put into a crock and thoroughly salted. The fish are incubated at room temperature until they have completely disintegrated. The remaining liquid or that which can be filtered off, is the fish sauce."

Today this method would be called *autolysis,* a process for which fish is particularly well suited because the gastrointestinal tracts of fish are rich in proteolytic en-

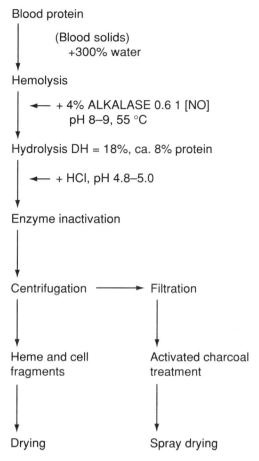

Blood protein

(Blood solids)
+300% water

Hemolysis

+ 4% ALKALASE 0.6 1 [NO]
pH 8–9, 55 °C

Hydrolysis DH = 18%, ca. 8% protein

+ HCl, pH 4.8–5.0

Enzyme inactivation

Centrifugation ⟶ Filtration

Heme and cell   Activated charcoal
fragments   treatment

Drying   Spray drying

Figure 5.70   Processing blood solids with alkaline bacterial protease.

zymes. Fifty years ago, Scandinavian researchers found that fish can be conserved by acid hydrolysis and salt. This process was called "fish silage."

In East Asia, fish sauces are of great importance. They are produced by natural lactobacilli fermentation or other microbes. It was found that this process could be enhanced by neutral bacterial protease or bromelain.

In the early 1960s, peptones from papain-hydrolyzed fish protein were developed as nutrients for fermentation. Hydrolysis was performed at pH 7.0 and 40°C (Sripathy et al., 1962). Fish hydrolysates were used in chicken feed and for producing synthetic calf milk.

Hale (1969) examined 20 technical protease preparations for their ability to hydrolyze fish protein. He found Pronase to be very effective but too expensive; however, a less active neutral bacterial protease is more economical. Papain, bromelain, pepsin, and pancreatin can also hydrolyze whole fish (Spinelli and Koury, 1974). The hydrolysates can be extracted with polar solvents to remove

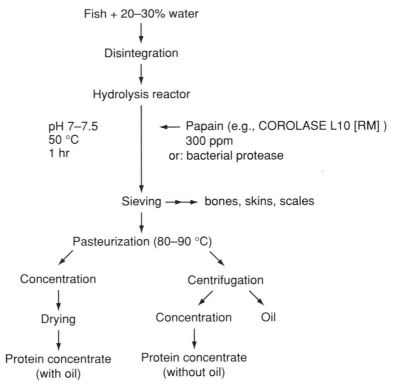

Figure 5.71 Converting fish or fish wastes to protein concentrates.

undesirable flavors (Fig. 5.71). Today, papain and alkaline bacterial proteases are employed for solubilizing fish wastes and to lower the viscosity of expressed fish fluids (stick water) in fodder manufacture.

5.8.9 Hydrolysis of Scrap Meat

In the meat processing industry, scrap meat remains attached to connective tissues and bones. O'Meara and Munro (1984) describe a process by which such meat residues can be solubilized. Alkaline bacterial proteases can solubilize 94% of the meat protein at pH 8.5 (with a pH-stat) and 55–60°C. Generally, meat hydrolysates should not be bitter. When bitterness occurs, cohydrolysis with gelatin can mitigate this effect.

5.8.10 Hydrolysis of Milk Proteins

Because of their significance in nutrition, milk proteins are among the most frequently studied proteins. The degradation of milk proteins, especially casein during cheesemaking, has been the subject of many scientific and technical investigations;

it was therefore logical to use the hydrolyzed milk proteins for other applications. A drawback is the characteristic bitterness of hydrolyzed milk proteins.

Studies by Clegg and McMillan (1974) seeking to develop a casein hydrolysate for dietary purposes were briefly mentioned. Casein was hydrolyzed at pH 6.2–6.3 and 40°C for 18 h. To suppress microbial growth, chloroform was added, a procedure totally out of favor in today's casein processing. The pH was adjusted to 7.8, then porcine kidney homogenate was added and incubated for another 24 h at 40°C; 50% of the product was made up of short-chain peptides and the other 50% consisted of amino acids. The product was not bitter and was successful in clinical trials. Technical peptidase preparations improve this procedure by providing a more rapid hydrolysis.

Nonbitter products for dietary purposes can be obtained from lactalbumin hydrolysis with a protease from *A. oryzae*. Generally, dairy products with the exception of casein have little tendency to become bitter. Trypsin and pancreatic protease appear to be best suited for whey protein hydrolysis.

5.8.11 Hydrolysis of Collagen and Gelatin

Many studies employing proteases were undertaken to shorten the time to manufacture gelatin, which usually requires several weeks. Many problems surfaced, including the difficulty in inactivating the proteases; it is most important that the hydrolytic process not proceed to the point of lowering the gelatin's molecular weight.

Gelatin can easily be degraded by various proteases. Pancreatic or neutral bacterial proteases readily degrade gelatin. Gelatin hydrolysates are not bitter and can, as mentioned earlier, be used in masking the bitterness in other acid fungal protease hydrolysates as described by Monsheimer and Pfleiderer (1981). Collagen hydrolysates are used today primarily in the cosmetic industry. The in-vitro hydrolysis of collagen is not restricted to a collagenase able to degrade native collagen; an alkaline or highly alkaline bacterial protease can hydrolyze collagen successfully at pH 9.0–10.

Some years ago, gelatin hydrolysates were promoted as special dietary weight-loss products. However, because gelatin does not contain the necessary amino acid composition essential for adequate nutrition, several deaths were reported as a consequence of chronic protein deficiency. Adler-Nissen (1986) reported that a gelatin hydrolysate per se presents no danger of malnutrition; but the problems reported are actually caused by unreasonable and extreme weight-loss programs.

5.8.12 Hydrolysis of Keratin

A number of studies have been conducted that examined keratin hydrolysates as a feed supplement or as a raw material for the production of cysteine. The cosmetics industry has a special interest in "soluble" keratin.

Because of the dense structure of keratin (i.e., feathers) and the great number of disulfide bonds, the raw material must be finely milled and pretreated with chemical reductants such as Na-sulfide, Na-hyposulfite, thioglycolic acid, or mercaptoethanol. After reduction of the disulfide bridges, keratin can be hydrolyzed with an alkaline or highly alkaline bacterial protease. For processing keratin wastes such as those generated by dewooling (see Section 5.9.4.5), a technical keratinase preparation has been developed from *Streptomyces fradiae*. However, in practice this enzyme presents no real advantage over the treatment of keratin with an alkaline bacterial protease subsequent to treatment with a reductant.

Linderström-Lang (1929) published an explanation for the ability of the wool moth to digest wool. Wool, treated with an enzyme preparation isolated from these moths will be digested in vitro only when the sample is pretreated with reducing agents. It was concluded that the moth's digestive tract must produce fluids with a strong reductive capacity. Lindström-Lang's photomicrographs clearly show how the wool fibers disappear in a short segment of the moth's digestive tract.

REFERENCES

Adler-Nissen, J., in Nagodawithana, T., and Reed, G. (eds.), *Enzymes in Food Processing,* 3rd ed., Academic Press, Inc., New York, 1993, p. 159.

Adler-Nissen, J., *J. Agric. Food Chem.* **24,** 1090 (1976).

Adler-Nissen, J., *Annales de la nutrition et de l'alimentation* **32,** (1978) 205.

Adler-Nissen, J., *Enzymic Hydrolysis of Food Proteins,* Elsevier, Amsterdam, 1986, p. 100.

Adler-Nissen, J., and Olson, H. S., *ACS Symp. Ser.* **92,** 125 (1979).

Arai, S., Watanabe, M., and Hirao, N.: in: Feeney, R.E. and Whitaker, J.R. (eds.): *Protein Tailoring and Reagents for Food and Medical Uses,* Marcel Dekker, New York, 1986.

Cheryan, M., and Deeslie, W. D., in Cooper, A. R. (ed.), *Ultrafiltration Membranes and Applications,* Plenum Press, New York, 1980, p. 591.

Clegg, K. M., and McMillan, A. D., *J. Food Technol.* **9,** 21 (1974).

Drepper, G., Drepper, K., Ludwig-Busch, H., *Fleischwirtschaft* **61,** 1393 (1981).

Feldman, J. R., Haas, G. J., Lugay, J. C., and Wiener, J., "Process for bland, soluble protein," U.S. Patent 3,857,966 (1974).

Fujimaki, M., Kato, H., Arai, S., and Yamashita, M., *J. Appl. Bacteriol.* **34,** 119 (1971).

Fujimaki, M., Kato, H., Arai, S., and Yamashita, M., Process for producing plastein, U.S. Patent 3,803,327 (1974).

Guigoz, Y., and Solms, J., *Chem. Senses Flavor* **2,** 71 (1976).

Güntelberg, A. V., and Ottesen, M., *Compt. Rend. Trav. Lab. Carlsberg ser. Chim.* **29,** 36 (1954).

Gunther, R. C., Vegetable aerating proteins, U.S. Patent 3,814,816 (1974).

Hale, M. B., *Food Technol.* **23,** 107 (1969).

Linderström-Lang, K., *Hoppe-Seylar's Ztschr. Physiol. Chem.* **182,** 151 (1929).

Löffler, A., *Food Technol.* **40,** 63 (1986).

Monsheimer, R., and Pfleiderer, E., Method for dissolving collagen-containing tissues, U.S. Patent 4,293,647 (1981).

Moore, S., and Stein, W. H., *J. Biol. Chem.* **176,** 367 (1948).

Murray, T. K., and Baker, B. E., *J. Food Agric.* **3,** 470 (1952).

Ney, K. H., *Z. Lebensm.-Untersuch. Forschg.* **149,** 321 (1971).

O'Meara, G. M., and Munro, P. A., *Enz. Microb. Technol.* **6,** 181 (1984).

Olson, H. S., and Adler-Nissen, J., *Process. Biochem.* **14,** 6 (1979).

Olson, H. S., and Adler-Nissen, J., *ACS Symp. Ser.* **154,** 133 (1981).

Plainer, H., and Sprössler, B., Poster, *Internatl. Conf. Biotechnol. and Food,* Univ. Stuttgart-Hohenheim, Feb. 20–24, 1989.

Pomeranz, Y., *Functional Properties of Food Components,* Academic Press, New York, 1985, p. 443.

Puski, G., *Cereal Chem.* **52,** 655 (1975).

Satterlee, L. D., *Food Technol* **35,** 53 (1981).

Schwimmer, S., *Source Book of Food Enzymology,* AVI, Westport, CT, 1981, p. 459.

Sörensen, S. P. L., *Biochem. Ztschr.* **7,** 45 (1908).

Spinelli, J., and Koury, B. J., Preparation of functional fish protein concentrates and isolates, U.S. Patent 3,826,848 (1974).

Sprössler, B., and Uhlig, H., Verfahren zum Abbau von Proteinen mit Proteinasen, *German Discl.* 3,003,679 (1981).

Sripathy, N. V, Sen, D. P., Lahiry, N. L., Sreenivasan, A., and Subrahamanyan, V., *Food Technol.* **16,** 141 (1962).

Umetsu, H., Matsuoka, H., and Ichishima, E., *J. Agric. Food Chem.* **31,** 50 (1983).

Watanabe, M., Fujii, N., and Arai, S., *Agric. Biol. Chem.* **46,** 1587 (1982).

Yamashita, M., Arai, S., Kokubo, S., Aso, K., and Fujimaki, K., *J. Agric. Food Chem.* **23,** 27 (1975).

5.9 ENZYMES IN THE TANNING INDUSTRY

References: Pfleiderer (1985); Herfeld (1976), *Magazine für die Lederindustrie* [Magazine for the Leather Industry (Röhm GmbH)]; Alexander (1988), Taeger (1988)

5.9.1 Introduction and Brief Overview

Since prehistory humans have obtained part of their nutrition from animals and processed the animal skins into products for their daily needs. The skins served as clothing and footwear, as a cover against rain and cold, and as protection against injury. Animal hides were also crafted into saddles and reins, belts and bags, and boats and tents. However, these articles were made of raw skins and were not very durable. Raw hides rot when wet as a result of rapid bacterial growth; alternatively, during periods of dryness and heat, the skins lose their malleability and become hard and stiff. Soon it was learned that when the hides were rubbed with fats or oils, they remained supple and became water-repellent. Further, smoking fresh hides could prevent their rapid disintegration. Treating the skins with salts and plant extracts produced an effect that resembled what is now called *tanning,* which gave the hides some resistance to microbial attack.

Figure 5.72 Egyptians processing and soaking pelts (Thebes, eighteenth dynasty, 1550–1295 B.C.).

Illustrations and sculptured reliefs from ancient Egypt provide evidence of the early technology of leather manufacture (Reisner, 1975). Wall paintings from burial chambers in Thebes show Egyptian tanners at work (Fig. 5.72).

Some leather articles found in burial chambers in the Valley of the Kings are of excellent quality and remained well preserved over the centuries because of the arid climate. Impressions of floral ornamentation and the names of kings were found on strips of fine-colored leather. Painted and gilded leather straps were found with the chariots in the tomb of King Tutankhamen. Scribes used a thin parchmentlike leather in addition to papyrus. That careful preparation and tanning of hides occurred can be concluded from the condition of these finds. Simple methods for pretreatment and cleaning of hides and pelts, using materials that were quite original and readily available but hardly pleasant (excrements, urine), were employed during Babylonian and Assyrian times. Therefore, it can be assumed that the vessels shown in the wall paintings were used for brine curing, soaking, liming, and bating hides. For the subsequent tanning an extract from the pods of the Nile acacia, or *Acacia nilotica,* was used.

Even the most ancient cultures along the Euphrates and Tigris rivers knew how to manufacture leather, but little is known about the methods they employed. Minoan art objects show persons wearing leather clothing and semihigh boots. For more information on this subject, the book by Bravo and Trupke (1970), entitled *100,000 Jahre Leder* (One Hundred Thousand Years of Leather), is recommended reading.

The practice of employing feces from animals such as dogs or pigeons as bating material (puering) to loosen hair and soften skin was still employed early in the twentieth century. Later, attempts were made to replace this "enzymatic" process with chemicals as bating agents. It was later found that the active material in the fecal matter was trypsin. Wood ash was used to loosen hair. The German term *Äscher* is derived from this activity. Lime and sodium sulfide were added for dewooling and dehairing, as well as for disrupting the hides' collagenous fiber network. However, further conditioning (bating) was required for cleaning the skins from residual epidermis, hair, and pigments.

In 1912, Wood and Law at Merck GmbH tested pancreatin as a bating agent; however, the results were unsatisfactory (Wood and Law, 1912). In 1904, Otto Röhm became interested in bating methods as part of the tanning process. Initially, he tried to replace the feces with a dilute ammoniacal solution; subsequently his interest turned to the proteolytic enzymes. Historically, the birth of the application of enzymes in industry commenced with Dr. Röhm's dairy entry on May 22, 1907, asking whether extracts from the pancreas whose enzymes affect the digestion of foods could also act as bating agents (Röhm, 1908; Herfeld, 1976; Fig. 5.73).

To employ pancreatic protease, it was essential to first empirically determine the effect of the reaction milieu and additives (activators). Addition of ammonium salts proved particularly favorable in that they could be adjusted for optimal effectiveness and also exhibited a deliming effect. In 1914, Röhm was granted a German patent for an enzymatic liming agent. This development involved the enzymatic removal of hair and wool along with their roots. The process is highly efficient and has been employed continuously; it is still being used with modifications at the present time. Its significance is borne out by the fact that the chemical environmental burden from leather manufacturing is virtually nil.

5.9.2 Skin Structure

Animal skin consists of three different layers (Fig. 5.74):

1. The epidermis, which constitutes about 1% of the total skin. It includes hair roots and sebaceous (oil) and sudoriferous (sweat) glands. It is subdivided into three additional layers: the *stratum corneum,* the outer epidermal layer or horny layer, followed by the *stratum granulosum,* or granular layer, and near the dermis or *cutis vera,* the *stratum mucosum,* or mucosal layer.

2. The dermis, *cutis vera,* or *corium* makes up 85% of the skin. It is composed of the *stratum papillare,* or papillary layer, making up 20–80% of the dermis, with the mass of collagen fibers. The *stratum reticulare,* or reticular layer, consists largely of connective tissue.

3. The *tela subcutanea,* subcutis, or hypodermis makes up about 15% of the skin, and is a loose fibrous layer containing interstitial fat, connective and muscle tissue, blood vessels, and nerves.

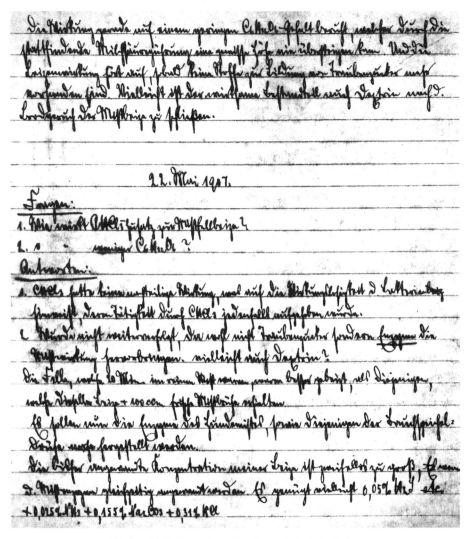

Figure 5.73   Page from the diary of Dr. Otto Röhm.

Chemically, the skin consists of 60–65% water, about 30–32% protein, approximately 10% fat, and 0.5–1% minerals. Figure 5.75 illustrates the main components of mammalian skin.

Leather is made from approximately 150 different skins and pelts. Each has a specific surface texture that the tanner calls *grain*. The grain of the most common hides and skins are illustrated in Figure 5.76.

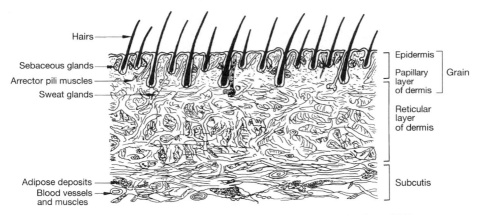

Figure 5.74 Histological cross section of the skin (after Stather, 1990).

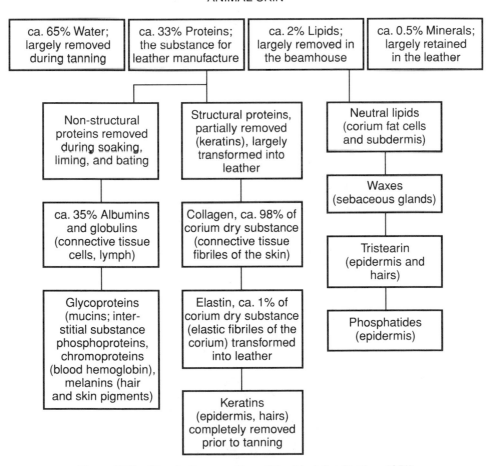

Figure 5.75 Chemical composition of the skin (after Statther, 1951).

(a) Cattle hide

(b) Calf hide

(c) Horse hide

(d) Sheep skin

(e) Goat skin

(f) Pig skin

5.9.3 Enzymes for Leather Processing

A large number of different proteases that have, over time, been used in the beamhouse in one way or another are cited in the literature. The most important are summarized by Pfleiderer and Reiner (1988). In addition to the pancreatic proteases, acidic, neutral, and alkaline bacterial and fungal proteases, as well as papain and bromelain of plant origin, are used. Applications using carbohydrases, lipases, and specific proteases such as keratinase and collagenase were also described by these authors.

Many experiments have been conducted to analytically define and standardize the enzymatic methods that could be correlated with the practical procedures in the beamhouse (Uhlig, 1968). Because this is not yet possible, the tannery chemists within the TEGEWA* have agreed to a traditional definition of proteolytic enzymatic activity. This is the Löhlein–Volhard unit (LVU) (Künzel, 1955; Reisner, 1971).

One Anson unit (AU) of a pancreatic protease preparation corresponds to about 116,000 LVUs. This is the specific activity of 1 g of enzyme preparation. Pancreatic trypsin with 3 AU/g corresponds to about 350,000 LVU/g. For definitions of LVU and AU, see Section 2.3.2.1.

Table 5.26 shows the relationship between the LV units and the data obtained from proteolysis with other methods and substrates. A comparison of the enzymes, standardized to 100,000 LVU/g protein, shows that the bacterial enzymes exhibit their highest activity with hemoglobin and gelatin substrates. The results obtained with the Kunitz method show large variations in activity between the various bacterial proteases. This is due to the different pH optima of the enzymes, a fact that is not addressed in the LVU method.

5.9.4 Leather Manufacture

The manufacture of aniline leather is shown in Figure 5.77.

5.9.4.1 *Raw-Stock Preservation*   Because skins and hides rapidly decompose after slaughter, they are immediately air-dried, or dehydrated and cured by packing with dry salt. During this first step of the hide-to-leather conversion process, the raw stock is extremely susceptible to microbial attack. Microbial decomposition of hide proteins results in a significant loss of leather quality. Adequate initial brine curing of the hides is especially important because the shipping distances for raw stock have increased over the years.

---

*Verband der Hersteller von Textilhilfsmitteln, Lederhilfsmitteln, Gerbereihilfsmitteln und Waschrohstoffen, e.V. (Association of Textile, Leather, Tannery and Laundry Additive Producers), Frankfurt, Germany.

Figure 5.76 Grain surfaces of hides and skins from different animals: (*a*) cattle hide; (*b*) calf hide; (*c*) horse hide; (*d*) sheep skin; (*e*) goat skin; (*f*) pig skin.

TABLE 5.26 Comparison of Various Proteases Using Different Analytic Methods (Uhlig, 1969)

| | Proteases[a] | | | | | | |
| | Bacterial | | | Fungal | | Pancreatic | |
Method (Substrate)	I	II	III	I	II	I	II
Anson hemoglobin	1.3	1.0	0.75	0.8	0.85	0.85	0.65
Modified Kunitz test							
pH 8 (PU method)[b]	190	90	180	185	250	100	170
pH 7	120	180	240	—	—	—	—
Gelatine viscosity test	18,000	—	25,000	3,000	3,800	7,500	4,500
Willstätter casein	26	21	30	—	—	20	—
Gelatin–formaldehyde titration	—	—	—	200	105	—	—
Determination of soluble N_2 after precipitation	29,000	19,000	26,000	22,000	30,000	24,000	22,500

[a]Enzyme source and preparation number from various microbial or pancreatic fractions; enzymes standardized to 100,000 LVU/g protein (LVU = Löhlein–Volhard unit).

[b]Protease unit (Section 2.3.1).

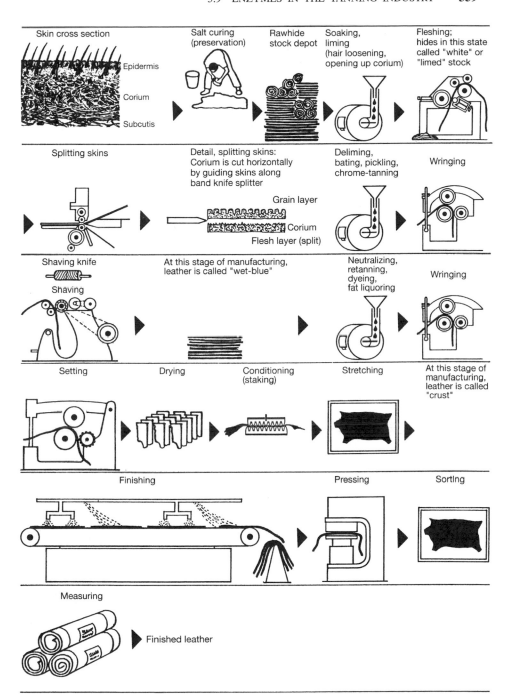

Figure 5.77 The process of manufacturing aniline leather.

5.9.4.2 Soaking   On arrival in the tannery, the hides are rehydrated in the so-called beamhouse. During the pretanning process, blood, dirt, salt, and fat residues are removed from the hides in an initial soaking operation. The main soak serves to restore the original moisture and hydration state of the hides. In addition, nonfibrous proteins such as globulins and albumins are removed during the main soak.

To improve and efficiently operate the complex soaking process, additives are introduced. They include surfactants (surface-active agents), reducing agents, preservatives, and enzymes. In 1914, O. Röhm was the first to recognize the advantages of employing proteases in the soaking process (Röhm, 1914, 1915).

Proteolytic enzymes effectively facilitate the soaking process. In addition, lipases have also been recommended for the soak. Most of the technical enzyme preparations can be combined with nonionic detergents. The synergistic effect observed with this combination is clear if one considers that the skin proteins are surrounded by lipids that must be removed in order to make the substrate (i.e., proteins) accessible to the proteolytic enzymes.

An important result of enzymatic soaking is the removal of undesirable (scud) matter from the skin. This consists of pigment cells and sebaceous and sweat glands. Such cleaning is a prerequisite for manufacturing distinctly grained high-quality aniline-dyed leather.

It should be mentioned that removal of lipids with surfactants during presoak and main soak is preferred to removal during the subsequent liming step. Penetration of the lime into the hide is greatly facilitated by previous enzymatic soaking, a procedure that eliminates the use of chemicals that may subsequently present an environmental pollution problem. Soaking will also stabilize the hides against microbial growth. This can be done quite economically by adjusting the pH to 9. In cases of extreme microbial contamination, a combination of preservatives and surfactants are recommended.

5.9.4.3 Enzymes for the Soaking Process   Pancreatic protease and pancreatin are employed along with neutral and alkaline bacterial proteases; in special cases, alkaline fungal proteases and lipases may also be used. In this case, as in other beamhouse operations, it becomes important that these proteases only hydrolyze the noncollagenous proteins, leaving the collagen essentially intact. Indeed, the highly substrate-specific proteolytic enzymes can selectively remove some proteins while leaving others unaffected.

Generally, only minimal enzymatic activity in the soak is required to produce satisfactory results; for instance, 50–200 g of a protease with 100,000 LVU/g protein is adequate for 1 t of hide. Pancreatic enzymes appear to be the enzymes of choice that can produce a satisfactory amount of skin structure (corium) loosening with a minor effect on collagen (Table 5.27). For barrel soaking, the volume required ranges from 100 to 400%, the pH from 7.5 to 8.5, and the temperature from 18 to 22°C.

The duration of soaking is 6 h for a salt-cured raw stock, and 8–12 h for flint-dried hides (Pfleiderer, 1975).

TABLE 5.27   Skin Loosening and Opening up Corium with Various Proteases (Pfleiderer, 1978)

Bating Enzymes	Löhlein–Volhard (units/g protein)	Loosening (pH 8.5)	Opening Corium (pH 8.5)
Pancreatic	500	228	95
Pancreatic	500	245	117
Bacterial	500	62	204
Bacterial	500	63	118
Bacterial	500	52	34
Fungal	500	77	120
Fungal	500	68	79

Example: Enzymatic Soak for Upper-Footwear Leathers

- Raw stock
  - Salted cattle hides with a weight class of 30–39.5 kg
- Presoak (vat)
  - 150% water at 32°C; duration 2 h at 4 rpm
  - Batch is drained
- Main soak (vat)
  - 150% water at 28°C
  - 0.5–0.7% soaking agent from *B. subtilis* with an activity of 1000 LVU/g (Pellvitt F [RM])
  - 0.4% soda (calcined); duration 4 h at 26°C
  - Density: 3.0°Bé (Baumé)
  - Batch pH 9.8

The hides rehydrated during soaking can be limed in the same batch. The percentage data shown are based on the weight of salt-cured hides.

Example: Enzymatic Soak and Dehairing of Goat Skins

- Presoak
  - 600% water, 4% soda (sodium carbonate), 0.1% preservative, and 0.8% nonionic surfactant
  - Duration: 12–18 h, pH 9.3–9.6 at 27°C
  - Batch is drained
- Enzymatic soak
  - 500% water at 28°C
  - 0.5% nonionic surfactant

- 0.6% soaking agent from *B. subtilis, A. sojae,* and pancreas with a total activity of 4000 LVU/g (Erhazym C [RM])
- 1.5–2% soda (calcined); agitated for 2 h, pause 1 h; then alternating cycle of 2 min agitation and 60-min pause
- Duration 12–18 h, pH 9.3–9.6 at 27°C
- Trimming
- Stretching
- Weighing
- Dehairing (vat)
  - 250% water at 28°C
  - 1.7–2.3% enzyme mixture of bacterial protease and protease from *A. sojae* with a combined activity of 6000 LVU/g (Arazym M 100 [RM])
  - 2.5–3.0% soda (calcined)
  - 0.5% nonionic surfactant
  - 0.05–0.1% preservative that must be enzyme-friendly
  - Agitation for 60 min, 60-min pause; then alternating cycle of 2-min agitation with 2-h pause
  - Duration 16–20 h, pH 9.3–9.5 at 27°C
  - Batch is drained and dehaired

5.9.4.4 Liming   The soak is followed by processes in which the hair and epidermis are removed and the corium is made accessible to tanning agents with an alkaline solution that causes the hide to swell. In addition, these processes also remove residual noncollagenous hide proteins after the soak. This is accomplished with a mix of substances containing lime, Na-hydroxide, and reducing agents. Otto Röhm had developed special enzyme preparations consisting of alkaline fungal proteases for this liming process. The proteases used for the liming process must be especially stable in alkaline solutions.

Example: Liming Mix for Furniture Leather

- Liming solution (vat)—in a soak batch that contains protease, the composition may be as follows:
  - 0.4% protease from *B. alkalophilus* with 500 LVU/g (Pellvitt KAB-P [RM])
  - 1.0% sulfide-free liming agent (Erhavit F [RM])
  - 1.0% of 72% Na-sulfhydrate
  - 1.0% Ca-hydroxide; agitated at 4 rpm for 45 min; then pause for 30 min
  - 0.8% of 60% Na-sulfide
  - 2.0% Ca-hydroxide; agitate at 4 rpm for 1 h, then every hour for 2 min at 2 rpm
  - Duration 20 h; final agitation for 15 min
  - Then the batch is drained

5.9.4.5 Dehairing and Dewooling The proteins of the epidermal basal membrane and the prekeratins of the hair root shaft are easily lysed by alkalis and sulfides. Therefore, traditionally, the dehairing and dewooling operation was conducted with lime and Na-sulfide. Pastes of lime and Na-sulfide are rubbed, mostly by hand, into the flesh side of the hides. The reaction is fast and the removal of the hairs complete. The hair roots remain in the skin.

In the lime drum the hairs are "gelled"; that is, they are largely dissolved by the sulfide. Subsequent alkaline treatment hydrates the skin, and the loose hair is discharged into the wastewater, thus giving rise to the following problems of today's tanneries:

1. The large biological oxygen demand for processing tannery effluents is largely due to the quantity (60–70%) of solubilized keratin.
2. The energy costs for oxidizing organic matter in wastewater is very high.
3. Sulfides are eliminated by either oxidation or precipitation.
4. Worldwide, serious accidents have been reported with hydrogen sulfide liberated from wastewater when acidified, such as by action of deliming agents. Also, some tanneries have been shut down because of the foul odor of hydrogen sulfide.

Presently, no ideal enzymatic dehairing process has been developed for all types of hides and skins; however, enzymatic procedures have been worked out that permit use of a fraction of the chemicals that were required earlier. One important consideration is the speed with which the enzymes penetrate the skin, where they impact the mesodermal hair papillae. The limiting factor during this process is enzyme diffusion, which can be influenced by pH changes, specific salt concentrations, and temperature regimes. Another consideration is the high level of enzymatic activity that has to be sustained at high pH. New proteases effective in the very high alkaline range (pH 10–11) could resolve this problem (see Section 3.2). By employing such enzymes, the amount of lime required can be cut in half. The amounts of enzymes needed depend on the type and quality of the hides.

Dehairing, however, is not exclusively the result of protease action. Combinations of proteases with carbohydrases are described in the patent literature. Proteoglycans present in the basal membrane area that surrounds the hair roots could be responsible for the synergistic effects reported for proteases in combination with carbohydrases.

Today one distinguishes between two dehairing processes: one in which the hair is preserved and one in which it is destroyed. Table 5.28 provides an overview.

Conventional reductive dehairing solubilizes proteins by complete reduction of all disulfide bridges in an alkaline milieu. Other reducing agents such as hydrosulfite, thioglycolic acid, mercaptoethanol, and amines can be used in concert for dehairing processes where hair is preserved (Taeger and Pfleiderer, 1986; Christiner, 1987).

Typically, the process of dehairing cattle hides proceeds as follows. For the initial treatment step, the hides are soaked in a bath of 1.5–10-fold water (150–1000%) with 1.5% soda and 0.1% dehairing enzyme, a neutral bacterial protease. For the second treatment step, dehairing is further promoted by the addition of a fungal pro-

TABLE 5.28   Various Dehairing Processes

Method	Dehairing Agent	Hide or Skin	Effect on Hair
Cleansing and liming	Sulfide <1 g Na$_2$S/liter Ca-hydroxide	Goat, sheep, cattle, calf	Intact
Liming	Sulfide >1 g Na$_2$S/liter Ca-hydroxide	All skins	Burned off
Sweating	Bacterial hydrolysis (bacterial proteases)	Goat, sheep, cattle, calf	Intact
Enzymatic dehairing	Technical proteases	Goat, sheep, cattle, calf	Intact
Dewooling	Technical proteases	Sheepskins	Intact
Debristling	Technical proteases	Pig	Intact
Pulling	Lime	Pig	Intact
Shearing	Mechanical	Sheep pelts	Intact

tease at pH 10. Treatment time is about 24 h; subsequently, the loosened hair is mechanically removed.

This process can be significantly accelerated if an enzyme solution is forced under pressure into the hide from the flesh side (Pauckner, 1982).

5.9.4.6   Deliming and Bating

Prior to tanning, the alkaline hydration of the skin has to be reversed. To accomplish this, the dehaired hides or skins must be delimed prior to or during bating. This requires partial neutralization during which the pH is reduced from 12 to 8. Commonly used deliming agents are acid salts such as ammonium sulfate, ammonium chloride, or organic acids.

Bating means to macerate, or to soften; it is done to purify the collagenous network from other proteins present, as it loosens the collagen fibers. Bating always involves the degradation of noncollagenous proteins such as the glycoproteins and proteoglycans of the interfibrillar matrix, which are undesirable substances in leather processing (Stirtz and Schröder, 1981). Individual collagen fibrils should become accessible to the tanning process that requires their cross-linking. This process also seeks to eliminate other skin components such as hair roots, pigments, and residual epidermis. Clean hides or pelts, ready for tanning, are an absolute requirement for aniline-dyed leathers.

5.9.4.7   Bating Enzymes

Starting in 1907, Otto Röhm produced bating compounds based on pancreatic proteases. It is important for satisfactory skin loosening that the complete pancreatic enzyme complex be available; thus, this requirement must be ascertained during its technical preparation. Other enzymatic activities from peptidases and carbohydrases also influence the bating results. A comparison of enzymatic efficacy of various enzyme preparations has shown that trypsin is least able to digest collagen. Stirtz and Schröder (1982) have shown that chymotrypsin can cause disruption of the cross-linkages between collagen molecules.

Today, pancreatic enzymes are combined with neutral and alkaline bacterial and fungal proteases. One problem encountered when bating with highly active alkaline bacterial proteases is that the bating effect is exaggerated; this will affect the grain and yield rough surfaces.

In addition to enzymatic bating in the alkaline range, bating with acid fungal proteases in the weak acidic pH range is also practiced. This bating mix is used predominantly for soft leathers.

Example: Bating Mix for Dehaired Pork Hides Used in Clothing Leather

- Partially processed hides
  - Limed, washed, and delimed hairless pork skins
- Bating mix (vat)
  - Deliming series (60% water)
  - +60% water at 32°C
  - +2% pancreatic proteases with 1200 LVU/g Oropon OO [RM]
  - 60-min agitation; check skin cross sections with phenolphthalein indicator
  - Drain batch when skin section remains colorless

(Dose percentages are correlated with the weight of the hides.)

The white stock (hairless hides, skins, or pelts) must be free of scud; the processed stock should be soft and pliable.

5.9.4.8 Enzymatic One-Step Process   The objective of this process, developed by Monsheimer and Pfleiderer (1974), is the optimization of tannery beamhouse operations. Soak, hair loosening, liming, and bating are streamlined into one operation. Such processing requires an intensive enzyme treatment in which hair loosening is achieved in a maximum of 12 h without damaging the grain.

Comparison of the conventional with the one-step process (Table 5.29) shows that treatment time and water consumption can be reduced by 50% (Fig. 5.78).

TABLE 5.29   Basic Outline for the One-Step Process

Washing: Water : Salt Ratio 1 : 1 (wt/wt) at 28°C)
One-step process: mixer, barrel or vat, tanning drums
50–100% water at 28°C
0.2–0.3% enzyme preparation
0.2–0.3% Na-hydroxide 1 : 10 (wt/v), 20 min agitation/h for 4 hrs
3.0–5.0% Ca-hydroxide
1.2–2.0% liming aids
0.3–0.5% 95% Na-sulfhydrate
0.3–0.5% 60% Na_2S
0.3–0.5% Na-hydroxide 1:10 (wt/v), 16 h, initial run 2 h, with short intermittent agitation

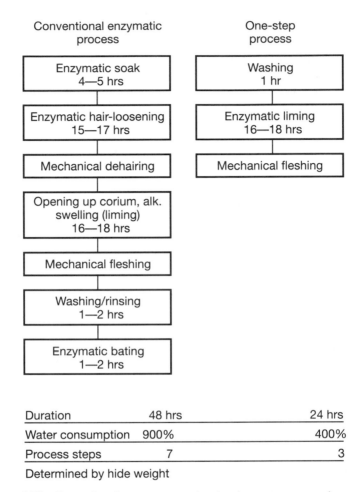

Figure 5.78   Comparison between conventional and one-step enzymatic processes.

5.9.4.9   Loosening Wet-Blue Stock
The starting material is chrome-tanned grain layer hides, chrome-tanned split (flesh-layer or blue drop) hides, or chrome-tanned pelts. The objective of the treatment is to loosen or soften the chrome-tanned stock for a more supple feel; better tensile strength, resiliency, and porosity; and a more even dyeing.

- Partially processed hides
  - Washed, chrome-tanned grain layer hides
- Loosening and softening
  - 150% water at 40°C
  - 2–3% enzymatic loosening (or softening) agent derived from fungal protease with an activity of 1000 LVU/g (Eropic WB [RM])

- 2-h agitation, then alternating a 30-min pause with 5-min agitation
- Duration 12–16 h, pH 3.8–4.2
- Then batch is drained

5.9.4.10 Processing of Tanning Wastes Tanning waste, consisting primarily of the so-called glue stock, cannot be deposited at waste dump sites but must be completely processed. To accomplish this, all proteins are solubilized at high-alkaline pH with bacterial proteases in order to partition the fat from the hydrolysate. After concentration of the protein solutes, the protein is dried and used as a nutritional supplement in animal feed or technical processes.

Table 5.30 reiterates the individual steps of enzyme-operated leather manufacture but focuses on economic considerations (Taeger and Pfleiderer, 1986).

REFERENCES

Alexander, K. T. W., "Enzymes in the tannery. The 1988 John Arthur Wilson Memorial Lecture," *J. Am. Leather Chem. Assoc.* **83**, 287 (1988).

Bravo, G. A., Trupke, J., *100,000 Jahre Leder,* Birkhäuser, Basel, Stuttgart, 1970.

Christiner, J., *Leather Manuf.* **105**, 6,7,19 (1987).

Herfeld, H., *Ledertechnologie—die Geburtsstunde industrieller Enzymanwendung. Dokumentation zum 100. Geburtstag von Dr. Otto Röhm* (leather technology—the birth of industrial enzyme application. Documentation on the occasion of the 100th birthday of Dr. Otto Röhm, Röhm GmbH, company publication, 1976.

Künzel, A., in *Gerbereichemisches Taschenbuch,* Steinkopff, Dresden, Leipzig, 1955, p. 86.

Monsheimer, R., and Pfleiderer, E., Verfahren zur Herstellung gerbfertiger Blössen aus tierischen Häuten und Fellen, German Patent 2 301 591 (1974).

Pauckner, W., in *Congr. Report Sci. Org. the Leather, Shoe and Leather Processing Industries,* Budapest, 1982, p. 309.

Pfleiderer, E., *Das Leder* **26**, 149 (1975).

Pfleiderer, E., *Röhm Spectrum* **35**, 9 (1985).

Pfleiderer, E., and Reiner, R., in Rehm, H.-J., and Reed, G. (eds.), *Biotechnology,* Vol. 6b, VCH, Weinheim, 1988, p. 730.

Reisner, W., *Das Leder* **22**, 121 (1971).

Reisner, W., *Röhm Spectrum* **1**, 56 (1975).

Röhm, O., "Verfahren zum Beizen von Häuten," German Reichspatent 200 519 (1908).

Röhm, O., "Verfahren zum Enthaaren und Reinmachen von Häuten und Fellen," German Reichspatent 268 873 (1914).

Röhm, O, "Verfahren zum Weichen von Häuten und Fellen," German Reichspatent 288 095 (1915).

Röhm GmbH, *Magazine für die Lederindustrie* (Magazines for the leather industry) (1985).

Stather, F., *Gerbereichemie und Gerbereitechnologie,* 2nd ed., Akademie, Berlin, 1951, p. 72.

Stather, F., in Herfeld, H. (ed.), *Bibliothek des Leders,* Vol. 1, Umschau, Frankfurt/M, 1990, p. 20.

TABLE 5.30 Individual Steps in Leather Manufacture (Taeger, 1988)

Process	pH	Substrates	Objectives	Enzymes	Alternatives
Presoak (rehydrating)	~7	Albumins globulins	Degrading proteins	Proteases, bacterial proteases	Surfactants Cl ions
Main soak	~9	Nonfibrillar proteins, prekeratins, lipid cell membranes	Solubilizing proteins, digesting cells loosening hair, opening up corium	Fungal, bacterial, and pancreatic proteases; papain	Surfactants, weak reducing agents, amines
Dehairing	~10	Keratin	Removing hair and epidermis	Keratinases, proteases	Inorganic and organic sulfides; strong reducing agents, OH ions
Liming	~12.5	Noncollagen protein, mucopolysaccharides	Separating fibers, opening up corium	Bacterial and fungal proteases, elastases	OH ions, hydroxylamines, ΔT
Deliming and bating	8–9	Pigments, sebaceous and lipid cells, hair-root-associated proteins	Cleaning grain surface, increasing grain elasticity	Pancreatic, fungal, bacterial, and plant proteases	Salts: $(NH_3)_2SO_4$ NH_3Cl; surfactants
Acid pickling	5–6	Collagen	Softening skin, increasing leather elasticity, improving dyeing	Fungal proteases, pancreatic amylases, papain, etc.	Prolonged acid hydrolysis, hydrotropic agents, ΔT
Wet-blue stock softening	6	Chrome- or vegetable-tanned leathers	Degrading proteins, softening skin, separating fibers	Fungal, bacterial proteases, special pancreatic extracts	Treatment with hydrotropic agents
Scrap leather processing	11	Glue leather	Hydrolyzing proteins for lipid separation	Bacterial proteases	ΔT, ΔP

Stirtz, T., and Schröder, I., *Das Leder* **32,** 155 (1981).

Stirtz, T., and Schröder, I., *Das Leder* **33,** 76 (1982).

Taeger, T., and Pfleiderer, E., *Hung. Bor-Cipo-Tech.* **36,** 406 (1986).

Taeger, T., Biotechnology Focus, Vol. 1, p. 289, Hanser, Munich, Vienna, New York, 1988.

Uhlig, H., *Das Leder* **19,** 28 (1968).

Wood, J. T., and Law, D. J., *Collegium Darmstadt* 121 (1912).

5.10   ENZYMES IN LAUNDRY DETERGENTS

References: Bahn and Schmid (1987), Trommsdorff (1976)

Enzymes have been used in laundry soaps for more than 75 years. The foundation for using proteases in laundry detergents was laid by Otto Röhm in 1913 (Röhm, 1913). His patent, on "Reinigung von Wäsche" (Process of laundry cleaning) initiated the development of incorporating proteases into detergents that today are currently used throughout the world. Röhm used porcine pancreatic proteases. As early as 1919, 65 t of porcine pancreases were processed for the proteases that were incorporated into the "Burnus" detergent. The uniqueness of this development and the role of Novo in Denmark in creating useful products for today's industry was discussed in Section 3.2. The success of these products stimulated the production of enzyme-containing laundry detergents so that, in some markets, 70% of the market has been tilled.

Until 1965 "Burnus" had a dominant share of the laundry detergent market, when Bio 40 entered the market and underwent a rapid development, especially in the United States, but later experienced a dramatic reversal. This was caused by the negative publicity regarding allergies associated with enzyme dust. Such problems were, however, observed in workers exposed to dusty enzyme concentrates during the manufacture of laundry detergents but not in consumers using the detergents in home laundering. Subsequently, several processes were developed to reduce enzyme dust during manufacturing.

5.10.1   Laundry Detergents

The purpose of detergents in laundering is to facilitate the removal of soil from clothes. Dirt and stains in clothes may consist of inert soil particles, the dust of inorganic and organic substances, lipids, and carbohydrates such as starches or pectic substances. In addition, proteins, tannins, and pigments cause stains that cling with varying degrees of tenacity to different textiles. Removal of dirt with water is bolstered by the laundry detergent. Washing starts with the soaking phase: the penetration of water into the fabric, wetting the fibers. Subsequently, the actual laundering processes involving emulsifying and suspending the dirt particles begins.

Laundry powders include nonionic and ionic detergents. They contain, in addition to soaps, sodium silicate or sodium bicarbonate, and sodium tripolyphosphates ("builders" or chelating agents) that bind calcium and magnesium ions, thereby eliminating soap scum, which results from the formation of insoluble salts of fatty acids during the wash cycle. Most recently, the phosphates have been replaced by other chelating agents to reduce phosphate in wastewater effluents. Laundry detergents may also contain bleaches, especially peroxyborate that act by liberating hydrogen peroxide when the washwater temperature is >50°C. Chlorine-containing components such as hypochlorite are also used as bleaches. Other laundry detergent additives are colloidal substances, including carboxymethylcellulose, optical brighteners, and perfumes.

5.10.2 Protein Stains

Household laundry soil consists primarily of lipids and carbohydrates that readily dissolve in an alkaline or highly alkaline pH range; these and other stains are removed by surface-active substances (surfactants), which break up and emulsify the dirt particles; surfactants also act as wetting agents which help to remove the soil from the fabric. Protein-containing stains are more difficult to remove. There are several reasons for this; proteins from milk, whey, meat juices, and blood can coagulate on the fabric and denature. These reductions are initiated not only by elevated temperatures but also by alkaline pH or oxidants. One such oxidation reaction occurs on contact of protein with laundry agents that contain perborate at temperatures >50°C, which stabilizes the protein. Glycoproteins present in bronchial mucus, or the fibrillar serum proteins, form solid films on textile fibers on drying and are very difficult to solubilize and remove during laundering. Protein residues must be completely removed from the textile fiber, because such protein films may also include pathogens that would, if not removed, be protected and stabilized on the fabric. In this manner, they could survive subsequent laundry disinfection. In the two-step wash cycle initially developed by Röhm GmbH, the laundry is first presoaked with an enzyme-containing laundry detergent (Enzymolin) at 35°C for 30 min. The second step involves only detergent at temperatures as high as 100°C. The warm prewash cycle, still practiced today, avoids setting blood, milk, and other protein stains.

5.10.3 Protease Requirements for Laundry Detergents

Laundry enzymes must meet the following requirements:

1. Possess high efficiency and little specificity
2. Exhibit activity at pH 9–11 for one hour at temperatures reaching 95°C
3. Be stable in the presence of chelating agents, perborates, and surfactants
4. Retain activity within the detergent without noticeable loss of activity for at least one year

These requirements are best met by the microbial serine proteases described in Section 3.2.4.1, although some laundry detergent components affect these enzymes' activity and stability adversely. For example, chelating agents, which are important as water softeners, reduce enzyme stability at high temperatures. The reason for this is that by complexing the Ca^{2+} ions they remove the ions that would otherwise act as enzyme stabilizers. Loss in enzymatic activity can be expected from perborate or other oxi-

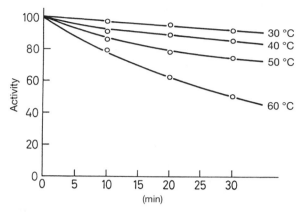

Figure 5.79 Sodium perborate stability of a laundry detergent protease (0.3% Na-perborate, pH 9.5, enzyme Optimase [SOEG]).

dants. Such agents can oxidize a methionine residue near the active site to a sulfoxide, thereby generating an activity-lowering steric blockage (Bahn and Schmid, 1986).

Figure 5.79 illustrates Na-perborate stability of an alkaline serine protease (Optimase [SOEG]) at various temperatures. A significant drop in activity can be observed during the wash cycle at 60°C (Richter and Konieczny-Janda, 1985).

An internationally approved test for demonstrating the cleansing power of a given enzyme is its ability to remove common stains from cotton fabrics. Test kits are commercially available under the name EMPA 116 (available from Swiss Material Test Institute, St. Gallen, Switzerland). The test sample is composed of blood, milk, and soot. Stained fabric samples are subjected, along with other laundry, to the standard wash cycle of a regular washing machine. After drying, the cleansing effect is determined by measuring the degree of light reflection. The tests are conducted in a standard wash system with the Terg-O-Tometer or the Launder-O-Meter. Dambmann et al. (1971) designed a laundry detergent standard for comparing the cleansing power of enzymes:

Enzyme	Cleansing Power (%)
Sodium lauryl sulfate (40%)	5.00
Alkyl phenol ethoxylate (11 EO)	3.00
Soap	3.00
Na-carboxymethylcellulose	1.60
Na-EDTA	0.18
Na-tripolyphosphate	38.00
Na-perborate $4 \cdot H_2O$	25.00
Na-metasilicate	1.00
Na-sulfate	≤ 100.00

5.10.4   Packaging Proteases into Detergents

Laundry detergent proteases are prepared by fermentation of *B. licheniformis, B. amyloliquefaciens,* or *Bacillus* sp. After completion of the fermentation step, the medium is purified by centrifugation or filtration; the purified culture medium is concentrated, if necessary, by thin-layer evaporation. Such enzyme concentrates are then stabilized for use in liquid laundry detergents. Stabilization is achieved by lowering the water content in detergent preparations by adding glycerol, glycols, or sorbitol. Many patents describe various stabilizing mixtures of glycerol, sugar, sugar alcohols, polysaccharides, proteins, polycarbonic acids, alkyloamines, ionic and nonionic surfactants, and inorganic salts.

Further concentration to a dry protease preparation is done by precipitating the enzyme with alcohol or salts, preferably with sodium sulfate. Dust from protease concentrates is particularly aggressive and readily causes inhalation allergies. Procedures for dust control have been developed.

Initially, binding the protease to the surface of an inert material, such as sodium tripolyphosphate, was attempted because the latter was already present in most detergents. For this and other procedures, it is important that the particle size of the enzyme-containing granules be similar or equal to the particle size of the other detergent components, to prevent separation of the components during transport or laundering. A disadvantage of the early granulations was frequent detachment of the enzyme from the surface of its carrier. Binding the enzyme to particle surfaces has been replaced by the prilling method, in which the protease is mixed with the molten detergent components and sprayed onto rotating disks; the droplets then solidify in the cold airstream into relatively homogeneous spherical granules that can be sifted for size. Such methods were described by Eygermans (1974) and Win et al. (1975).

Another method of packaging enzymes is by extrusion granulation (see Table 5.31). The enzymes, mixed with detergents and other inert substances, are extruded, and the extrusion strands are then cut into cylinders and shaped into spheres with a "Marumerizer" (a machine with rapidly rotating corrugated disks), to receive granulated powders.

After drying and sifting, the granules are coated with wax. Other types of granules can be prepared with fluidized-bed granulation. This procedure is especially suited for more thermally labile enzymes. Some enzyme producers granulate laundry detergent enzymes in systems that are also used in preparative food technology.

5.10.5   Wash Cycle

The wash cycle in Europe usually consists of a prewash at 20–30°C for 10–20 min duration. The main wash cycle follows at 40–50°C or higher for 30–60 min. This process is more enzyme-friendly than one-step cycles in which hot (50–60°C) water is added at the onset of the wash cycle. The need to save energy promoted the development of enzymes that have optimum activity at temperatures of about 40°C. Recent experimental products have shown such properties (Nielsen et al., 1981).

TABLE 5.31   Procedures for Incorporating Proteases into Detergents (after Bahn and Schmid, 1986)

Prilling	Extrusion Granulation	Mixed Granulation
Basic ingredients and enzymes	Basic ingredients and enzymes	Basic ingredients and enzymes
↓	↓	↓
Melting	Mixing	Mixing
↓	↓	↓
Prilling	Extrusion	Drying
↓	↓	↓
Sieving	Marumerization	Wax coating
	↓	↓
	Fluidized-bed drying	Cooling
	↓	↓
	Sieving	Sieving
	↓	
	Wax coating	
	↓	
	Cooling	
	↓	
	Sieving	

Typical compositions of enzyme-containing prewash laundry detergents are:

Components	Composition (%)
Anionic components	10–15.00
Nonionic components	2–3.00
Na-tripolyphosphate	20–45.00
Na-carboxymethylcellulose	0.5–1.00
Alkaline protease with 1.5 AU (see Section 2.3.2.1)	0.5–1.00
Na-sulfate	≤100.00

5.10.6   Liquid Laundry Detergents

Liquid detergents present no dust problem and can be dispensed and dosed more easily than powder. Liquid detergents have already achieved a market share of more than 25% in the United States. Liquid detergents containing powerful enzymes are also of interest in stain pretreatment. A presoak with concentrated liquid detergents has become a customary household procedure today. The application of enzymes as spot-removal agents has been practiced in dry cleaning for years.

The enzyme is stabilized by reducing the water component in the formulations. Liquid detergents contain about 30–50% of anionic and nonionic surfactants and 10–25% of builders; the overall water content is 45%.

According to Weber-Meyer (1979), proteases in liquid detergents can be stabilized with glycerol, propylene glycol, and a number of other polyfunctional alcohols.

5.10.7   Other Enzymes for Detergents

5.10.7.1   Bacterial Amylases   These enzymes are often mentioned in formulations for presoak or prewash detergents. While these additives have little importance in laundry detergents, they are of interest for dishwashing detergents. The effects are significant when removing sticky, starchy, food residues, such as those of noodles and oatmeal. The thermally stable bacterial α-amylases (see Section 3.1.2.4) maintain their activity beyond pH 9.0. It would be interesting to integrate amylases into hand soaps; however, there are problems with enzyme stability in such applications.

5.10.7.2   Cellulases   An unexpected discovery found cellulases to have the ability to soften recycled cotton fabric without measurably damaging the fiber. Employing cellulases in laundry detergents has been described in a Japanese patent and a European patent by Murata et al. (1981, 1982) listing 23 bacterial strains and 126 fungal strains as suitable sources for the cellulases.

An enzyme presently in use is the cellulase from *Humicola insolens* [NO]. This enzyme is active under mild alkaline conditions (pH 8.5–9.0) and at temperatures of ≥50°C. The softening optimum is pH 8.0. In addition to the softening effect, cellulase can impart color freshness to colored, frequently laundered cotton fabrics. This phenomenon is related to the fiber-straightening effect of cellulase as shown in Figures 5.80*a*–5.80*c*.

The amount of enzyme required represents only ~ 0.4% of the total detergent; this, however, still constitutes a considerable expense.

5.10.7.3   Lipases   It has been argued for some time that the use of lipases in the laundry industry may have little value because surfactants at higher washing temperatures adequately remove lipids.

Until recently, no lipases were available that exhibited the necessary activity and stability in the pH range 9–10. However, if the quantity of surfactants could be reduced and the cleansing conducted at lower temperatures, then the new lipases could be of commercial interest. A fungal lipase (Lipolase [NO]) produced by recombinant DNA techniques is effective under alkaline conditions (pH ≤ 12) and temperatures of ≤60°C. Numerous cleansing tests were carried out with Lipolase in the Terg-O-Tometer with various basic laundry detergent formulations. Significant effects were reported for European formulations under cold-wash conditions as well as for U.S. laundry detergents at higher temperatures (40°C). Lipase (0.2–0.6%) with 100 KLU[3]/g was added to the detergent.

Cornelissen et al. (1988) have described the use of lipases from *Humicola languinosa* or *Thermomyces languinosus* in laundry detergents. Also, patent rights were filed for the use of a bacterial lipase from *Chromobacter viscosum,* which exhibits special stability in the presence of bleaching agents.

(a)   (b)   (c)

Figure 5.80   Effect of cellulose containing detergent on cotton fibers [NO]: (a) cotton fibers of new fabric; (b) cotton fibers of used and laundered fabric; (c) cotton fibers of used and laundered fabric after Celluzyme treatment.

Thus, recombinant DNA technology has already produced suitable lipases for detergent application. Another lipase was isolated from *Pseudomonas fragi* and its gene transferred to *E. coli,* a more acceptable source organism (Aoyama et al., 1988).

Cutinases can also act as general lipases having a useful high pH optimum. A cutinase from *P. putida* has a high lipase activity within a pH range of 6–11 and, therefore, is also a candidate for detergents. Presently, genetically engineered lipases are being actively studied in the detergent industry.

An enzymatic system that could hydrolyze tannins and plant phenols present in fruit stains, which are still extremely difficult to remove by laundering, would be particularly valuable.

REFERENCES

Aoyma, S., Yoshida, N., and Inouye, S., *FEBS Lett.* **242,** 36 (1988).

Bahn, M., and Schmid, R.-D., *Labor* **2000,** 16 (1986).

Bahn, M., and Schmid, R.-D., *Biotec* **1,** 119 (1987).

Cornelissen, J. M., Lugkist, J., Lagerwaard, C. A., Swarthoff, T., Thom, D., "Enzymatic detergent and bleaching composition," Eur. Patent 271 152 (1988).

Dambmann, C., Holm, P., Jensen, V., and Nielsen, M. H., *Develop. Industr. Microbiol.* **12,** 11 (1971).

Eygermans, P. J., "Preparation of enzymes in particulate form," U.S. Patent 3,801,463 (1974).

Murata, M., Chiba, J. P., and Suzuki, A., "Detergent compositions containing cellulase enzyme capable of converting cotton to reducing sugars," Jpn. Patent Appl. 81/31674 (1981).

Murata, M., Chiba, J. P., Suzuki, A., and Funabashi, J. P., "Reiningungsmittelzusammensetzung," German Patent 3 207 847 (1982).

Nielsen, M. H., Jepsen, S. J., and Outtrup, H., *J. Am. Oil Chem. Soc.* **58,** 644 (1981).

Richter, G., and Konieczny-Janda, G., *Seifen. Öle, Fette. Wachse* **15,** 455 (1985).

Röhm, O., "Verfahren zum Reinigen von Wäschestücken aller Art," German Reichspatent 283 923 (1913).

Trommsdorff, E., *Dr. Otto Röhm, Chemiker und Unternehmer,* Econ, Düsseldorf, 1976.

Weber-Meyer, M., "Liquid cleaning composition containing stabilized enzymes: Protease," U.S. Patent 4,169,817 (1979).

Win, M. H., Desalvo, W. A., and Kenney, E. J., "Enzymatic detergents," U.S. Patent 3,858,854 (1975).

5.11   ENZYMATIC DESIZING OF TEXTILES

Weaving textile fibers exposes the fabric's warp and weft strands to substantial mechanical stress. To prevent breakage of these strands, they are almost always reinforced with sizing. Sizing can consist of various materials. Gelatin, guar gum or carbo bean meal, polyvinyl alcohol, methacrylate, water-soluble cellulase derivatives, and last—but not least—starch have been employed as sizings. The essential requirement to qualify as a suitable sizing material is good adhesion to the threads and easy removal after weaving.

Despite the many advantages of synthetic sizings, starch remains the sizing of choice because of its low cost. In Europe potato starch is used predominantly, in the United States cornstarch, and in the Far East, rice starch. For imported raw fabrics that still contain sizing, the substances used for sizing must be identified. Sometimes combinations of starch with plant gums or egg white have been used, which can present unexpected difficulties during desizing.

Coating the fibers with sizing causes changes in the fabric. Thus, the sizing must be removed prior to further processing because subsequent treatment such as dyeing, bleaching, or finishing can be compromised.

Many sizings, primarily synthetic ones, can be removed with warm water. Sizing with starch must be executed in such a way that the sizing is water-soluble. This can be achieved by steam heating with sodium hydroxide or by treatment with oxidants. Both processes, however, strongly attack the cellulose fibers and contribute to the wastewater burden.

The safest and environmentally least hazardous process of desizing starch is enzymatic.

5.11.1   Desizing Enzymes

Earlier, malt α-amylase or pancreatic amylase were primarily employed. The drawback of amylases is their limited temperature stability. To obtain a rapid starch degradation, starch must first be gelatinized. For this, the fabric is briefly boiled in water. When treating with malt or pancreatic amylase, the fabric is cooled to 60 or 50°C.

Generally, the bacterial amylases will permit starch degradation above the gelatinization temperature, at 85–90°C, or even higher, at 105–115°C, which will accelerate the desizing process.

Selected amylases for starch desizing have been described in Section 3.1.1.2. Table 5.32 summarizes their application.

TABLE 5.32   Enzymes Used in Desizing Textiles

Enzymes	pH Range	Maximum Operating Temperature (°C)	Activators, Stabilizers
Malt α-amylase	4.5–5.5	55–65	Ca^{2+} ions
Pancreatic amylase	6.5–7.5	45–50	NaCl, Ca^{2+} ions
Fungal α-amylase	4.5–5.5	55–65	Ca^{2+} ions
Standard bacterial α-amylase	5.5–7.5	75–85	NaCl, Ca^{2+} ions
Thermostable bacterial α-amylase	5.0–7.0	90–105	NaCl, Ca^{2+} ions (with very soft water)

The maximum operating temperature can be increased by ~5°C under special conditions. The system then operates at optimal activity but also with the least enzymatic stability.

5.11.2   Desizing Procedures

The process can be divided into four steps:

1. *The Prewash:* The prewash with boiling water and surfactants promotes acceleration of the entire process, especially with very heavy fabrics. This operation removes lipids, waxes, and nonstarch sizings. Surfactants must be selected for enzyme compatibility. This is the case for most nonionic surfactants, but anionic wetting agents must first be tested for enzyme compatibility.

2. *Soaking:* During the prewash or during soaking, the fabric should absorb 90–100% of its weight in water. Before the enzyme is added, the saturation bath is prepared by dissolving the wetting agent and the substances that stabilize and activate the enzyme in water and control the pH. When a standard bacterial α-amylase is used, about 300 g of sodium chloride, 50 g of calcium chloride, and about 50 g of nonionic surfactant are dissolved in 100 liters of water at 65–70°C. Then about 100–200 g of an amylase preparation with an activity of 3000 SKB/g is mixed into the brine. If the thermostable α-amylase is used, the temperature can be increased to 70–80°C. A 100-liter batch should contain 30 g of calcium chloride and 400 g of sodium chloride with a pH of 6–8.

3. *Starch Degradation:* Degradation commences with the soaking step and lasts, depending on the process, from 2 min to 16 h. The variable parameters are the enzyme, enzyme concentration, temperature, and reaction time. Short reaction times can be obtained with a thermostable amylase at high concentration and high temperature. To prevent the fabric from drying, which would halt starch degradation during batch desizing, the rollers are covered and rotated. Progress in starch degradation is monitored by the reaction of iodine with starch. A 0.1 N iodine/

potassium iodide solution is diluted with water (1:19) and used as the test reagent. Degradation is complete when the iodine test is negative, indicated by a light yellow spot remaining on the fabric after a drop of the solution is added.

4. *Rinsing:* The degradation products are removed by rinsing with water at temperatures of 95–100°C. The rinse is supplemented by surfactants or detergents and conducted with vigorous mechanical agitation. Unusually heavy particulates are removed with a 5–10-g NaOH/liter rinse.

5.11.3   Desizing Processes

5.11.3.1   Desizing in a Pad Roller Unit or Foulard   In this process (see Fig. 5.81) the prewash and soaking steps proceed continuously. For starch degradation, the material is put on a slowly rotating roller or alternatively, put into a trough. Hydrolysis is carried out with a standard bacterial α-amylase (e.g., 250 g Degomma HB [RM] or 250–500 g Aquazyme 120 L [NO] per 100 liters of soaking bath) at 70–75°C for 2–4 h, or at temperatures < 50°C for 10–14 h. A rinsing process follows.

5.11.3.2   The Jigger Desizing Process   This is a simple process requiring only two rollers and a trough. The material passes through the bath, which must be changed for each step.

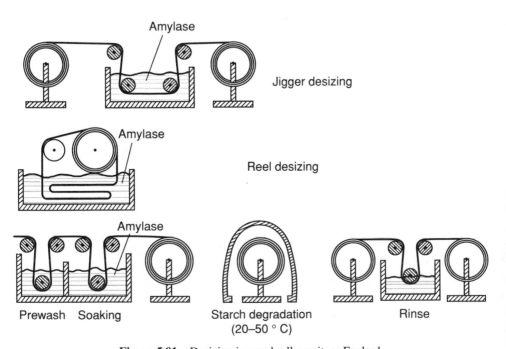

Figure 5.81   Desizing in a pad roller unit, or Foulard.

For the prewash, two passages in boiling water containing a mild surfactant capable of emulsifying the oils and waxes are required. For soaking, 50–150 g of enzyme with an activity of about 3000 SKB units/g of a standard bacterial α-amylase (e.g., Optisize [SOEG]) is required to hydrolyze the starch. Starch degradation at 65–70°C requires four passages. For the last two passages, the temperature is increased to about 90°C.

5.11.3.3 The Reel Desizing Process

This is also a relatively simple process in which the bath must be changed as well. Prewash and postwash proceed in the same way as that in the jigger process. The amount of enzyme needed is about 25–100 g of a preparation with an activity of about 3000 SKB units/g (e.g., Aquazyme [NO]).

5.11.3.4 Continuous Desizing Process with a Standard Bacterial α-Amylase

The process requires approximately 30 min. The material is, as described earlier, continuously prewashed and soaked. Starch hydrolysis proceeds continuously in a J-tank, which permits controlled dwell times at elevated temperatures. The starch degradation process is concluded by continuously passing the material through a steam chamber. Finally, a vigorous hot-water rinse containing 20–30 g of sodium hydroxide per liter follows ending with cold water rinses (Fig. 5.82).

A 100-liter soaking solution consists of the following:

- 300 g of sodium chloride
- 200 g of calcium chloride
- 50 g of surfactant
- 250–500 g of a standard bacterial α-amylase (e.g., Degomma HB 20 [RM])

5.11.3.5 Continuous Desizing Process with a Thermostable Bacterial α-Amylase

The process time is reduced to 2–5 min. Removal of sizing is performed in a steam chamber at 105–115°C requiring only 15–120 s. A vigorously boiling alkaline wash significantly improves the cleansing process.

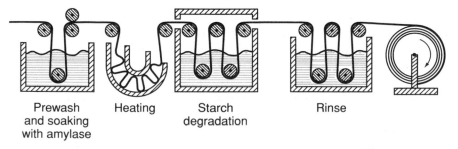

| Prewash and soaking with amylase | Heating | Starch degradation | Rinse |

Figure 5.82 Continuous desizing process.

A 100-liter soaking solution consists of the following:

- 400 g of sodium chloride
- 300 g of calcium chloride
- 50 g of surfactant
- 50–300 g of thermostable bacterial α-amylase with an activity of ~2600 SKB units/g (e.g., Thermamyl [NO]).

5.11.3.6   Stone Wash with Amylase and Cellulase   Recently stonewashing was introduced as a means of giving blue jeans a worn, rugged appearance. To enhance the stoning effect, bacterial amylase were introduced. About 1–1.5 liters of a standard bacterial α-amylase with 2000–2500 SKB units/g (e.g., Aquazyme [NO]) is added to a 1000–liter batch.

Since 1992, it has become standard procedure to use cellulases for finishing jeans. Cellulases are especially suitable for the stonewashing of denim dyed with indigo, as this dyestuff hardly penetrates the fibers, but stays mainly on the surface of the yarn. Depending on the process, neutral/alkaline or acid cellulases from *Humicola insolens* [NO] or *Trichoderma reesei* [MJ] are used, also in the combination with bacterial amylase (Videbaek et al., 1994).

The photographic industry produces large amounts of gelatin–silver salt emulsions. To remove the silver from exposed films, such as exposed X-ray films, the gelatin layer must be separated from the carrier material. This allows recovery of both the silver and the carrier, such as celluloid, which can be recycled. Bacterial protease as well as pancreatic protease (e.g., Degomma PNL [RM]) are used to remove the emulsions. For gelatin, which is difficult to separate, film fragments are briefly presoaked at 90 or 70°C for several hours. The gelatin is then digested with a dilute protease solution, such as 2–3 g of Degomma PNL in 100 liters with nonionic surfactants at 20–45 for 1–2 h. With color film, the process requires 24–28 h in an agitated reaction bath. The treatment may have to be repeated to remove any residual color layer.

5.11.4   Enzymes for Degumming Silk

The raw-silk filament has a gray–yellow-colored coating consisting of protein called *sericin* or *silk gum,* which envelops and cements the fibroin of the silk filament. Sericin must be partially removed prior to further processing of the silk filament. Chemically, sericin is composed of 90% protein and 5% wax, fats, salts, and ash. It must be softened to permit unwinding of the filament into a continuous thread. Raw silk consists of about 20% sericin and 80% fibroin. Only about 1% of the sericin is removed because some silk gum is required for further handling of the delicate filament. Earlier, degumming was accomplished with alkaline solutions and soap. Degumming of raw silk with enzymes was discovered by O. Röhm and a patent was filed on January 3, 1915, and granted in 1917 (Röhm & Haas Chem. Fabrik, 1917).

Of historical interest is a letter written by Otto Röhm on June 16, 1910, to his brother in Lyon, then an established center of silk manufacturing:

Please procure a sample of raw silk for me that has not yet been degummed. I shall try to degum it. I note that it is done with ordinary soap. Ask your contact if this was always done with soap and if he knows of attempts to replace soap and if so, possibly with what.

In 1914, Röhm marketed a pancreatic protease preparation for silk degumming under the name Degomma S. Today, degumming is also performed with solutions of alkaline bacterial protease.

A typical degumming solution includes

- Alkaline bacterial protease from *B. licheniformis* with 1 AU/g, ~1.0–2.5 g/liter or pancreatic protease with 2 AU/g, 0.5–1.5 g/liter
- Sodium bicarbonate, 5.0 g/liter
- Nonionic surfactant, 0.5–1.0 g/liter

The temperature for protease from *Bacillus* can be 55–60°C. Pancreatic protease should be heated only to 55°C. The degumming process requires about 0.5–1 h with gentle agitation. For manufacturing crepe de chine (technical information, Novo Industri), a nonenzymatic wash at 95°C for 30–60 min is recommended as a first step in the degumming process. Weight loss of the silk after this process should be about 16–20%. Further degumming is performed with alkaline bacterial protease, bleaching with hydrogen peroxide follows, and finally, the silk is thoroughly rinsed.

REFERENCES

Novo Nordisk, technical information [NO].

Röhm GmbH, technical information [RM].

Röhm & Haas Chem. Fabrik, Darmstadt, Germany, "Verfahren zum Entbasten von Seide," German Reichspatent 297 394 (1917).

Solvay Enzymes GmbH, technical information [SOEG].

Videbaek, T., Fich, M., and Screws, G., "The jeans effect comes into being," *ITB Dyeing/ Printing/Finishing*, **1**, 1–4 (1994).

5.12 LIPASE-CATALYZED HYDROLYSIS AND MODIFICATION OF FATS AND OILS

References: Godtfredsen (1993), Tombs (1995)

In the past the use of technical enzyme preparations was confined to the hydrolysis of proteins and carbohydrates. Employing enzymes for lipolysis remained in the domain of medicine, where lipases still play an important role as medicinal aids for lipid digestion. However, lipids are important raw materials for industrial processes and in the food industry, which has an annual lipid production of about 60 million metric tons.

Although the utilization of technical lipase preparations is still in its infancy, there are a number of current and potential applications. Other chapters in this book discuss some of these areas; examples are the use of lipases in cheesemaking and for the production of aroma and flavor. Another use is found in the leather industry for fat removal. A very recent application is the addition of lipase to detergents.

The following areas of lipase utilization are discussed in this chapter:

1. Partial and total hydrolysis of lipids into fatty acids and glycerol.
2. Interestification and transesterification reactions
3. Ester synthesis

5.12.1 Complete Hydrolysis

The complete hydrolysis of lipids and oils to free fatty acids and glycerol is, technologically, a mature process. Almost all lipids and oils are hydrolyzed without difficulty with superheated pressurized steam at 250°C and 60 bar. Although the use of castor bean lipase for lipolysis was employed at the turn of the century, the process did not become popular because of the high costs involved. In more recent years, however, it became possible to drive the lipolytic process with suitable microbial lipases at significantly lower costs; this created renewed interest in an enzyme-driven system. In addition, energy considerations favor this development in that an enzymatic process requires less energy.

This renewed interest is evidenced by the numerous publications and patent applications, largely from Japan, that have appeared recently. Bühler and Wandfrey (1987) have thoroughly studied the kinetics of complete hydrolysis, especially the hydrolysis of coconut oil. Their results show that the following factors play important roles:

1. *Use of a Nonspecific Lipase:* Among the 34 technical preparations studied, the authors found the enzyme isolated from *Candida cylindracea* to be the most suitable because all three ester bonds in the triglycerides are cleaved rapidly, almost at the same speed.

2. *Importance of Phase Interface for Activity:* The lipase must be absorbed at the lipid–water interface. Therefore, the surface area of this interface has a significant role in enzymatic activity. Increasing the speed of a stirrer will increase the surface area of the interface. This will also increase the reaction velocity of a given enzyme concentration. The velocity can be accelerated to the point where it is limited by substrate concentration. Thus, the hydrolytic rate of coconut oil is doubled with the doubling of the stirrer's rotational speed; with olive oil, however, the hydrolytic rate increases by only about 10%.

3. *The Lipid–Water Relationship:* A quantitative conversion is possible only with a high proportion of water. If the water content in the reaction mixture is decreased, the glycerol concentration in the water phase increases, and a back reaction

can be expected. Bühler and Wandrey (1987) reported that, given an optimal 70:30 lipid:water ratio, only 3% of the enzymes were found in the water. Only traces (~2%) of lipase were found in the lipid, meaning that most of the lipase is located at the lipid–water interface.

4. *Influence of the Substrate:* The significance of the substrate was determined at constant reaction conditions, with a 70:30 ratio of lipid:buffer at pH 5.6 and 30°C. Lipase OF-360 from *Candida cylindracea* [SA] was used for the lipolysis of several oils (Fig. 5.83).

According to the results shown above, 95% of soybean oil is hydrolyzed after 6 h, while beef tallow and coconut oil hydrolyze at a much lower rate under the same reaction conditions.

Nielsen (1985) studied complete hydrolysis with various lipases. The maximum degree of hydrolysis (DH) with lipases from *Rhizopus, Aspergillus,* and *Mucor miehei* were compared with that obtained with the *Candida* lipase. The results showed that *Mucor* lipase was found to be the most effective enzyme for hydrolyzing olive oil; a maximum DH of 98–99% was reached after hydrolysis for 48 h at pH 7–8 and 45°C.

From an economic point of view, the costs for an enzyme-driven process are significantly less. Lipase carries the greatest process cost. In 1987, the costs for the enzymatic process were still twice those of conventional methods. According to

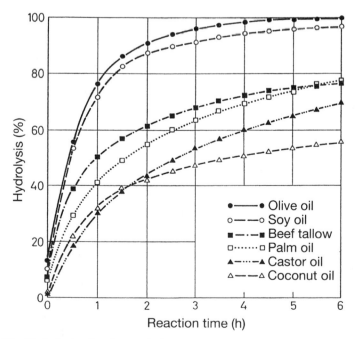

Figure 5.83 Hydrolysis of various technical fats with lipase from *Candida* [Meito Sangyo] (Bühler and Wandrey, 1987).

Schmid (1986), enzymatic lipolysis has an advantage only where complete lipolysis is not required, such as in soap production. In 1983, the Miyoshi Oil and Fat Company was routinely hydrolyzing about 1000 t of fat per month making soap employing a lipase from Meito Sangyo. Up to 92% cleavage was obtained; the process was conducted at 90°F and and at standard atmospheric pressure rather than the 480°F and 700 psi (lb/in.²) used in the conventional process. It was reported that an annual saving of $850,000 was achieved. The enzymatic process is, however, also of interest for generating pure unsaturated fatty acids; this process requires lipolysis under very gentle conditions.

Furthermore, in many specialized applications, albeit of minor economic significance, the use of lipases is advantageous. These areas include food processing technology, removing fat from fat traps in commercial kitchens, and cleaning pipe assemblies, filters, and surfaces. Uhlig at Röhm GmbH successfully removed fat from printing plates with pancreatin along with nonionic detergents, replacing the earlier method that used trichloroethylene.

These processes do not require complete lipolysis. A protease-free lipase from *A. niger* [AM] can be used to improve the aerating (whippability) of chicken egg whites because egg-white contains up to 0.04% fat, which adversely affects foam stability.

5.12.2   Specific Hydrolysis

Several specific lipases are presently available for the hydrolysis of fats and oils. Among these are the lipases from *Aspergillus* which primarily split off short-chain fatty acids and are used to ripen cheese and for the preparation of enzyme-modified cheeses (EMCs). The lipase from *Geotrichum candidum* possesses a particular specificity that splits off long-chain fatty acids from triglycerides that have a cis double bond in the C-9 position.

Examples of the stereo-, regio-, and group-specific hydrolysis of carbohydrate acylesters were described by Kloosterman et al. (1988).

Quite a different application of lipase, or more generally esterase activity, is in organic chemical synthesis. Schneider and Reimerdes (1987) report on the possibility of synthesizing chiral carbonic acids and alcohols via specific hydrolysis, esterification, and transesterification.

5.12.3   Interesterification and Transesterification

References: Macrae (1983), Nielsen (1985), Schneider and Reimerdes (1987)

Interesterification of triglycerides (or triacylglycerols) is a means of modifying the triglycerides' properties by changing their fatty acid composition. This process is accomplished with chemical catalysts such as sodium or Na-methylate. These stimulate acyl-group migration between the individual glyceride molecules. New triglycerides arise from mixtures in which the fatty acid moieties are randomly distributed. This procedure can also be applied to mixtures of triglycerides having different melting points with free fatty acids (transesterification–acidolysis) or fatty acid methylesters (interesterification). In this manner, a triglyceride can be enriched with a specific fatty acid that will change its melting point. This process also allows the hardening of plant oils avoiding high-pressure hydrogenation. If a nonregiospecific lipase is used for an interesterification, then the triglyceride mix will yield the same number of isomers that a purely chemically driven interesterification would generate. In an interesterification (ester interchange) reaction of a trigylceride with the composition A–B–A with another triglyceride with the composition A–B–A with another triglyceride composed of C–B–C, 18 different triglycerides may be formed. The same situation applies to the transesterification of a triglyceride with a free fatty acid (acidolysis).

However, when a 1,3 regiospecific lipase is employed for the interesterification reaction with the same triglyceride mixture described in the preceding paragraph, it produces a new mixture with only three new triglycerides (1); such a mixture cannot be generated by random chemically catalyzed ester interchange. This is also true for the transesterification of a triglyceride with a free fatty acid C (2). The reaction scheme is as follows:

$$
\begin{bmatrix} A \\ B \\ A \end{bmatrix} + \begin{bmatrix} C \\ B \\ A \end{bmatrix} \rightarrow \begin{bmatrix} A \\ B \\ A \end{bmatrix} + \begin{bmatrix} A \\ B \\ C \end{bmatrix} + \begin{bmatrix} C \\ B \\ C \end{bmatrix} \quad (1)
$$

or

$$
\begin{bmatrix} A \\ B \\ A \end{bmatrix} + C \rightarrow \begin{bmatrix} A \\ B \\ A \end{bmatrix} + \begin{bmatrix} A \\ B \\ C \end{bmatrix} + \begin{bmatrix} C \\ B \\ C \end{bmatrix} \quad (2)
$$

The opportunity to generate triacylglycerols with "new" properties by the action of regiospecific lipases is of great interest to the manufacture of edible fats. The development of a triglyceride mixture with the properties of cocoa butter has received special attention (Macrae, 1985). The reaction proceeds as follows:

$$
\text{Palm oil} + \text{methyl stearate} \xrightarrow[\text{lipase}]{\text{1,3-regiospecific}} \text{cocoa butter} + \text{methyl oleate}
$$

Another way of producing cocoa butter is the interesterification of palm oil with glycerol tristearate (tristearin):

$$
\begin{bmatrix} P \\ O \\ P \end{bmatrix} + \begin{bmatrix} S \\ S \\ S \end{bmatrix} \rightarrow \begin{bmatrix} P \\ O \\ P \end{bmatrix} + \begin{bmatrix} P \\ O \\ S \end{bmatrix} + \begin{bmatrix} S \\ O \\ S \end{bmatrix} + \begin{bmatrix} P \\ S \\ P \end{bmatrix} + \begin{bmatrix} P \\ S \\ S \end{bmatrix} + \begin{bmatrix} S \\ S \\ S \end{bmatrix}
$$

where: O = oleic acid (18:1)
         P = palmitic acid (16:0)
         S = stearic acid (18:0)

According to Tanaka et al. (1986), a cocoa butter substitute can be obtained in this manner with the following composition: S–O–S, 45–70%, P–O–S, 25–45%; and P–O–P, <10%.

Another application for a lipase-driven transesterification is that of enriching oil with highly unsaturated fatty acids. This yields products of particular interest to nutritionists and the health food industry.

For hardening, soybean oil and tallow are routinely transesterified as an alternative to hydrogenation. Fats suitable for margarine production are treated in this manner.

5.12.4   Immobilized Lipase

Immobilized lipases are most significant in transesterification. These products are commonly prepared by adsorption of the enzyme to a carrier or support, such as silica gel or kaolin (aluminum silicate). The enzymes are precipitated onto the carrier or are mixed wet with the carrier and then dried.

Immobilization usually means that the enzyme is attached to the carrier or support by strong and generally covalent bonds. But lipases must remain free to enter the interface. Covalent bonds would result in loss of activity. Therefore, supported lipases are contained in gels, water droplets, and so on. Several immobilization methods have been used quite well for lipases (see also Section 4.4.5.2).

In addition, the flow characteristics of the support media are key determinants in choosing enzyme supports. The ability of depositing the enzyme while retaining a high activity will ultimately govern the usefulness of lipases, and that of other enzymes, for industrial applications.

Macrae and How (1988) developed a tonnage-scale pilot plant run at 40°C for 600 h. The rate of lipase-catalyzed interesterification was adequate for a continuous pumped operation with a 10-min dwell time. Presently, this plant is the closest approach to a commercial operation.

In earlier work by Macrae (1983), a lipase from *A. niger,* bound to silica gel and activated with a small amount of water, was able to transesterify (acidolysis) the intermediate palm oil fraction (1 part) with stearic acid (0.5 part) (Macrae, 1983). The change in product composition is illustrated in Table 5.33.

An analysis of the triglycerides formed after 16 h (40°C) indicates that the stearic acid moieties were selectively inserted into the 1- and 3-positions.

Yet another application of the transesterification reaction is the regiospecific acylation of glycols (alcoholysis), catalyzed by pancreatic lipase in organic solvents. Pancreatic lipase shows a preference for a primary hydroxyl group over the secondary hydroxyl or the second primary site (Cesti et al., 1985), as illustrated in Table 5.34.

In addition to immobilizing lipases by adsorption to an inert carrier matrix, these enzymes can also be bound to a macroporous ion-exchange resin. Hansen and Eigtved (1986) describe a 1,3-regiospecific lipase from *Mucor miehei* that was bound to such a

TABLE 5.33 Products from Transesterification of Palm Oil Midfraction with Stearic Acid Using *Aspergillus* Lipase (Macrae, 1983)

Triglyceride[a]	Palm Oil, Midfraction (%)	Transesterified Product (%)
P–O–P	58	19
P–O–S	13	32
S–O–S	2	13
S–S–O	7	2
S–L–S	9	7
S–O–O	4	11
S–S–S	5	13

[a]L, linoleic acid; O, oleic acid; P, palmitic acid; S, stearic acid.

carrier (Lipozyme, [NO]). This immobilization improved temperature stability so that the half-life of the enzyme is now approximately 1600 h at 60°C.

5.12.5 Unexpected Reactions in Nonaqueous Systems

Reference: Klibanov (1986)

In recent years new knowledge has indicated that enzymes may also be active in essentially nonaqueous systems. Enzymes can function in organic solvents provided a small amount of water is present. This observation, with credit due to Klibanov and his collaborators, opened new approaches for preparative organic chemistry.

Many substances are only soluble in organic solvents. Water is detrimental for many reactions because, for instance, anhydrides must by hydrolyzed. Further, the recovery of end products from the water phase is frequently quite difficult.

The alternative use of enzymes is simple and has several advantages:

- Powdered enzyme preparations can be employed, eliminating expensive and inefficient carrier-binding procedures.
- The enzymes are more stable in an organic solvent than in water. No enzyme inactivation by proteases can occur as is often the case when operating in dilute aqueous solutions.
- The enzymes can be recovered from the solvent by simple filtration.

TABLE 5.34 Regiospecific Transesterification of Glycols (Cesti et al., 1985)

Glycol	Ester	Monoester (%)	Nonesterified Glycol (%)	Diester (%)
1,2-Butanediol	Ethyl acetate	96	3	1
1,2-Hexanediol	Ethyl acetate	96	2	2
1,3-Butanediol	Ethyl caprylate	91	9	0

Also, a minimum amount of water is always required.

Water is indispensable for enzymatic catalysis because water participates directly or indirectly in all noncovalent reactions. This includes hydrogen bonding, hydrophobic bonds, and binding by van der Waals forces. Only a monomolecular water layer is required (less than 1%) to coat the enzyme. The rest of the solution can be replaced with an organic solvent. The solvent should not be miscible with water. Other reactions can be expected with methanol, ethanol, or acetone as solvents.

To minimize the interaction with water, hydrophobic solvents are chosen preferentially. When using more hydrophilic solvents such as ethyl acetate, the solvent should be saturated with water. The influence of the solvent during the hydrolysis with palm oil and beef tallow was studied by Takahashi et al. (1984); some of their data are presented in Table 5.35. The optimum activity was found in solutions containing 20% solvent.

Enzymatic activity is also dependent on pH in nonaqueous systems. The question arises as to how the enzyme recognizes the pH in an organic solvent. The enzyme is conditioned to the pH it experienced in its previous solvent; in other words, the ionogenic groups of the enzyme that have a given degree of ionization at a defined pH retain this condition when dry. For example, the transesterification reaction (alcoholysis) between N-acetyl-L-phenylalanine ethylester and propanol in octane is accelerated 300-fold if, instead of a commercial subtilisin preparation, subtilisin precipitated from a pH 7.8 solution is used. The pH optimum for subtilisin in an aqueous solution is 7.8 for both cleaving the ester and transesterification.

When planning such reactions, an enzyme solution should be adjusted to its optimal pH and converted into a powder by freeze drying. Klibanov (1986) demonstrated the remarkable degree of stability of these enzymes in such a desiccated state. Pancreatic lipase, which is quite labile in aqueous solution, acquires a half-life of 10 h in a reaction system having 0.1% water and a temperature of 100°C. According to Coleman and Macrae (1980), enzymatic transesterification with all known lipases can be conducted over a period of several days in a reaction milieu containing 0.2–1% water at 40–60°C.

TABLE 5.35   Influence of Solvents on Lipid Hydrolysis

Solvent	Degree of Hydrolysis for Palm Oil
Control, H_2O/buffer	50
n-Hexane	61
n-Heptane	70
n-Octane	78
n-Nonane	86
i-Hexane	79
i-Octane (2,2,4-trimethylpentane)	99

5.12.6 Other Esterifications

During the past few years, enzyme-catalyzed esterification of many substances have been published, patent applications filed, and patents granted; among these substances are compounds affecting aroma and flavor. This largely involves esters that cannot be readily obtained with chemical methods, or that can be obtained only in poor yields. In addition, legal considerations may play a role in the production of flavor compounds; because enzyme-catalyzed reactions can be viewed as "natural" conversions such as fermentation, their products can be considered safe.
The following may serve as examples:

1. A patent concerning the enzyme-catalyzed production of esters and lactones (Gatfield and Sand, 1982) describes the production of esters that find application as raw materials in the perfume and fragrance industry. Lipase from *Mucor miehei* yields significant amounts of geranyl-*n*-butyrate from geraniol and butyric acid (reaction time 72 h); in the same way, geranyl caprylate, ethyl oleate, and *n*-butyl oleate can be generated. A mixture of fatty acids from butter, prepared by lipolysis of butter fat, was esterified with ethanol or *n*-propanol with a ~60% yield. Performing these reactions is quite simple. Acid and alcohol are mixed; ~1–5% (wt/wt) of the selected dried enzyme is added and stirred for 48–72 h at room temperature. After the enzyme is removed by filtration, the reaction products are recovered by distillation.

2. For the synthesis of various waxes, 2 g of carrier-bound lipase (Lipozyme [NO]) is incubated with 0.05 mol of fatty alcohol and fatty acid at 70°C and stirred for 6 h. The following compounds with a 75–87% yield are formed: oleyl palmitate, oleyl stearate, octadecan–caprylate, –stearate, –oleate, and other esters (Eigtved et al., 1986).

Genetic manipulation of selected microorganisms may lead to lipases of increased fatty acid specificity, which most certainly will increase their commercial usefulness. Because lipase-catalyzed interesterification–transesterification of triglycerols is not yet fully commercialized, progress in genetic technology will lead to improvement of this process.

In a review on lipases, Godtfredsen (1993) has compiled a well-referenced list of the most important lipase-producing microorganisms covering synthesis and characterization of lipases. Lists of important industrial sources of lipases and their commercial producers and industrial applications are also provided.

REFERENCES

Bühler, M., Wandrey, C., *Fett, Wissenschaft, Technologie* **89**, 156 (1987).

Cesti, P., Zaks, A., and Klibanov, A. M., *Appl. Biochem. Biotech.* **11**, 401 (1985).

Coleman, M. H., and Macrae, A. R., "Fat process and composition; interesterification of oils and fats by treating with a microbial lipase enzyme," U.S. Patent 4,275,081 (1980).

Eigtved, P., Hansen, T. T., and Huge-Jensen, B., Characteristics of immobilized lipases in ester synthesis and effects of water in various reactions. Technical information, Novo Industri, Copenhagen, 1986.

Gatfield, I., and Sand, T., "Verfahren zur enzymatischen Herstellung von Estern und Laktonen," German Discl. 31 08 927 (1982).

Godtfredsen, S. E., in Nagodawithana, T., and Reed, G. (eds.), *Enzymes in Food Processing,* 3rd ed., Academic Press, San Diego, London, New York, 1993, p. 205.

Hansen T. T., and Eigtved, P., in Baldwin, A. R. (ed.), *World Conference Emerging Technologies in the Fat and Oil Industry,* Cannes, 1985; *Am. Oil Chem. Soc.* **63,** 365 (1986).

Klibanov, A. M., *Chemtech* 354 (June 1986).

Kloosterman, M., Schoemaker, H. E., and Meijer, E. M., "Lipase catalyzed reactions on carbohydrates," poster, XIV Internatl. Carbohydr. Symp., Stockholm, 1988.

Macrae, A. R., *J. Am. Oil Chem. Soc.* **60,** 291 (1983).

Macrae, A. R., in Tramper, J., Van Ter Plast, H. C., and Linko, P. (eds.), *Studies in Organic Chemistry,* Vol. 22, Elsevier, Amsterdam, 1985, p. 195.

Macrae, A. R., and How, P., "Rearrangement process," U.S. Patent 4,719,178 (1988).

Nielsen, R., *Fette, Seifen, Anstrichmittel* **87,** 15 (1985).

Schneider, M., and Reimerdes, R. H., *Forum Microbiologie* 302 (1987).

Schmid, R. D., *Fette, Seifen, Anstrichmittel* **88,** 555 (1986).

Takahashi, K., Nischimura, H., Yoshimoto, T., Saito, Y., and Inada, Y., *Biochem. Biophys. Res. Commun.* **125,** 761 (1984).

Tanaka, Y., Omura, H., Irinatsu, H., and Kobayashi, T., "A cocoa butter substitute composition," Eur. Patent Appl. 196 210 A2 (1986).

Tombs, M. P., in Tucker, G. A., Woods, L. F. J., (eds.), *Enzymes in Food Processing,* 2nd ed., Blackie Academic & Professional (imprint of Chapman & Hall), New York, 1995, p. 258.

5.13   ENZYMES IN ANIMAL NUTRITION

References: Spring (1992), Simon et al. (1994)

Raw materials used as animal feed contain not only carbohydrates and proteins that are nutritionally complete but also fiber-rich components consisting of cellulase and hemicellulase. Because of the limited digestion of cellulose and hemicellulose in the gastrointestinal tract of monogastric animals, the feed is not completely utilized. This holds true especially for young animals whose digestive potential for digesting certain raw foodstuffs has not yet fully developed. However, it is not only the underutilization of the fiber components but also the indigestibility of some of the protein that has been denatured with extreme heat that is of concern. In addition, starch sequestered by β-glucans and pentosans is not degradable in the digestive process.

Use of certain raw materials in animal nutrition is further restricted by the presence of antinutrients. In whey, the high lactose content limits its unrestricted use. Further, as discussed earlier, whey cannot be concentrated without a prior lactose hydrolysis. The keeping quality is limited, and transportation costs for such a dilute nutrient are exorbitant. The classic antinutritive substances are raffinose and stachyose, which are the oligosaccharides that limit the amount of soybean meal that can be put into animal feed. These problems may be resolved in the future by enzymatic treatment. Specific α-glucosidases exist in various microorganisms that can cleave oligosaccharides; however, the use of such enzymes is still too expensive.

Hemicellulases also qualify as antinutritive substances, which, because of their high water-binding capacity, can cause digestive problems.

In feed manufacture, enzymes are of great help in preserving feed components. The protein portion of fish and meat wastes and blood solids can be enzymatically hydrolyzed and the hydrolysates subsequently concentrated. This is a prerequisite for energy-saving drying. Another way to preserve feed components is by silaging. In Europe and in the United States, the nutritive value of grasses, maize, and beets can be extended in this manner. Because silaging is influenced by many factors, silaging aids have always been used; these are now being increasingly replaced by enzymes.

The following enzyme applications in animal nutrition can be cited:

1. Using enzymes for preserving fodder, such as in silaging
2. Using enzymes for hydrolyzing antinutritive substances or for macerating raw materials used in pelletized feed
3. Using enzymes as feed additives—to support the digestive process—in the same way as aiding the digestive process in humans

5.13.1 Silage Enzymes

References: McDonald (1981), Van Belle (1986)

The objective of silaging green fodder is to maintain the nutrient quality present at the time of harvest. Silaging is a very ancient technique of fodder preservation, Van Belle (1986) refers to Egyptian wall paintings from the period between 1500 and 1000 BC. These illustrate that fodder at that time was already stored in some kind of stone silos. The principle of fodder preservation by silaging is fermentation by lactobacilli. Generating lactic acid by fermentation inhibits acetic acid formation and should completely inhibit butyric acid formation.

The microflora associated with grass or maize depends on the season, location, and climate. It consists of numerous aerobic bacteria and facultative anaerobic organisms. The microflora during silaging changes rapidly within the first few hours. The lactic acid bacteria take over. The number of cells increase from 10^6 to 10^8 per gram of silage. *Lactobacillus plantarum* is considered the most important bacterium in silaging and is responsible for the rapid and intensive lactic acid formation. During the initial phase of an efficiently performing silaging process, more than 50% of the microorganismal population consists of *L. plantarum* (Beck, 1968), considered to be the principal factor in acidification.

5.13.1.1 Silage Starter   The technical production of *L. plantarum* employs fermentation that produces an excellent starter organism for silaging. The starter dose is adjusted so that about 10^6 bacteria are available for each gram of plant material to be silaged. In practice, inoculation cell densities of 10^5 are usually sufficient. The starter marketed by Hüls AG (Germany) contains 2.5×10^{11} cells and is inoculated at 10–20 g per metric ton of plant material. This corresponds to an inoculation density of 2.5×10^6 cells/g plant material.

5.13.1.2 Enzymes for Improving Silage   An extensive literature exists on this topic; namely, many studies describe the effects of enzyme additives such as pectinases, cellulases, hemicellulases, proteases, amylases, and glucose oxidase. The general consensus is that enzymes are added because the polysaccharides of plant cell walls cannot be directly converted by lactobacilli. Addition of appropriate hydrolases is thought to disrupt the cell wall polymers and thus provide added nutrients for the lactobacilli.

In 1969, Röhm GmbH performed tests on grass silaging with high doses of a cellulase–hemicellulase mix from *A. niger* (Rohament CW [RM]). The preparation contained activities of carboxymethylcellulase, pectinases, β-glucanase, arabinofuranosidase, xylanase, arabinase, and other enzymes. Silaging could be improved significantly, but the commercial utilization of such enzyme preparations was not feasible economically at that time. Today, these earlier results are being reconsidered because it is now possible to work with higher doses of very efficient cellulases that are cost-effective. Bertin (1978) did extensive experimentation with enzyme preparations from various sources and with different substrates such as alfalfa, rye grass, and corn. He found that the combined activity of carboxymethylcellulase (CMCase), xylanase, salicinase, and amyloglucosidase was the most effective. Glucose oxidase, combined with cellulase, lowers acetic acid and butyric acid fermentation during silaging. In this case, the butyric acid content was lower than that in silaging with formic acid as a silaging aid (Hiyama, Jpn. Pat., 1985).

Today, cellulase preparations are used for grass silaging and amylase preparations for corn silaging, each dosed together with silage starter microorganisms (Hiyama, 1986).

Figure 5.84 illustrates grass silaging in which a bacterial starter preparation [RM] is combined with an enzyme preparation [RM].

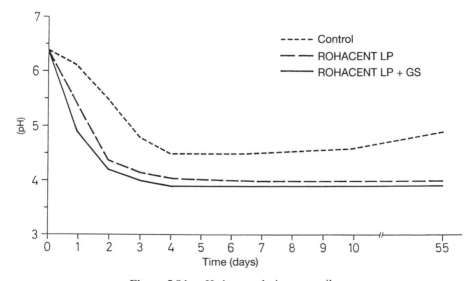

Figure 5.84   pH changes during grass silage.

The drop in pH caused by cellulase is minor; the titratable acidity has, however, markedly increased.

In addition to an increase in fermentable sugar, the enzyme additives (Rohacent GS [RM]) elevate the convertible energy and digestibility increases to 3%. Further, a 6% increase in net energy of cow lactation is achieved.

5.13.2   Enzymes for Fodder Production

Enzymatic pretreatment of raw materials facilitates their degradation during fodder production. The costs encountered with such enzymatic degradative procedures lie in the fact that water is required for enzymatic activity, and that optimal maceration can succeed only in a dilute aqueous solution in which the pH and temperature can be precisely controlled. Such wet maceration requires subsequent drying, which is expensive. Provided enzyme and substrate can be brought into intimate contact, and the system can be agitated under high pressure, only about 15–20% water would be necessary to effect hydrolysis. Such conditions can be met with an extruder. However, here the thermostability of the enzymes becomes the limiting factor. Thermostable bacterial α-amylase is a good candidate for such an application. On the other hand, cellulase and β-glucanase preparations are less thermostable and will be inactivated during pelletizing.

5.13.3   Enzyme Additives for Fodder

Reference: Scott et al. (1987)

Enzyme preparations are new and very important fodder additives for the future development of animal nutrition. Today, they are available to both the animal breeder and the fodder manufacturer.

A large number of experiments were conducted to enzymatically improve fodder or, more precisely, to improve the nutritive value of animal fodder. The results vary. In many cases, no effect or only very minor improvements were noted. Other cases presented a more favorable picture. Positive results were most often observed in cases where the fodder quality was low and the enzyme doses high. Just a few years ago, the high enzyme doses needed to yield results were not feasible economically. Now, this has drastically changed. New studies show that enzymes added to fodder under special conditions can be highly effective.

Much remains to be done to unravel the precise conditions for effective enzyme use. The types and ages of the animals, the type of fodder, and the type of husbandry are parameters to be considered when selecting enzymes and their specific activities. Precisely such experiments are now being conducted by many research groups to learn which enzyme system and what dose is required to improve fodder utilization for animal growth. For certain animal species, various countries have already introduced specific practices.

It is known that proteins, carbohydrates, and fats must be digested by native digestive-tract enzymes. Proteins are hydrolyzed to peptides and amino acids; carbohydrates to simple sugars such as maltose and glucose; and lipids to glycerol and fatty acids. Some raw materials such as proteins and carbohydrates are coated with substances that cannot, or can only with difficulty, be removed or digested by the mammalian digestive system. These include the β-glucans and pentosans, which are extremely hydrophilic and bind water. Enzymatic additives should support the native mammalian digestive system to enable the animals to utilize fodder rapidly and more efficiently. The nutritive effect will be more dramatic if fodder of initially low caloric value is enzymatically digested, or if indigestible feed proteins, denatured by heat treatments, will again become more digestible with enzyme treatment.

5.13.4   Enzymes for Chicken Feed

As early as 1957, tests were performed in the United States to replace corn-based chicken feed with enzymatically digested barley. Success was closely dependent on the barley variety fed to the chickens. Today, barley is a preferred feed component in chicken husbandry in many northern countries such as Scandinavia, Finland, Poland, and Canada. However, the amount of barley in the feed is critical. A feed with only 10–20% untreated barley reduces poultry growth and results in moist or liquid droppings. More recently, according to Scott et al. (1987), chicken feed can contain more than 70% barley if treated with an enzyme preparation derived from *Trichoderma reseii* that contains cellulase and β-glucanase activity (Avizyme [FR]), added to the feed at 1 kg per metric ton. The high viscosity and water binding caused by the barley's β-glucan in the chicken intestines is reduced by a cellulase from *T. viride* (cellulase Onazuka [YT]) (White et al., 1981). Experiments by Broz and Frigg (1985) indicate that supplementing the barley feed ration with purified β-glucanase from *A. niger* [NO] significantly increases the growth rate by 6.4–10.5% in newly hatched chicks. In addition, the enzyme increased the dryness of the chicken droppings. Supplementing a purified α-amylase showed no significant effect. The addition of β-glucanase to corn feed had no impact on the fattening of broiler chicken. In tests conducted by Scholtyssek and Knorr (1987), about 30% of a total ration of triticale (*Triticum*) or rye was supplemented with a cellulase preparation from *T. viride* carrying the added activities of xylanase and pectinase. Feed utilization was improved for both cereals by 5.5% for chicks up to 29 days old. The raw fiber of triticale feed and the NFE (nitrogen-free extracts, such as starch and sugar) were digested better with a rye ration, sugar utilization and nitrogen retention were improved.

A positive influence of β-glucanase on growth and feed utilization and quality of droppings was also obtained with feed mixes containing a high proportion of oats (Elwinger and Saeterby, 1987). In Japan, feed tests with white leghorns were conducted in 1987 by Y. Isshiki at the University of Kagawa. The feed contained 58% corn, 10.5% wheat bran, 12.5% rice bran, and 10% soy flour. The enzyme preparation used contain a mix of pectinase and xylanase (Aspellase [SA]). The weight at the age of 40 days improved by 7%, and the feed uptake, by about 2%. The feed utilization for the entire test period was 31.7% in the controls, and 33.7% in the group in which enzymes had been added to the feed.

The composition of the enzyme preparation must be adapted to the feed components. If the feed mix contains a higher proportion of rye that possesses soluble and insoluble pentosans, a strong binding of water occurs. Then, the enzyme preparation should contain xylanase activity in addition to the β-glucanases and cellulases mentioned earlier.

5.13.5 Some Feed-Test Results

Feed tests conducted jointly with technical enzyme manufacturers [RM], [MY], and at the University of Stuttgart-Hohenheim by Scholtyssek et al. (1989) employed complex cellulase preparations with the following results:

1. About a 10% improved digestibility of barley feed; feed utilization (wt/wt in kilograms) improved from 2.0 to 2.14%. Egg laying improved in laying hens fed enzyme-supplemented barley instead of corn.

2. The meat quality (breast muscle) could be markedly improved when feed containing 57–70% corn and milo together with 17–32% fish meal was supplemented with a 0.5% cellulase preparation and a cellulase–protease mix [MY]. This yielded a profit of U.S. $4.00 per 100 broilers in the cellulase-fed group, and U.S. $5.00 for those in the cellulase–protease group. In all cases the droppings had improved and, concomitantly, the hygienic conditions in the chicken coops.

3. On the addition of Roxazyme G [HLR], an enzyme complex from *T. viride,* the convertible energy of two different wheat strains could be increased by 7 and 8.1%, respectively and barley, by 13.4%. The added utilization of normally undigestible feed components not only improves the economics of animal husbandry but also contributes to a lowering of the environmental burden by the reduced amount of animal manure produced (Spring, 1992). Similar enzyme mixes can be employed for the husbandry of young pigs when the fodder mix is programmed to contain various kinds of grains.

4. Rye can be used only to a limited extent for poultry fattening. The high content of pentosans reduce the assimilation of nutrients. Jeroch (1991) tested rations with 31% of rye with the multienzyme preparation Bergazyme [BS]. The result was a higher feed absorption rate and an improved feed conversion. The mortality rate was reduced.

5.13.6 Enzymes for Pig Husbandry

In 1955/56, Lewis, Carton, and co-workers (Carton, 1956) reported piglets < 5 weeks old to be deficient in intestinal proteases. At this age, they are unable to digest soy protein. After an early weaning, enzyme secretion in piglets declines (Makkink, 1989). Because amylase and maltase are present in inadequate amounts, the feed for these animals must be formulated for easy digestion. This is accomplished with high-quality raw materials such as milk powder, whey powder, grain silage, and amino acid mixes. Alternatives are being sought because such feeds are quite expensive.

When young pigs weighing 5–10 lb were fed with fodder containing soy protein, their growth was greatly impeded. However, if fodder mixes were supplemented with bacterial proteases and bacterial α-amylase, the growth improved and fodder utilization normalized.

In addition to proteolysis, the degradation of carbohydrates plays an important role in the husbandry of young pigs.

A rapid disintegration of complex polysaccharides permits a rapid hydrolysis of oligosaccharides in the upper gastrointestinal tract. If sugars can be absorbed rapidly and continuously, diarrhea in young pigs can be prevented. Such diarrhetic conditions are associated with an uneven substrate fermentation by the intestinal flora (Inborr and Isshiki, 1987).

5.13.7   Enzymes for Calf and Cattle Fodder

Burroughs et al. (1959) report that supplementary enzymes such as amylases and proteases in calf fodder result in the following:

- Increased live-weight gain average of 12%
- Increased fodder uptake of ~5%
- Improved fodder utilization of ~7%
- Equal or improved meat quality
- Added cost of $8 per calf compared to standard fodders

The enzyme supplements consisted of bacterial proteases and amylases. The fodder ration contained a high proportion of maize that was partially substituted during experimentation by barley or sorghum.

Although these outstanding results were strongly dependent on fodder composition, they do show that it is possible for enzymes to generate positive results in calf nutrition or husbandry.

5.13.8   Use of Phytase in Animal Nutrition

The main part of feedstuffs for poultry is derived from plants such as wheat, corn, and rye. Up to 80% of the grain phosphorus is bound in the phytic acid. With Ca^{2+}, Mg^{2+}, and zinc, nearly unsoluble salts—phytin—are formed. The phosphorous of phytin can be resorped by the animal only after hydrolysis to inositol and inorganic phosphate (Nelson et al., 1968).

Despite improved phosphorus uptake by the animals, the quantity of excreted phosphorus increases environmental pollution, especially eutrophia of rivers and lakes in countries with high animal populations. A 30% reduction of phosphorus excretion seems to be possible (Simon et al., 1994). Phytase preparations from *Aspergillus* and yeast have been developed by several enzyme producers (e.g., Alko [AO] and Gist-Brocades [GB]).

REFERENCES

Beck, T., *Das Wirtschaftseigene Futter* **14**, 177 (1968).

Bertin, M. G., *Dissertation,* Univ. Paul Sabatier, Toulouse, 1978.

Broz, J., and Frigg, M., *Arch. Geflügelk.* **50**, 41 (1985).

Burroughs, W., Woods, W., Culbertson, C. C., Grieg, J., and Theurer, B., *J. Animal Sci.* **18**, 1524 (1959).

Carton, D. V., *Proc. Distillers Feed Conf.* **11**, 27 (1956).

Elwinger, K., and Saeterby, B., *Swed. J. Agric. Scand.* **37**, 175 (1987).

Hiyama, K., "Production of silage," Jpn. Patent 61/209 554 (1986).

Huhtanen, P., Hissa, K., Jaakkola, S., and Poutiaien, E., *J. Agric. Sci. Finl.* **57**, 285 (1985).

Inborr, J., and Isshiki, Y., "Enzymes in pig feed," *38th Annual Meeting European Association Animal Production,* Sept. 28–Oct. 10, Lisbon, 1987.

Isshiki, Y., Kagawa University, personal communication to Meiji Kasei Co., 1987.

Jeorch, H., "Verbesserung des Futterwertes von auf Gerste und Roggen basierendem Broiler-futter durch Enzymsupplementierung," Internatl. Scientific Conf.: Industrial Enzymes, Probiotics and Biological Additives, Kaunas/Litauen, 1991.

Lewis, C. J., Carton, D. V., Liu, C. H., Speer, V. C., and Ashton, G. C., *Agr. and Food Chem.* **3**, 1047 (1955).

Makkink, C., *Kraftfutter* **12**, 482 (1989).

McDonald, P., *The Biochemistry of Silage,* Academic Press, New York, 1981.

Nelson, T. S., Shieh, T. R., Wodzinski, R. J., and Ware, J. H., "The availability of phytate phosphorus in soya bean meal before and after treatment with a mold phytase," *Poultry Sci.* **47**, 1842–1848 (1968).

Scholtyssek, S., Swiercczewska, E., and Niemiec, J., *Kraftfutter* **4**, 116 (1989).

Scholtyssek, S., and Knorr, R., *Arch. Geflügelk.* **51**, 10 (1987).

Scott, D., Hammer, F. E., and Szalkucki, T., in Knorr, D. (ed.), *Bioconversions: Enzyme Technology in Food Biotechnology,* Marcel Dekker, New York, 1987, p. 413.

Simon, O., Ruttloff, H., Klappach, G., "Einsatz von Enzympräparaten in der Tierernährung und in der Futtermittelindustrie," in H. Ruttloff (ed.), *Industrielle Enzyme,* Behr-Verlag, Hamburg, 1994, pp. 961–986.

Spring, W. G., in Heinemann, R., and Wolnak, R. (eds.), *Opportunities with Industrial Enzymes,* Wolnak, Chicago, 1992, p. 82.

Spring, W. G., Technical information, F. Hoffmann-La Roche Ltd., 1993.

VanBelle, M., *L'ensilage—Nouveaux Aspects Biologiques.* Ed. Sanofi Publishers, Paris, 1986.

White, W. B., Bird, H. R., Sunde, M. L., Prentice, N., Burger, W., and Marlett, J. A., *Poultry Sci.* **60**, 1043 (1981).

5.14 LYSIS OF MICROBIAL CELL WALLS

Reference: Andrews and Asenjo (1987)

A large number of compounds of interest to both industry and medicine are produced by microorganisms. Many of these compounds are released into the surrounding culture medium passing through the plasma membranes and cell walls of the microorganism. These include many industrial enzymes. Other proteins, carbo-hydrates, and enzymes remain in the cell. To obtain these substances, the cell walls have to be made permeable to such substances. Thermal and mechanical methods are employed in industrial technology. Mechanical processes such as pressurized homogenization or treatment with pebble mills require an extensive financial outlay for cooling if one wishes to extract active enzymes from the cells. The production of yeast extract by autolysis is the most widely used cell disruptive method. This enzymatic method will be referred to later; however, a few references regarding products obtained from lysing cell walls are presented in Figure 5.85.

The lysis of cell walls is also of interest for other processes that involve killing microorganisms by destroying their cell walls, or controlling microbial contamina-tion in products. In a review article, Scott (1988) refers to such lytic enzymes as an-timicrobial enzymes.

Nature evolved several substances for building cell walls. The number of basic building blocks is relatively limited. Cell walls are composed of amino acids and various sugars; the plasma membranes consist primarily of lipids. A sophisticated cross-linking and weaving of polymers composed of these building blocks yield different, firm, yet elastic structures within which other reinforcing polymers are embedded. The structure reminds one of today's steel-reinforced concrete in which steel rods are wired together into mats that are then bound to each other and onto which the cement is poured.

5.14.1   Yeast Cell Lysis

The cell wall of yeast has two principal layers. The outer layer consists of a man-nan–protein complex; the inner layer is composed of glucan; the sketch shown in Figure 5.86 illustrates these structures (Hunter and Asenjo, 1986).

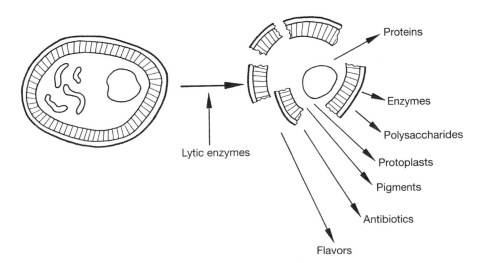

Lytic enzymes

Proteins

Enzymes

Polysaccharides

Protoplasts

Pigments

Antibiotics

Flavors

Figure 5.85   Products obtained from lysing microbial cell walls.

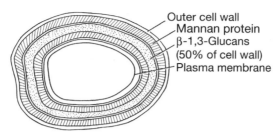

Figure 5.86   Cell wall structure of yeast (Hunter and Asenjo, 1986).

Enzymes used in disrupting the yeast cell wall must, therefore, be composed of a mixture of several enzymes such as β-1,3-glucanase, protease, β-1,6-glucanase, mannanase, or chitinase. These enzymes act synergistically in the degradation of both layers. However, the cell walls of various yeast species differ in their composition; therefore, not all yeast respond equally well to specific enzyme mixtures. Generally, for *Saccharomyces* species, mannanases with and without proteases act in concert with β-glucanases. Commercial galactomannanase preparations can be employed for this purpose. According to Knorr et al. (1979), a combination of lysozyme and endo-β-1,3-glucanase can digest the outer mannan–protein layer quite easily. Proteases, particularly papain, are able to support the autolytic processes involved in the manufacture of yeast extract.

5.14.2   Yeast Extract

Reference: Kelly (1983)

Yeast is produced on a grand scale for the baking industry. Large quantities of yeast are by-products in the brewing and distilling industries, but only small amounts can be employed in the subsequent fermentation processes. About 75% remains as surplus, some of which is used in fodder production. Another portion is used for making yeast extract. In 1983, about 25,000 t of yeast extract was produced worldwide. Yeast extracts are used as flavoring agents in the food industry. Yeast extracts are integrated with dietary products and thus find use in the pharmaceutical industry.

Yeast extract manufacture involves several process steps. First, the yeast cells are killed. This is accomplished by plasmolysis-inducing substances; for example, salts or organic solvents such as isopropanol (e.g., 0.5% for ~5 h duration) or ethyl acetate can be used; these are subsequently removed from the extracts by distillation. Plasmolysis can also be effected by heat treatment (for ~48 h at 58°C, pH 5.5). Heat plasmolysis requires careful monitoring to prevent thermal denaturation of the enzymes that are sought.

The next step is autolysis: the destruction of the cell by its own enzyme systems. Autolysis is conducted at temperatures between 30 and 55°C for about 20 h. Papain sometimes is added to support autolysis (~0.04% of a standard papain preparation at pH 5.0 and 50–60°C); this will improve the yield of solutes (Kelly, 1983). The next step is pasteurization followed by clarification of the extract with centrifugation and filtration. The extracts are concentrated to 70–75% of dry substance.

Making yeast extract with cell-wall-lysing enzymes from *Arthrobacter* sp. was recommended by two Japanese manufacturers [AM, DA]. At pH 7.5, *Arthrobacter* glucanase and alkaline bacterial protease are added to a 20% cell suspension at 45–50°C, which was previously plasmolyzed with 3% NaCl or 1% ethyl acetate and incubated for 10–13 h.

In still another method, the yeast cell suspension is heated to 90°C for 10 min prior to enzyme treatment. These enzymatic experiments (Tunicase [DA]) resulted in the reduction of autolysis time (from 48 to 16 h) and increased extract yield from 55 to 70%.

5.14.3   Fungal Cell Wall Lysis

According to Rockem et al. (1986), the cell wall of *Aspergillus niger* is composed of chitin:glucan at a ratio of about 1:1.4. Thus, the chitinase system, consisting of endochitinase, exochitinase, and chitobiase, can, together with other enzymes, lyse fungal cell walls.

The enzyme cocktail produced by *Trichoderma harzianum* (Novozyme 234 [NO]) is well suited for lysing the cell walls of *Aspergillus, Saccharomyces,* and *Penicillium.* This preparation is used to isolate protoplasts from fungal organisms. The key activity of such an enzyme preparation is directed against an insoluble glucan called *mutan;* in addition, it exhibits cellulase, laminarinase, xylanase, chitinase, and protease activities.

5.14.4   Bacterial Cell Wall Lysis

The cell wall of gram-positive bacteria is composed of several layers; the peptidoglycan layer constitutes the largest portion and is responsible for cell wall strength. This layer is interspersed with lipopolysaccharides, teichoic acid, and proteins. The latter form an outer membrane that can be removed by different substances, thus providing access for a lysozymic attack on the peptidoglycan layer. Such substances include ethylenediaminetetraacetic acid (EDTA) and Tris buffer. Polycations, together with anionic detergents such as Na-dodecyl sulfate, can lyse the lipopolysaccharide layer.

To lyse a bacterial cell suspension in one hour, approximately 250 mg/liter of ovolysozyme with an activity of about 20,000 units/g protein is required.

Gram-negative bacteria have a cell wall consisting of a peptidoglycan layer (also called *muramic acid* or *murein*), located between the lipopolysaccharides of the outer and inner membranes.

The outer membrane must be removed with detergents before the lytic enzymes can exert their action.

Lysozyme is currently the only technical enzyme specific for lysing bacterial cell walls.

Its main use in the food industry is for making hard cheeses such as Parmesan, Grano, Provolone, Asiago, and Emmentaler. Organisms such as *Clostridium tyrobu-*

tyricum must be killed because they can cause cheese swelling. These organisms enter the process via the fodder silage fed to the cows. This is the reason for the prohibition of feeding silage in some areas.

REFERENCES

Andrews, B. A., and Asenjo, J. A. *Tibtech,* Vol. 5, Elsevier, Cambridge, UK, 1987, p. 273.

Hunter, J. B., and Asenjo, J. A., in Asenjo, J. A., and Hong, J. (eds.), *Separation, Recovery, and Purification in Biotechnology: Recent Advances and Mathematical Modelling,* ACS, Washington, DC, 1986, p. 9.

Kelly, M., in Godfrey, T., and Reichelt, J. (eds.), *Industrial Enzymology,* The Nature Press, New York, 1983, p. 457.

Knorr, D., Shetty, K. J., and Kinsella, J. E., *Biotechnol. Bioeng.* **21,** 2011 (1979).

Rockem, J. S., Klein, D., Toder, H., and Zomer, E., *Enzyme Microbiol. Technol.* **9,** 588 (1986).

Scott, D., "Antimicrobial Enzymes," Internatl. Symp. Enzymes in the Forefront of Food and Feed Industries, June 17, 1988, Tkk, Espoo/Otaniemi, Finland.

5.15   TECHNICAL APPLICATIONS OF GLUCOSE OXIDASE

References: Hammer (1993), Hamer (1995), Law and Goodenough (1995), Woods and Swinton (1995)

Glucose oxidase (GO) is used in food technology for stabilizing various foods and beverages to prevent color and taste changes. Such changes that reduce food quality can be the result of oxidative processes mediated by atmospheric oxygen or the Maillard reaction. The latter occurs between aldehydes and amino groups in foods. The aldehyde groups are associated primarily with glucose, while the amino groups are derived from amino acids or, indirectly, from proteins. In some situations, the Maillard reaction may even be desired where browning of baked goods, roasted meat, or potatoes is required.

On the other hand, this reaction is undesirable for egg products, for canning fruits and vegetables, or in fruit juices and wine.

One Maillard reaction partner, the aldehyde group, can be decreased in number or eliminated from the reaction by the glucose oxidase–catalase system.

5.15.1   Glucose Removal

An important application for the glucose oxidase–catalase system was developed mainly by Scott (1953) for desaccharification in order to stabilize commercial liquid egg white.

To create a stable product with extended shelf life, the concentration of free glucose must be reduced to less than 0.1%. Egg white is normally concentrated about eightfold; whole egg, about fourfold; and egg yolk, only about twofold. Therefore, the glucose concentration in fresh egg white will be reduced the greatest.

The reaction is performed with excess oxygen. The reaction time for desaccharification to a specific glucose concentration is inversely proportional to the enzyme concentration. Generally, hydrogen peroxide is added, producing excess foaming. The problem can be controlled with incremental dosing of hydrogen peroxide (Scott, 1953).

Richter (1975) designed the process as follows. Egg white is heated to a temperature of approximately 32°C and the pH adjusted with granular citric acid to pH 7.0–7.5. Prior to addition of the enzyme, about 1.2 liters of 35% hydrogen peroxide per metric ton of egg white is added in incremental aliquots. Then, about 100–150 g of glucose oxidase with an activity of 1500 GO units/ml is introduced into the reaction mixture. Desaccharification requires about 8 h. This process can be monitored with a glucose oxidase test. After desaccharification, the product is pasteurized.

The method for whole-egg desaccharification differs somewhat from that for egg white. The pH is adjusted and the amount of enzyme should be about 200,000 units per metric ton; this is somewhat higher than that for egg white. Desaccharification requires about 6–8 h.

5.15.2   Oxygen Removal (Deaeration)

References: Szalkucki (1993), Lea (1995)

5.15.2.1   *Citrus Concentrates and Citrus Beverages*   Both products contain solubilized oxygen, which yields peroxide on exposure to sunlight. These peroxides cause significant flavor changes in citrus beverages. But orange and grapefruit juice concentrates also experience oxidative changes, so it seems desirable to remove oxygen.

The liquid enzyme preparation should be added to the beverage immediately prior to bottling or canning. The bottled or canned beverage must not be sterilized because the enzyme will be destroyed on pasteurization.

The rule for dosing the enzyme is as follows: 3–7 ml of glucose oxidase (e.g. Asgo 1500 [FS]) with an activity of 1500 units can replace ~100 g ascorbic acid (Scott and Craig, 1966). With decreasing pH, the enzyme dose must be elevated considerably; this is also true for highly acidic concentrates.

If, at pH 3.6, about 3500 units of enzyme per 100 liters are required, then, at pH 2.8, about 10,000 units will be needed; for juice concentrates, 10 times this amount is required. Glucose oxidase, added to freshly expressed orange juice, will cause a significant decrease in the potential for yeast growth. As a result, the shelf life (half-life) is doubled (Sagi and Mannheim, 1988).

5.15.2.2   *Beer Stabilization*

Reference: Power (1993)

Beer changes taste and color significantly under the influence of oxygen. The possibility of beer stabilization with glucose oxidase was, therefore, studied from several points of view. Generally, 15–40 GO units per liter should suffice to stabilize beer against changes in taste. In practice, however, this kind of beer stabilization has yet to be realized.

Hartmeier (1982) studied beer stabilization with immobilized glucose oxidase and catalase. The oxygen in beer could be deaerated more rapidly and completely in the presence of a coimmobilized amyloglucosidase. Cells of *A. niger* (carrying glucose oxidase) were coimmobilized with an amyloglucosidase in solution. Coupling was performed with tannin and glutaraldehyde.

5.15.2.3   Wine Stabilization   Glucose oxidase–catalase can also stabilize white wine against both browning and changes in taste (Ough, 1960).

5.15.2.4   Mayonnaise Stabilization   The product—an emulsion of oil and water—contains oxygen that, on exposure to light, reacts at room temperature and undergoes a considerable change in taste. On addition of 15–50 GO units per kilogram of product, mayonnaise can be stabilized (Bloom et al., 1956). One must, however, employ an enzyme preparation that is completely amylase-free; otherwise, the viscosity will be rapidly reduced and the emulsion disrupted.

5.15.2.5   Gluten Stabilization   Frequently, oxidants are used for strengthening the gluten in flour. A similar but weaker effect was achieved with a catalase-free glucose oxidase (Luther, 1957). Glucose oxidase, in the presence of catalase and ascorbic acid, has a positive influence on the baking properties of flour (Maltha, 1961).

5.15.2.6   Cheese Stabilization   Previously glucose oxidase was used in cheese packaging as a seal against oxygen. A film of glucose oxidase was sprayed on the packaging material to provide an oxygen barrier. Vacuum packaging, however, has since been shown to be superior to glucose oxidase packaging. Scott (1988) reported on a new application of this enzyme for cheese powder stabilization. A mixture of glucose and powdered glucose oxidase replaces about 50% of the customarily used anticlumping agent. The enzyme is activated by residual moisture in the cheese powder, thereby completely removing the oxygen from the sealed package.

REFERENCES

Bloom, J., Scofield, G., and Scott, D., *Food Packer* **37,** 16 (1956).

Hamer, R. J., in Tucker, G. A., and Woods, L. F. J. (eds.), *Enzymes in Food Processing,* 2nd ed., Blackie Academic & Professional (imprint of Chapman & Hall), New York, 1995, p. 215.

Hammer, F. E., in Nagodawthana, T., and Reed, G., *Enzymes in Food Processing,* 3rd ed., Academic Press, New York, 1993, p. 251.

Hartmeier, W., in Dupuy, P. (ed.), *Utilisation des enzymes en technologie alimentaire, Internatl. Symp.,* May 5–7, Techn. Doc. Lavoisier, Paris, 1982, p. 205.

Law, B. A., and Goodenough, P. W., in Tucker, G. A. and Woods, L. F. J. (eds.), *Enzymes in*

Food Processing, 2nd ed., Blackie Academic & Professional (imprint of Chapman & Hall), New York, 1995, p. 114.

Lea, A. G. H., in Tucker, G. A., and Woods, L. F. J. (eds.), *Enzymes in Food Processing,* 2nd ed., Blackie Academic & Professional (imprint of Chapman & Hall), New York, 1995, p. 246.

Luther, H. G., "Treatment of flour with glucose oxidase," U.S. Patent 2,783,150 (1957).

Maltha, P. R. A., "Verfahren zur Verbesserung der Backfähigkeit von Mehl oder Teig," German Patent 1 050 703 (1961).

Ough, C. S., *Mitt. Klosterneuburg* **10A,** 14 (1960).

Power, J., in Nagodawithana, T., and Reed, G., *Enzymes in Food Processing,* 3rd ed., Academic Press, New York, 1993, p. 455.

Richter, G., technical information, Miles-Kali-Chemie, 1975.

Sagi, I., and Mannheim, C. H., "Shelf life of enzymatically deaerated fresh chilled citrus juice," *Internatl. Citrus Congress,* March 6–12, 1988, Tel Aviv.

Scott, D., *J. Agric. Food Chem.* **1,** 727 (1953).

Scott, D., and Craig, T., *Proc. Soc. Soft Drink Technol.* 25 (1966).

Scott, D., in Whitaker, J. R., and Sonnet, P. E. (eds.), *Biocatalysts in Agricultural Biotechnology,* p. 176. ACS Symp. Series 389 (3rd Chem. Congr. North America, 195th ACS Meeting, Toronto, 1988), ACS, Washington, DC, 1989.

Szalkucki, T., in Nagodawithana, T., and Reed, G., *Enzymes in Food Processing,* 3rd ed., Academic Press, New York, 1993, p. 285.

Woods, L. F. J., and Swinton, S. J., in Tucker, G. A., and Woods, L. F. J., (eds.), *Enzymes in Food Processing,* 2nd ed., Blackie Academic & Professional (imprint of Chapman & Hall), New York, 1995, p. 260.

6 Legal Considerations

Enzymes are primarily ingested from food. All fresh fruits and vegetables contain active enzymes. Products of fermentation, including cheese, kefir, yogurt, raw sausage, sauerkraut, pickles, beer, and wine, contain both the native enzymes and the microorganisms that produced the enzymes. These include a number of non-pathogenic microbes such as lactobacilli, micrococci, yeasts, streptomycetes, and molds. Generally the fermentation products are not considered health hazards; many food chemistry studies have ascertained that such products contain no detectable amounts of toxic substances such as, for example, mycotoxins. A good example of the concept that enzymes used in food processing are nothing other than normal cell constituents was provided by Frank (1988). Curing young Matjes herrings involves the native muscle enzymes and microbial exoenzymes. The latter are not native to the fish, but cannot be avoided because the fillets are not sterile. Thus, they become auxiliary agents in processing the Matjes herring or the current substitute, the Matjes-style herring filet.

The active enzymes that are consumed are proteins and as such are equally harmless. Boiling food inactivates the enzymes, and the enzyme protein is denatured in the same way as other proteins. This may be the reason why some regulatory groups list the enzymes with food products. Enzyme technology utilizes enzymes from animals, plants, and microorganisms. In most of the processes discussed, the enzymes are heat-inactivated at the time the reaction is terminated. Traces of denatured protein that remain are catabolized during gastrointestinal passage. On the other hand, enzyme processes have been described during which a carrier-bound enzyme comes in contact only with the substrate, and does not remain as part of the product. In these instances, the enzyme is considered a technical processing aid. However, other views hold that enzymes are technical aids even when they are denatured and remain in the food product in small quantities (~ 1 ppm).

Regardless of whether enzyme preparations are defined as food additives or technical aids, it must be established that they are not health hazards. Naturally, it could be possible to purify the enzymes to such an extent that only the "pure enzyme protein" would be used. Such purified enzymes, used in medicine and laboratory analysis, are very expensive, and therefore the cost is prohibitive for processing food. Denner (1983) describes an example: "It would be a terrible misuse of public funds if one would salt the streets in winter with reagent-grade table salt, and vice versa an extreme violation of public health to add rock salt that contains significant amounts of heavy metal to food." Thus, specifications for food and food additives that consider the health factors first and the economics second will have to be developed.

405

6.1 REGULATORY REQUIREMENTS FOR ENZYME PREPARATIONS

The purpose of this brief review on the legal aspects of enzyme utilization is to inform the reader about the measures that the scientific community and legislators have taken to provide maximum safety for both the user and the consumer of enzyme preparations. It should also be pointed out that the regulations for enzyme preparations and their use are by no means uniform throughout the world or even within the countries of the European Community (EEC*); therefore, much work is still needed to institute uniform regulations.

Employing recombinant DNA (rDNA) technology for the production of enzymes initiated discussion on the potential health hazards and the safety of such enzymes in the consumer marketplace. The EEC (1990a,b,c) has published safety guidelines on the contained use of genetically modified microorganisms, their deliberate release into the environment, and risk management of workers exposed to such organisms. Toet (1992) reported on the effects of rDNA technology relative to the safety requirements for enzyme preparations. According to AMFEP[†] (1988), it is necessary to regulate only those microorganisms—genetically engineered or not—that serve as a source of enzymes used in food production and processing with no limitation other than good manufacturing practice (GMP).

6.2 WHO IS INVOLVED WITH ENZYMES?

Public health agencies in several countries have made an effort to establish regulatory guidelines for the use of enzymes in concert with legislation on food additives. The regulations vary greatly and often are not published in detail. Therefore, it was most commendable that the FAO/WHO[‡] Joint Expert Committee on Food Additives (JECFA[¶]) dealt with the problems associated with the safety of enzymes. A group of toxicologists discussed questions of toxicity and purity for enzymes used in food production. Reports from the Committee sessions held in 1971 (15th), 1974 (18th), and 1977 (21st) [FAO/WHO, 1972, 1974, 1978, 1981] became the guidelines for national laws or regulations. Session 21 generated the criteria for evaluating the health risks of enzymes according to their origin (FAO/WHO, 1978).

It was determined that enzymes from animal organs or plant parts require no toxicological studies for approval. This primarily concerns the industrial enzyme preparations from the pancreas, and pepsin, rennet, papain, and bromelain. Some toxicological

---

*European Economic Community, Research and Advisory Consortium, 48 rue du Cardinal, B-1040 Brussels, Belgium.

[†]Association of Microbial Food Enzyme Producers (EC), 172 ave de Cortenbergh, B-1040 Brussels, Belgium.

[‡]FAO, United Nations Food and Agriculture Organization, Viale delle Terme di Caracalla I-00100 Rome, Italy; WHO, World Health Organization, 20 Avenue Appia, CH-1211 Geneva 27, Switzerland.

[¶]Joint Expert Committee on Food Additives of the United Nations Food and Agricultural Organization (FAO; see address under this listing).

testing may be required for enzyme preparations originating from microorganisms that occur as contaminants in food but generally are viewed as harmless. However, enzyme preparations from cultures of *Bacillus subtilis, Aspergillus niger*, and *A. oryzae* must be examined for acute toxicity as well as in a short-term toxicological test. Long-term toxicological tests are required for enzymes from less well-known organisms. All pathogenic microorganisms have been excluded from use in production systems. Furthermore, the Committee has established guidelines for purity of industrially used enzyme preparations [15th Session, JECFA; see FAO/WHO (1972)].

The Code Committee on Food Additives (CCFA) is concerned itself with food additives. An open-ended list of additives and technical aids is continually examined by the Committee's study groups. If a new substance is added to the list, the CCFA can ask the Expert Committee (JECFA) for further testing.

The enzyme preparations are then listed with their general purity requirements, specifications, and analytic methods in the Food Chemical Codex (NAS, 1981).

The General Directorates (GD) in the EEC coordinate the various enzyme regulations in member countries. Responsibilities lie with GD III (internal market and industry), GD IV (agriculture), and GD XII (science, research, and technology). The latter oversees the developments in biotechnology, an important future field in the European Economic Community (EEC).

Currently, no horizontal EEC directive regarding food enzymes exists; there are, however, a series of vertical directives that include requirements for the use of enzymes (e.g., Council Directives 79/337/EEC, 90/219/EEC, 90/220/EEC, 90/679/EEC).

One of these is the Fruit Juice Directive (EEC 726/75), which states in Article 4 that only pectinases, proteases, and amylases are allowed in juice production. This restricted enzyme use is based on an approved list that was adopted by legislation in Denmark and Germany. In contrast, the United Kingdom (UK) did not adopt Article 4, and enzymes can be used in juice production that are not contained in Article 4 of the EEC Directive. Cellulase, for example, can be employed in juice production in the United Kingdom.

The EEC Wine Regulation (EEC 79/337) permits treatment of grapes and grape must (unfermented or partially fermented) with pectolytic enzymes. An application for approval of β-glucanases in winemaking has been submitted.

The Association of Microbial Food Enzyme Producers (AMFEP), founded in 1977, represents an organization of enzyme producers in western Europe. It has 12 members, including Germany (2), the United Kingdom (2), Finland (2), Belgium (1), the Netherlands (1), France (1), Denmark (2), and Switzerland (1). Several study groups deal with questions of toxicity testing, microbial safety, and approval in general. Guidelines for the production and application of enzymes were published in 1983 in a brochure entitled *Regulatory Aspects of Microbial Food Enzymes* (AMFEP, 1988) and presented to the EEC and the individual governmental agencies.

The Association of Manufacturers of Animal-derived Food Enzymes (AMAFE*) is an organization with members in Austria, Denmark, France, Germany, Holland,

---

*Association of Manufacturers of Animal Derived Food Enzymes, c/o [CH], Boege Allee 10-12, DK-2970 Hørsholm, Denmark.

Italy, and Sweden. A proposal that will regulate enzymes from animal organs and edible plant parts for use in the food industry has been drafted.

6.3   TOXICITY AND ALLERGIES

Pure enzymes per se are not toxic, but as all proteins they can cause allergies. There is no reason why the consumption of enzyme proteins should result in more allergies than the consumption of any other protein. However, inhaled enzymes, especially protease concentrates, are highly effective allergens. Allergies from enzymes are in most cases inhalation allergies and rarely contact allergies. A sensitization to enzymes that have been used as food additives is rare because

1. The amounts added are small, in the range of a few parts per million of enzyme protein.
2. Only in a few cases are the enzymes consumed with food not thermally denatured. In breadmaking, fruit juice, and glucose production, all the enzymes are denatured on heating. It is a known fact, however, that consumption of cooked protein can trigger allergic reactions.

Enzyme producers themselves have taken the initiative to avoid or minimize these problems. Enzyme dust can be avoided by having the enzymes granulated, enclosed in microcapsules, and used as liquid preparations.

6.4   TOXINS

6.4.1   Toxic Activities

It has been known for some time that microorganisms produce toxic metabolites. One group of mycotoxins produced by fungi, the aflatoxins, have drawn attention recently because of severe cases of poisoning. Worldwide, many people die annually from poisoning by the botulism toxin in spoiled food. It is also known that these toxins are carcinogenic.

A large number of such toxins have been detected over the past few years by highly refined analytic methods. Their chemical structures have been determined and sensitive analytic methods for their detection have been developed. The level of detection has been lowered to the part-per-billion range, that is, to the level of 1 mg per metric ton of foodstuff.

What must enzyme producers do to ensure that their products are free of such toxins? Initially, the producer will exclude all strains of microorganisms known to generate toxins from the production process, even though such toxins are not formed during exponential growth of the microorganisms. However, this is the phase during which the enzymes are generated. It is standard procedure to test for toxins in exploratory laboratory experiments when screening microorganisms.

Currently, tests for mycotoxins are conducted during product quality control [see AMFEP (1988) recommendations].

6.4.2 Antibiotic Activities

It is well known that microorganisms synthesize antibiotics. To assure the absence of antibiotic activity in an enzyme preparation, JECFA has recommended series of tests employing specific microorganisms. These organisms and the associated test methods are listed in the AMFEP brochure (AMFEP, 1988).

6.4.3 Documentation and Classification

The standards for the classification of commercial enzyme products were, as mentioned earlier, developed by FAO/WHO committee experts. The various enzyme producers did, in turn, draft procedures for documentation and classification of enzymes that will be used in the approval process by the national regulatory agencies. The AMFEP has derived its *General Standards for Enzyme Regulations* from this information. These standards serve as guidelines for licensing enzymes in the EC member countries.

The standard describes industrial enzymes and their nature and properties. It further contains suggestions for the requirements of free and carrier-bound enzymes and includes a list of microorganisms that produce the enzymes in question. This list is based on JECFA recommendations. It groups the organisms into three categories (see Sections 6.4.3.1–6.4.3.3).

6.4.3.1 a) Microorganisms Traditionally Used in Food Production

- *Bacillus subtilis*, including strains such as *Bacillus mesentericus, B. natto, B. amyloliquefaciens* (amylases, proteases, β-glucanases, hemicellulases)
- *Aspergillus niger*, including strains of *A. aculeatus, A. awamori, A. ficuum, A. foetidus, A. japonicus, A. phoenicis, A. saitoi,* and *A. usamii* (α-amylases, amyloglucosidase, cellulase, catalase, β-glucanase, lipase, pectinases, proteases, pentosanases, lactase, α-glucosidases, β-glucosidases, glucose oxidase, etc.)
- *Aspergillus oryzae*, including *A. effusus* and *A. sojae* (α-amylases, amyloglucosidase, glucanases, cellulase, pectinases, proteases, lactase, etc.)
- Various other microorganisms:
  - *Mucor javanicus* (lipase)
  - *Rhizopus arrhizus* (lipase, amyloglucosidase)
  - *Rhizopus oligoporus*
  - *Saccharomyces cerevisiae* (invertase)
  - *Kluyveromyces fragilis* (lactase)
  - *Kluyveromyces lactis* (lactase)
  - *Leuconostoc oenus* (maleic acid decarboxylase)

This group also includes the innoculation cultures used in the production of fermented food (DFG, 1987).

b)   Microorganisms Recognized as Safe Based on Scientific Evidence from Toxicological Testing

- *Bacillus stearothermophilus* (proteases)
- *Bacillus licheniformis* (proteases, amylases)
- *Bacillus coagulans* (glucose isomerase)
- *Bacillus megaterium* (β-amylase)
- *Bacillus circulans* (β-glucanase)
- *Klebsiella aerogenes* (pullulanase)

c)   Microorganisms Not Listed in Sections a) and b)   The recommendations of the Deutsche Forschungsgemeinschaft* (DFG, 1987) refer to "enzyme preparations from lesser known organisms which commonly are not known to be associated with food." Regarding this statement, the authors caution that this classification is unclear from a food microbiologist's point of view and, therefore, unlikely to be used.

Bacillus acidopullulyticus
Bacillus cereus (protease, β-amylase)
Mucor miehei (rennet, lipases)
Mucor pusillus (rennet)
Endothia parasitica (rennet)
Actinoplanes missouriensis (glucose isomerase)
Streptomyces albus (glucose isomerase)
Trichoderma reesei (T. viride) (cellulase, β-glucanase)
Trichoderma harzianum (β-glucanases)
Penicillium lilacinum (dextranase)
Penicillium emersonii (β-glucanase)
Penicillium funicolosum (dextranase)
Penicillium simplicissimum (pectinase)
Sporotrichum diamorphosporum (cellulase, hemicellulase)
Streptomyces olivaceus (glucose isomerase)
Streptomyces raserubiginosus (glucose isomerase)

On the basis of this list, recommendations were made for the toxicological examination of enzyme preparations from microorganisms in groups a) and c); in essence, these represent a more precise definition of the JECFA recommendations. The study of the DFG (DFG, 1987) has incorporated these recommendations.

Address: Kennedy Allee 40, D-53175 Bonn, Germany.

6.5   DOCUMENTATION

The documentation prepared in support of an enzyme approval application should include the following:

1. Information on the enzyme source (referring to the preceding list of micro-organisms).
2. Major enzyme activities (EC numbers); the most common coenzymes should also be listed.
3. Data obtained from toxicological tests.
4. Methods used for production and product quality control.
5. Information about the area of application and quantities required.
6. Information about the activity and stability of the product.
7. Information about contaminants, based on the JECFA recommendations.
   a. Arsenic, not exceeding 3 mg/kg
   b. Lead, not exceeding 10 mg/kg
   c. Heavy metals, not exceeding 40 mg/kg, calculated on the basis of lead
   d. Hygiene requirements; microbes
      i. *Salmonella*, not present in a 25-g sample
      ii. *E. coli*, not present in a 25-g sample
      iii. Coliforms, not exceeding 30/g
      iv. Total live bacteria, not exceeding $5 \times 10^4 \text{ g}^{-1}$

   The Society of German Chemists (GDCH, 1984) standards provide a practical supplement; specifically, enzyme preparations must be devoid of viable pathogens that could damage human health. Because the total permissible bacterial counts in food are defined, only those enzyme preparations that do not exceed the maximum bacterial count permitted are employed. Determining this is important because enzymes can be blended into food for which bacterial counts have not been established.

   e. Mycotoxins. The German standard permits not more than 10 μg of total aflatoxins per kilogram and not more than 5 μg of aflatoxin B_1 per kilogram.
   f. Antibiotic activity. Test negative.
8. Data on carrier material, and blending agents, as well as antioxidants, stabilizers, and preservatives.

In addition to information about the enzyme manufacturer, items 1 through 8 should also be part of the enzyme specification for enzyme users.

6.6   SOME NATIONAL REGULATIONS

Argentina   There is no specific legislated regulation for enzymes. Pectinases are permitted in winemaking. For licensing, data on activity and purity must be submitted.

Australia   Only papain is included in the list of food additives. Enzymes are authorized as technical aids in winemaking. Use of amylases in the starch industry is permitted as well.

Belgium   No special legislation exists for food enzymes; yet there is a general regulation for food additives that includes enzymes.

Denmark   An approved list exists for food additives (1982). The registry format includes data on enzyme characteristics, source, and production methods and requires submission of toxicological data. Enzymes from animal organs and edible plant parts, as well as immobilized enzymes, require no toxicological data. Tests of enzymes from other sources follow JECFA recommendations.

Germany   Enzymes are additives according to paragraph 2 of the current "Food and Consumer Goods" law. Paragraph 11, Section 3 of this law specifies, however, that Subsection 1, number 1.a, is not applicable to enzymes or innoculation cultures. Therefore, enzymes can be used in food without additional approval. Some rules and regulations provide exceptions and permit the use of only very specific enzymes.

Restricted applications are specified in the Dairy Act of December 16, 1986, and in the cheese regulation of April 14, 1986. For cheese production, only rennet from animal and microbial origin is permitted; and for lactose hydrolysis, only lactase.

For fruit juice production, the fruit juice regulation permits the use of pectinases but not cellulases and hemicellulases, which, however, can be used in vegetable processing. The enzymes employed in winemaking are set forth in an EC regulation and at present include pectinases and β-glucanase preparations.

Special regulations apply to beer. The well-known "Reinheitsgebot," or purity law, decreed by the Bavarian State Parliament in 1516, stipulates that no substances other than malt, hops, and water are permitted in brewing beer. Today, the beer tax law of March 14, 1952, regulates the German brewing industry.

The purity requirements for enzyme preparations correspond to those for food additives.

The enzyme standards (GDCH, 1984) and the DFG study (DFG, 1987) were developed to protect the consumer and inform the legislature.

The AMFEP recommendations detailed above are largely identical to the requirements defined in the enzyme standard.

The Netherlands   Enzymes are not specifically regulated, but their purity requirements correspond to JECFA recommendations. Their application must agree with the individual food standard.

France   Enzymes are considered as additives (Decree of 1983). Applications of certain enzyme preparations are regulated by decrees. All enzyme preparations must be registered with the "Repression des Fraudes" agency. Testing of new enzyme preparations from organismal sources that have not yet been registered will be conducted by the expert committee "Conseilles Supérieur de la Hygiène." There is an approved list giving the source and application of microbial enzymes.

At the "Centre National de Coordination des Etudes et de Recherches sur la Nutrition et l'Alimentation" (CNERNA), science, industry, and regulatory agencies cooperate in an effort to establish enzyme regulations. Preparations currently registered are presented in Table 6.1.

TABLE 6.1   Enzymes Approved in France

Enzyme	Source	Application Food or Beverage
1. Function: Glycosidic Bond Hydrolysis in Polysaccharides		
α-Amylase	*Bacillus subtilis*	Starch products
	Bacillus licheniformis	Cookies, pastries, viennese pastries, rusks
	Aspergillus niger	Bread production
	Aspergillus oryzae	Bread production, beer, fruit juices and concentrates, vegetable juices, fruit nectars, syrups
Amyloglucosidase	*A. niger*	Starch products, cookies, pastries, viennese pastries, beer
2. Function: Sucrose Hydrolysis		
Invertase	*Saccharomyces*	Invert sugar, sweets
3. Function: Peptide Bond Hydrolysis in Proteins		
Papain	*Carica papaya*	Beer
Proteases	*B. subtilis*	Cookies, pastries, viennese pastries Beer
	A. oryzae *A. wentii*	Fruit juices and concentrates, nectars
Acid proteases	*Mucor pusillus* *Mucor miehei* *Endothia parasitica*	Cheese production from cow milk only
4. Function: α-1,4-Galacturonic Acid Bond Hydrolysis in Pectins		
Pectinases	*A. niger* *A. wentii*	Fruit juices and concentrates Fruit pulps, apple and pear cider
5. Function: Glucose Isomerization		
Glucose isomerase (free, or immobilization by glutaraldehyde)	*Streptomyces violaceoniger* *Streptomyces olivochromogenes* *B. coagulans*	Glucose syrup with increased fructose content

Italy Enzymes are not specifically regulated. Enzymes are addressed in some governmental ordinances. In the regulation on fruit juice production, pectinases, proteases, and amylases are mentioned. Rennet, microbial rennet, and lipase are permitted in cheese production; pectinases, in winemaking. Malt can be treated with proteases. Papain and microbial proteases are permitted in brewing, and amylases, in breadmaking.

The United Kingdom Enzymes are not specifically regulated. An independent group of experts, the "Committee on Food Additives and Contaminants" (FACC) issued a report dealing with enzyme preparations in 1982. They recommended that an approved list be developed to regulate enzymes. Special identifying test data must be presented for each enzyme. Future regulations will include not only the JECFA recommendations but also the suggestions of the FACC.

The use of enzymes in fruit production is permitted and not limited to the enzymes listed in the EEC directive.

The regulation for cheese production permits the use of milk-clotting enzymes as well as the enzymes used for ripening cheese and making processed cheese.

Canada A long list of authorized enzymes and their applications is available. Because this list provides a compendium of enzyme applications, it is excerpted. It contains the name and source of the enzyme, and the authorized use or area of application, and the maximum dose recommended for the given application. The dose was excluded from this list (Table 6.2) because good manufacturing practice will not administer a given enzyme in larger quantities than absolutely necessary.

The United States of America Certain substances in the United States are classified as "generally recognized as safe" (GRAS). Such substances can be used in the production of food, provided that "good manufacturing practice" is applied. The GRAS status for an enzyme preparation can be granted by the Food and Drug Administration (FDA) on the basis of the documentation presented. This status, however, can also be issued by an expert panel of scientists that is qualified to judge the safety of food additives. For example, the use of papain and several fungal proteases as meat tenderizers was approved by the U.S. Department of Agriculture.

GRAS status pertains to enzymes that are obtained from plants and animals and also those from microorganisms that have been used as enzyme sources for a long time. This includes carbohydrases, lipases, catalase, and glucose oxidase from *Aspergillus niger* and the carbohydrases, lipases, and proteases from *A. oryzae*. Enzymes without GRAS status can be approved for a specific application by means of a special food additive petition (GRASP).

"Standards for Identity" have been developed in the United States for many foodstuffs in which the number and the type of additives have been specified for the production of a given food. Such food products are flour, bread, cheese, soft drinks, liquid egg, and egg white.

TABLE 6.2   Enzymes Approved in Canada

Enzymes	Source	Application
Amyloglucosidase	*Aspergillus niger* var. *A. oryzae* var. *Rhizopus oryzae* var.	Ale, beer, light beer, malt beer, porter, malt liquor Bread, flour Chocolate syrup Distillery mash Precooked instant cereals Starch for producing dextrins, glucose, glucose syrup, powdered glucose Nonstandardized baked goods
Amyloglucosidase	*Rhizopus niveus* var. *R. delemar* var.	Distillery mash Vinegar mash
Bromelain	*Ananas comosus* *Ananas bracteatus*	Beer, light beer, malt extract, porter, malt liquor Bread, flour, whole wheat flour Edible collagen sausage casings Hydrolyzing meat, milk, plant proteins Tenderizing cuts of meat Meat tenderizing and processing Beef marinades
Catalase	*Aspergillus niger* Calf liver	Lemonades Egg whites
Cellulase	*Aspergillus niger* var.	Distillery mash Liquid coffee concentrates Spice extracts
Ficin	Latex of *Ficus* sp.	See "Bromelain"
Glucanase	*Aspergillus niger* var. *Bacillus subtilis* var.	Ale, beer, light beer, malt beer, porter, malt liquor Distillery mash Vinegar mash Degerming Nonstandardized baked goods
Glucose isomerase	*Bacillus coagulans* var. *S. olivochromogenes* *S. olivaceus* *Actinoplanes missouriensis* var.	Glucose isomerization
Glucose oxidase	*Aspergillus niger* var.	Lemonades Liquid whole egg, egg white
Hemicellulase	*Bacillus subtilis* var.	See "Cellulase"
Invertase	*Saccharomyces* sp.	Confections with soft or liquid centers Nonstandardized baked goods
Lactase	*Aspergillus niger* var. *Saccharomyces* sp.	Lactose reducing enzymes Milk used in ice cream
Lipase	*Aspergillus niger* Pregastric Pancreatic	Flavors from milk products Liquid and powdered egg, or egg white Romano cheese

continued

TABLE 6.2   *(Continued)*

Enzymes	Source	Application
Milk-coagulating enzymes	*Mucor miehei*	See "Rennet"
	M. pusillus	Swiss cheese
	Enthia parasitica	
Papain	*Carica papaya* (fruit)	See "Bromelain"
Pectinase	*Aspergillus niger* var.	Cider, wine
	Rhizopus oryzae var.	Distillery mash
		Fruit juices
		Extracting pigments and flavors
		Treating citrus peels for jams, marmalades, and candied fruits
		Utilizing vegetables for soups
Pentosanase	*Aspergillus niger* var.	Ale, beer, light beer, malt extract, porter, malt liquor
	Bacillus subtilis var.	Degerming
		Distillery mash
		Vinegar mash
		Nonstandardized baked goods
Pepsin	Pig gasters	See "Protease": "Beer," etc.
		See "Rennet": "Cheese," etc.
		Fat-free soy flour
		Precooked instant cereals
Protease	*Aspergillus oryzae* var.	Beer, light beer, malt extract, porter, malt liquor
	Aspergillus niger var.	
	Bacillus subtilis var.	Bread and flour
		Flavors from milk products
		Distillery mash
		Edible collagen sausage casings
		Hydrolyzing proteins
		Spray-dried powdered cheese
		Meat tenderizer
		Precooked instant cereals
		Nonstandardized baked goods
Rennet	Rennet, stomach of calves, sheep, goats (aqueous extracts)	Cheese, cottage cheese, cheese spread, cream cheese

In breadmaking, malted wheat, malted barley, malt extract, *A. oryzae* amylase, or protease, papain, and bromelain can be utilized. It is interesting to note that while the protease from *A. niger* has GRAS status, it is not included in the "Standard for Identity" discussed earlier.

Table 6.3 lists enzymes that have qualified for and obtained GRAS status. Also included are enzymes for which a GRAS petition (GRASP) has been filed and others, which, because of their acceptance in the Code of Federal Regulations (CFR, June 1983), have been approved.

TABLE 6.3 Enzymes with GRAS, GRASP, and CFR Status in the United States

Enzyme	Source	Status[a]
α-Amylase	*Aspergillus niger*	GRAS
	Aspergillus oryzae	GRAS
	Bacillus licheniformis (mixed amylase–protease)	21 CFR 184.1027
	B. subtilis	GRAS
	Barley malt	GRAS
β-Amylase	Barley malt	GRAS
Bromelain	*Ananas comosus*	GRAS
	Ananas bracteatus	GRAS
Catalase	*A. niger*	GRAS
	Beef liver	GRAS
	Micrococcus lysodeikticus	21 CFR 173.135
Cellulase	*A. niger*	GRAS
	Trichoderma reesei	GRASP 9G0260
Ficin	*Ficus* sp.	GRAS
α-Galactosidase	*Morteirella vinacea*	21 CFR 173.145
β-Galactosidase	*A. niger*	GRAS
	Candida pseudotropicalis	GRASP 2GO282
	Kluyveromyces fragilis	GRAS
	K. lactis	GRASP 6GOO88, 7GOO88
Glucoamylase	*A. niger*	GRAS
	A. oryzae	GRAS
	Rhizopus oryzae	21 CFR 173.145
	Rhizopus niveus	21 CFR 173.110
Glucose isomerase	*Actinoplanes missouriensis*	21 CFR 184.1372
	Arthrobacter globiformis	GRASP 7GOO87
	Streptomyces olivaceus	21 CFR 184.1372
	S. olivochromogenes	21 CFR 184.1372
		GRASP 1GO277
Glucose oxidase	*A. niger*	GRAS
Invertase	*S. cerevisiae*	GRAS
Lipase	Gaster of calf, sheep, and goat	GRAS
	Pancreas	GRAS
	A. niger	GRAS
	A. oryzae	GRAS
Lipase esterase	*Mucor miehei*	21 CFR 173.140
Papain	Papaya	GRAS
Pectinase	*A. niger*	GRAS
Pepsin	Gaster of pig and cattle	GRAS
Protease, generic	*A. niger*	GRAS
	B. licheniformis (mixed amylase–protease)	21 CFR 184.1027
Rennet	Gaster of ruminants	GRAS
	Endothia parasitica	21 CFR 173.150
	B. cereus	21 CFR 173.150
	Mucor miehei	21 CFR 173.150
	Mucor pusillus	21 CFR 173.150
Trypsin	Pancreas	GRAS

[a]GRAS, generally recognized as safe; GRASP, GRAS petition; CFR, Code of Federal Regulation.

Japan   Enzymes are considered natural products and are subject to existing regulations for food additives. The legislative and regulatory bodies strongly promote innovation. The Japanese enzyme producers are presented in the "Japanese Enzyme Organization."

REFERENCES

AMFEP (Association of Microbial Food Enzyme Producers) (ed.), *General Standards for Enzyme Regulations*, 3rd ed., Brussels, 1988.

Council Directive 90/219/EEC, April 23, 1990, "On the contained use of genetically modified micro-organisms (1990a)," in *Official Journal of the European Communities*, Vol. 33, No. L 117, May 8, 1990, p. 1.

Council Directive 90/220/EEC, April 23, 1990, "On the deliberate release into the environment of genetically modified organisms (1990b)," in *Official Journal of the European Communities*, Vol. 33, No. L 117, May 8, 1990, p. 15.

Council Directive 90/679/EEC, Nov. 26, 1990, "On the protection of workers from risks related to exposure to biological agents at work (7th individual Directive within the meaning of Art. 16(1) of Directive 89/391/EEC), (1990c)," in *Official Journal of the European Communities*, Vol. 33, No. L 374, Dec. 31, 1990, p. 1.

Council Regulation 79/337/EEC, Feb. 5, 1979, "On the common organization of the market in wine (1979)," in *Official Journal of the European Communities*, Vol. 22, No. L 54, March 5, 1979.

Denner, W. H. B., "Legislation and regulation in industrial enzymology," in Godfrey, T., and Reichelt, J. (eds.), *Industrial Enzymology: The Applications of Enzymes in Industry*, The Nature Press/Macmillan, New York 1983, p. 111.

DFG Senate Committee (FRG), Starterkulturen und Enzyme für die Lebensmitteltechnik, in *XIth Comm. Examining Food Additives and Food Substances*. VCH, Weinheim 1987.

FAO/WHO, FAO Nutrition Meeting Report Series 50B, "Specifications for the identity and purity of some enzymes and certain other substances," in *WHO Food Additives*, Series 2, Geneva 1972.

FAO/WHO, FAO/WHO Joint Expert Committee, "Evaluation of certain food additives," in *18th Report on Food Additives*, WHO Technical Report, Series 557, Geneva, 1974.

FAO/WHO, FAO/WHO Joint Expert Committee, "Evaluation of certain food additives," in *21st Report on Food Additives*, WHO Technical Report, Series 617, Geneva, 1978.

FAO/WHO, FAO/WHO Joint Expert Committee, "Evaluation of certain food additives," in *25th Report on Food Additives*, WHO Technical Report, Series 669, Geneva, 1981.

Frank, H. K., "Enzyme und Lebensmittelrecht," in *Enzyme in der Lebensmittelverarbeitung*, Bauer, W. (Director), Behr's Seminars, Munich, Feb. 25–26, 1988.

GDCH (Society of German Chemists), Committee Food and Judicial Chemistry; Subcommittee on Enzyme Chemistry (eds.), "Enzympräparate. Standards für die Verwendung in Lebensmitteln," in *Lebensmittelchemie, Lebensmittelqualität*, Vol. 5, Behr, Hamburg, 1984.

NAS [National Academy of Sciences (USA)], *Food Chemicals Codex*, 3rd ed., Washington, DC, 1981.

Toet, D. A., in Casper, R., and Landsmann, J. (eds.), *The Biosafety Results of Field Tests of Genetically Modified Plants and Microorganisms. Proc. 2nd Internatl. Symp.*, May 11–14, 1992, Goslar (publ. by Biol. Bundesanst. Land- und Forstwirtsch.), Braunschweig, 1992, p. 201.

7 Economic Considerations for the Use of Technical Enzymes

The economic benefit of using technical enzyme preparations lies in lowered process costs, often increasing product quality, reducing environmental impact, and making use of renewable resources in technology and food possible (see Table 7.1). The significance of these considerations can be demonstrated with a few examples. There will, of course, be some overlapping between those situations in which utilization of natural products go along with cost reduction.

7.1 INCREASING YIELDS AND IMPROVING RAW-MATERIALS UTILIZATION

7.1.1 Improving Product Yields

In food and beverage manufacturing many examples of enhanced product yields can be cited such as processing fruits and vegetables into juices and canned foods, or producing glucose syrup from starch. Glucose yields could be increased from about 75% to 90% upon substituting enzymatic starch hydrolysis for acid hydrolysis and, in addition, process costs could be lowered.

7.1.2 Raw-Materials Utilization

Various fruit and vegetable varieties can, except for small residual amounts, be completely hydrolyzed with the combined use of pectinases and cellulases. This represents optimal utilization of a staple commodity.

It has been shown that the action of enzymes can generate useful food industry products from whey, which previously was considered only a waste product. Because some countries still emit large quantities of whey into wastewater systems, processing whey into food or fodder will reduce the environmental burden.

Raw materials for fodder can be digested by enzymes so that the nutrient value of food is improved. Ethanol for fuel can be derived from starchy raw materials. An objective of modern enzyme technology is to replace starch with cellulose for ethanol fuel.

TABLE 7.1   Economic Considerations for the Use of Technical Enzymes

Objectives	Measures	Products/Processes
Cost reduction	Increase yields	Glucose, glucose–fructose syrup, fruit processing
	Upgrade raw-materials utilization	Fruit and vegetable juices, meat processing
	Process costs	Glucose and fructose syrups, removing sizing from textiles
	Filtration costs	Beer, wine, fruit juices
Quality improvement	Preservation	Fruit juice concentrates, lemonade stabilization, silage
	Alter technical properties	Protein modification flour, baked goods, transesterification of lipids
	Flavor	Milk products, cheese, flavor concentrates
	Cleansing power	Laundry and cleansing detergents
	Process control	Leather
Utilization of raw materials	Whey, hydrolysis of starch and cellulose	Beverages, baking aids
		Ethanol
		Animal feed
Lowering environmental impact		Leather production, whey utilization

7.2   LOWERING COSTS

7.2.1   Process Costs

Starch hydrolysis with amylases can again be mentioned. Even greater advances can be possible when raw starch can be hydrolyzed without prior temperature treatment. In fact, all processes in which energy can be conserved through the use of alternative methods that employ enzymes can be included, such as lipid hydrolysis with lipases rather than with high temperatures and pressures.

7.2.2   Filtration Costs

These process costs are listed separately because they are essential for certain procedures. Filter clarification and sterilization of beverages or fermentation broths are processes that entail considerable costs. These include

1. Expenditures for filter material, filter plates, membranes, or other filtering devices

2. Expenditures for labor in servicing and operating the facilities
3. Loss of liquids in the filters

Highly hydrophilic substances such as pectins, β-glucans, pentosans, or glyco-proteins form very viscous solutions in water and make filtration difficult. Applied technology has demonstrated that enzymes can be employed here to great advantage. Often, filtration efficacy can be doubled, if not increased 10-fold.

7.3 ALTERING TECHNICAL PROPERTIES

This section includes methods in which the characteristics of natural products are altered to fit nutritional or technological needs, for example, changing the properties of wheat gluten to produce various kinds of baked goods or improving a protein's ability to be whipped and emulsified on enzymatic partial hydrolysis. Similarly, altering the properties of lipids discussed earlier (see Section 5.12) should also be mentioned in this context.

7.4 IMPROVING PRESERVATION, FLAVOR, AND CLEANSING ACTION

7.4.1 Preservation

Preservation makes a significant impact on quality, especially in fruit juice production. As shown earlier, modern processes convert juices into concentrates that, except for aroma, can be stored for a long time without loss in quality. Stabilizing flavor and color in beverages with glucose oxidase is also an example of improved preservation.

7.4.2 Flavor

Aroma formation aided by lipases and peptidases has been discussed in Section 5.6.2; however, flavor improvements can also be obtained on partial hydrolysis of proteins or digestion of cell material.

7.4.3 Cleansing Action

This topic, which includes enzymes in laundry detergents, has become especially relevant today because new efforts are required to reduce the environmental burden. Enzyme-powered detergents are especially significant for removing stubborn, ground-in dirt, which in the past could be removed only with strong alkali, surfactants, or solvents. Enzymes should be considered for such future applications, although current detergent manufacturing processes utilizing enzymes are still expensive in Europe. However, laundry detergents with enzyme power are successfully marketed in the United States and Canada.

Numerous examples of this topic can be cited from the previous chapters. Leather manufacturing, as an example, is one area in which process control can be accomplished by introducing enzymes. The advantages of enzyme use are reducing losses, significant quality improvement, and last, but not least, a lowered environmental impact by limiting chemical effluents into wastewater.

7.5 COSTS OF TECHNICAL ENZYMES

The costs of technical enzymes are determined primarily by microorganismal productivity, energy, and investment requirements. Because enzymes are employed in relation to their activity, it is not the price per kilogram but rather the price per unit activity that determines the cost. Unfortunately, the activity data from the various enzyme producers are, with some exceptions, not comparable. If users wish to compare enzyme products from various suppliers, they must conduct their own comparison test for their specific applications.

Users require information on how enzyme costs will affect their process costs as compared to the costs presently incurred with a conventional enzyme-free process. Table 7.2 lists enzyme costs for specific processes or applications; however, it does not show the costs of the processes themselves. Some of the data presented in Table 7.2 were taken from Poldermans and Roels (1986).

The data presented are certainly subject to large fluctuations and depend on enzyme quantity purchased, shipping costs, customs duty, and other factors. The strong competitive marketing over the past 30 years has resulted in more concen-

TABLE 7.2   Cost of Technical Enzymes in 14 Specific Processes (data from 1989)

Process or Application	Cost (U.S.$)/Unit of Measure
Starch hydrolysis	~2/t starch
Saccharification	3.5/t starch
Isomerization	6/t starch
High-fructose corn syrup in United States (1986)	6–7/t starch
Alcohol production (1984)	1/100 liters alcohol
Brewing	0.1/100 liters beer
Baked goods	
United States	0.05–0.1/100 kg flour
Europe	0.05–0.5/100 kg flour
Fruit juice	0.05–0.1/100 liters juice
Wine	0.05–0.1/100 liters wine
Carbonated fruit drinks	
Glucose oxidase stabilization	0.26–0.79/100 liters
Cheese production	0.05/100 liters milk
Laundry detergents	0.04–0.06/kg laundry detergent
Leather production	1.2–3/t hide

TABLE 7.3   Quantities and Cost of Enzymes in Relation to Product Value

Product	Amount of Enzyme[a] as ppm of Product	Enzyme Cost as Percentage of Product Value
Laundry detergent	150–200	1
Glucose	150–200	1
High-fructose corn syrup	150–200	3
Ethanol	300–400	3
Cheese	3–6	0.1–0.3

[a]Pure enzyme.

trated, purer, and less expensive enzyme preparations. For example, the enzyme costs required to produce 1 t of high-fructose corn syrup were approximately \$12–\$14 in 1980, and only \$6–\$7 in 1986. Another aspect may be of interest as well (Aunstrup, 1984) in which costs of enzyme preparations are related to the value of the product generated (Table 7.3).

Because questions are continuously being raised about the enzyme price per kilogram, and further, about the percent of enzyme needed for this or that product, another cost analysis is offered. Answers to questions such as how much a kilogram of pure enzyme will cost and what amount of pure enzyme is present in one kilogram of a technical-grade enzyme preparation are provided in Table 7.4 for some enzymes (Aunstrup, 1984).

Amyloglucosidase has become a surprisingly inexpensive enzyme after a long developmental phase involving expensive strain, culturing, and process optimizations. Enriched technical-grade bacterial α-amylase contains approximately 50% pure enzyme. Because microbial rennet is based on technical enzyme protein, it costs only one-tenth as much as animal-based rennet.

TABLE 7.4   Cost of Enzymes in Relation to Amount Present in Technical Preparations

Enzyme	Price in \$/kg of Pure Enzyme	Percent Enzyme in Technical Preparation[a]
Amyloglucosidase	40	5–30
Bacterial α-amylase	314	5–60
Laundry detergent protease	200	3–6
Glucose isomerase	471	5–10
Fungal α-amylase	114	1–25
Chymosin	8000	0.1–0.2
Microbial rennet	1029	0.2–0.4

[a]From Aunstrup (1984) and other sources.

REFERENCES

Aunstrup, K., *Third European Congress on Biotechnology*, Vol. III, Verlag Chemie GmbH, Weinheim, 1984, p. 143.

Poldermans, B., and Roels, J. A., in Mark, H. F., Bikales, N. M., Overberger, C. G., and Menges, G. (eds.), *Encyclopedia of Polymer Science and Engineering*, 2nd ed., Vol. 6, J. Wiley, New York, 1986, p. 132.

8 The Market for Technical Enzyme Preparations

This market emcompasses the so-called food enzymes and enzyme preparations for industrial processes that include detergent enzymes. In both areas, the market can be divided into two sectors: (1) enzyme concentrates sold by the enzyme producer to manufacturers who produce custom formulations or (2) preparations for specific consumers. But enzyme producers now increasingly formulate such products themselves.

Well-known marketing studies were conducted by Hepner (1978) or by Frost and Sullivan (1983) who presumably evaluated data from genuine enzyme producers.

Formulations are prepared for specific applications in which more than one enzymatic activity is required for the desired effect. Furthermore, use of enzyme concentrates is often difficult in practical applications, such as when introducing small enzyme quantities into large substrate volumes. This is the case in leather manufacturing, where excess dosage can result in great damage. Also, when treating flour with enzymes at the mill, it is important to work with premixes to ascertain homogeneous distribution. Companies that produce wine- and juice-making aids offer their customers diluted enzyme combinations tailored to specific processes.

Limited information is available about market development. In 1970, $40–$50 million in sales were registered for enzymes worldwide; in 1977, the figure was $180 million (estimate by Gist Brocades); in 1981/82, it may have been as much as $310 million; and professional estimates for 1988 are about $571 million.

The author has attempted to derive an estimate for the enzyme market in 1990 from personal communications of experts familiar with market trends; this estimate includes both the major markets and the most important areas of enzyme application or formulations. Total sales of $700 million could be distributed as shown in Table 8.1.

More recent marketing reports on the EC market (1992) and western Europe in general (1993) on industrial enzymes have been published by Frost & Sullivan, who specialize in commodity market research.

TABLE 8.1   Market for Technical Enzymes in U.S. $ Million (1990)

Application	United States	EU,[a] Japan, and Others
Starch processing, dextrins, glucose	20	14
High-fructose corn syrup	95	3
Brewing	12	13
Fruit processing	10	10
Wine	8	6
Flour and baked goods	15	20
Animal rennet	20	12
Microbial rennet	20	22
Cheese ripening and flavors	3	1
Various candy confections	1	1
Silage and animal feed	8	8
Paper and textiles	4	4
Leather	—	70
Detergents, proteases, lipases, and cellulases	—	300

[a]European Union, formerly known as EEC (European Economic Community).

REFERENCES

Frost & Sullivan Market Research Report, *Study of Enzymes in Europe*, Frost & Sullivan, New York, 1983.

Frost & Sullivan Market Research Report 1431, *The EC Market for Industrial Enzymes*, Frost & Sullivan, London, 1992, p. 315.

Frost & Sullivan Market Research Report 1944, *West European Industrial Enzyme Markets*, Frost & Sullivan, London, 1993, p. 166.

Hepner, L., *Enzyme Using Industries toward 2000*, L. Hepner & Associates, Tavistock Square, London, 1978.

APPENDIX A
List of Enzyme Producers

ABM ABM (Group of Rhone) Poulenc Chemicals Ltd., Poleacre Lane Woodley, Stockport, Cheshire, SK 6 1PQ, UK

AM Amano Pharmaceutical Co. Ltd., 1-21-chrome, Nishiki Naka-Ku, Nagoya, Japan
Amano Enzyme USA Co., Ltd., 1157 North Main Street, Lombard, IL 60148 (USA)

AO Alko Ltd., SF-05 200 Rajamäki, Finland

BN Biocon Ltd., now Quest International NL B.V., P.O. Box 2, N1-1400, CA Bussum, the Netherlands

BÖM Boehringer Mannheim GmbH, P.O. Box 51, Mannheim, Germany

BÖI Boehringer Ingelheim Backmittel GmbH, Mainzer Strasse 152–160, 55411 Bingen, Rhein, Germany

BF BASF AG, D-67056, Ludwigshafen, Germany

BS Biocatalysts Ltd., Main Avenue, Treforest Industrial Estate, Pontypridd, Mid Glamorgan, CF 37 5UT, UK

B&S Berg und Schmidt, Feed Additives, An der Alster 81, D-20 099 Hamburg, Germany

CH Christian Hansen Laboratories A/S, 3 Sankt Annae Plads, DK-1250 Copenhagen, Denmark

DFL Dairyland Food Laboratories, now Sanofi Bio Industries Inc. 20 Progress St., Waukesha, WI 53186 (USA)

EB Enzyme Bio-Systems Ltd., 2600 Kennedy Drive, Beloit, WI 53511-3969 (USA)

FR Finish Sugar Co., Ltd., Enzymes: Genencor International Kyllikiportti 2, P.O. Box 105, Fin-00 241 Helsinki, Finland

GB Gist-Brocades, Food Ingredients Division, P.O. Box 1, 2600 Delft-Holland, the Netherlands

GO Godo Shusei Co., Ltd., 2-10, Ginza 6-chome, Chuo-Ku, Tokyo

GR Genencor International, Inc. (North & South America), 4 Cambridge Place 1870 South Winton Road, Rochester, NY 14618 (USA)

GT Danisco Ingredients (Grindsted Division), Edwin Rahrs Vej 38, DK-8220, Brabrand, Denmark

GX Glaxo Operations UK Ltd., Ulverston, Cumbria LA12 9 DR, UK

HÖ Höchst AG, D-65926 Frankfurt, Germany

HU   Hankyo Kyoei Bussan Co. Ltd., Enzyme Division (see "UA")

KK   Kakem Chemical Co. Ltd., 2-28-8 Honkomagome Bunkyo-ku, Tokyo 113, Japan

KI   Kumiai Chemical Industry Co. Ltd., 1-4-26 Ikenohata Taitoku, 110 Tokyo, Japan

MC   Mühlen-Chemie GmbH, Kornkamp 40, D-22926 Ahrensburg, Germany

MJ   Meiji Seika Kaisha Ltd., 4-16 Kyobashi, 2-chome, Chuo-Ku, Tokyo

MK   E. Merck, Frankfurter Strasse D-64, 293 Darmstadt, Germany

MS   Miles Laboratories Inc., Elkhart, IN 46514 (USA); see SOE

MKC   See "SO"

MZ   Megazyme Co., 2/11 Ponderosa Parade Warriewood (Sydney), NSW 2102 Australia

NG   Nagase CO. Ltd., Konishi Bldg., 2-2-chome Honcho Nihonbashi Chuo-Ku, Tokyo

NO   Novo Nordisk A/S, Novo Allee, DK-2880 Bagsvaerd, Denmark

OY   Oriental Yeast, Nihonbashi Fukawa Bldg., Chuo-Ku, Tokyo

RM   Röhm GmbH, Kirschenalle, D-64 293 Darmstadt, Germany

RH   Röhm and Haas (see "GR")

SA   Sankyo Company Ltd., No. 7-12, 2-Chome, Ginza Chuo-ku, 104 Tokyo, Japan

SN   Shin Nihon Chemical Company Ltd. Anjyo, Aichi, Japan

SE   Sturge Biochemicals, Denisson Road, Selby, North Yorkshire Yo8 8EF, UK

SO   Solvay Enzymes GmbH & Co., Hans Böckler Allee 2, D-30 173 Hannover, Germany

SOE   Solvay Enzymes Inc. (formerly Miles Laboratories) Elkhart, IN 46514 (USA)

ST   Stern-Enzym GmbH & Co. KG, Kornkamp 40, D-22926 Ahrensburg, Germany

UA   Ueda Chemical, now Hankyu Bioindustries Co., Ltd., 2-6, Tatayanagi 1-Chome Neyagawa-City, Osaka 572, Japan

YT   Yakult Pharmaceutical Industry Co., Ltd. 5-13-15 Shinbashi Minato-ku Tokyo, Japan

APPENDIX B
List of Some Industrial Enzymes

Name	Type of Enzyme	Origin*	Producer
50:50	Rennet	Animal	CH
Agidex	Amyloglucosidase	*Asp. niger*	GX
Alfamalt VT	Fungal α-amylase	*Asp. oryzae*	MC
Alfamalt PRO	Bacterial protease	*Bac. subtilis*	MC
Alkalase	Bacterial protease	*Bac. licheniformis*	NO
Ambazyme	Amyloglucosidase	*Asp. niger*	ABM
Amylase P	Fungal α-amylase	*Asp. oryzae*	GB
Amyliq	Bacterial α-amylase	*Bac. subtilis*	GB
Amylolysin	Bacterial α-amylase	*Bac. subtilis*	HU
Anthocyanase YA 2	Anthocyanase	*Asp. niger*	YT
Arazyme	Protease	Microbiell	RM
Aquazyme	Bacterial α-amylase	*Bac. amyloliquefaciens*	NO
Asgo 1500	Glucoseoxidase	*Asp. niger*	FR
Avizyme	Cellulase	*Trich. reesei*	FR
BAN	Bacterial α-amylase	*Bac. subtilis*	NO
Beta-Glucanase	β-Glucanase	*Arthobacter sp.*	DA
Beta-Glucanase	β-Glucanase	*Rhiz. solani*	KI
Bergazyme H	Cellulase, hemicellulase	*Trich. reesei*	B&S
Biobake	Fungal α-amylase	*Asp. oryzae*	BN
Biocellulase A	Cellulase	*Asp. niger*	BN
Biocellulase T	β-Glucanase	*Trich. reesei*	BN
Biodiastase	Fungal α-amylase	*Asp. oryzae*	AM
Bioinvert	Invertase	*Sacch. sp.*	BN
Biokleistase	Bacterial α-amylase	*Bac. subtilis*	DA
Biopectinase	Pectinase	*Asp. niger*	BN

Bac. = Bacillus
Asp. = Aspergillus
Trich. = Trichoderma
Pen. = Penicillium
Rhiz. = Rhizopus
Strep. = Streptomyces

Bioprase	Bacterial protease	*Bac. subtilis*	NG
Biotamylase	Bacterial α-amylase	*Bac. subtilis*	NG
Biotex	Bacterial α-amylase	*Bac. subtilis*	NG
Gabo	Rennet	Animal	CH
Canalpha	Bacterial α-amylase	*Bac. amyloliquefaciens*	BN
Capalase	Lipaesterase	Goat	DFL
Capalase KL	Lipaesterase	Goat, Lamb	DFL
Catalase LA 3	Catalase	*Asp. niger*	SE
Catalase U 5L	Catalase	Micrococcus Lysodeikticus	UA
Celluclast	Cellulase	*Trich. reesei*	NO
Cellulase Onozuka	β-Glucanase, cellulase	*Trich. viride*	YT
Cellulase CP	Cellulase	*Pen. funiculosum*	SE
Cellulosin AP	Cellulase	*Asp. niger*	HU
Cellulosin HC	Hemicellulase	*Asp. niger*	HU
Celluzyme	Cellulase	*Humicola insolens*	NO
Cereflo	β-Glucanase	*Bac. subtilis*	NO
Ceremix	β-Glucanase	*Bac. subtilis*	NO
Clampzyme	Cellulase	*Fungal* sp.	FR
Collupulin	Papain	Papaya	GB
Corolase L 10	Papain	Papaya	RM
Corolase N	Bacterial proteinase	*Bac. subtilis*	RM
Corolase PK	Pancreatic protease	Animal	RM
Corolase PN	Fungal alkal. protease	*Asp. sojae*	RM
Cyclodextrinase	Glucanotransferase	*Bac.* sp.	AM
Degomma	Bacterial amylase and protease	*Bac. subtilis*	RM
Denapsin	Fungal acid protease	*Asp.* sp.	NG
Deoxin	Glucose–oxidase	*Pen. amagasakiense*	NG
Dex-Lo	Bacterial α-amylase	*Bac. subtilis*	GB
Dextranase 50 L	Dextranase	*Pen. lilacinum*	NO
Dextrozyme	Glucoamylase, pullulanase	*Asp. niger* *Bac. acidopullulyticus*	NO
Diase	Amyloglucosidase	*Asp. niger*	GB
Emporase	Rennet	*Mucor pusillus*	DFL
Esperase	Bacterial alkal.protease	*Bac.* sp.	NO
Extractase	Pectinase	*Asp. niger*	FR
Fermcolase	Catalase	*Asp. niger*	FR
Filtrase B	β-Glucanase	*Bac. subtilis*	GB
Filtrase NL	β-Glucanase	*Asp.* sp.	GB
Finizym	β-Glucanase	*Asp. niger*	NO
Flavourzyme	Protease, peptidase	*Asp. oryzae*	NO
Fromase	Rennet	*Mucor miehei*	GB
Funcellase	β-1,3 Glucanase	*Trich. viride*	YT
Glucanase GV	β-Glucanase	*Asp. niger*	GT

Glucanex	β-Glucanase	*Asp. niger*	NO
Glucoseoxidase L	Glucose–Oxidase	*Asp. niger*	SO
Glucoseoxidase P	Glucose–Oxidase	*Asp. niger*	SO
Glucox	Glucose–Oxidase	*Asp. niger*	SE
Glucozyme	Amyloglucosidase	*Rhiz.* sp.	NG
Gluczyme NL 3	Amyloglucosidase	*Rhiz.* sp.	AM
Glutase	Amyloglucosidase	*Rhiz.* sp.	HU
Godo-AGI	Glucoseisomerase	*Strep. griseofuscus*	GO
Godo-BSP	Bacterial neutral protease	*Bac. subtilis*	GO
Godo-MSD	Dextranase	*Asp. niger*	GO
Godo-YNL	Lactase	*Kluyveromyces lactis*	GO
Grindamyl A	Fungal α-amylase	*Asp. oryzae*	GT
Grindamyl S	Xylanase, hemicellulase	*Asp. niger*	GT
Hannilase	Rennet	Microbiell	CH
Hazyme	Fungal α-amylase	*Asp. niger*	GB
Hitempase	Bacterial α-amylase	*Bac.* sp.	BN
Hydrolact	Lactase	*Kluyveromyces lactis*	SE
Inulinase MJ	Inulinase	*Asp.* sp.	MJ
Italase	Lipaesterase (pregastric)	Calf	DFL
Kitalase	β-1,3 Glucanase	*Arthrobacter* sp.	DA
Kleistase	Bacterial α-amylase	*Bac. amyloliquefaciens*	DA
Klerzyme	Pectinase	*Asp. niger*	GB
Lactase Y-AO	β-Galactosidase	*Asp. niger*	YT
Lactase Y-L	β-Galactosidase	*Kluyveromyces lactis*	AM
Lecitase	Phospholipase A 2	Hog pancreas	NO
Lipase AP	Lipase	*Asp.* sp.	AM
Lipase CE	Lipase	*Humicola lanuginosa*	AM
Lipase MAP	Lipase	*Mucor* sp.	AM
Lipase N	Lipase	*Rhiz.* sp.	AM
Lipase P	Lipase	*Pseudomonas* sp.	AM
Lipase Seiken	Lipase	*Rhiz.* sp.	NG
Lipolact K1	Lipaesterase	*Asp. niger*	RM
Lipolase	Lipase	Microbiell	NO
Lipozyme L	Fungal lipase	*Mucor miehei*	NO
Lysozyme	*N*-Acetylmuramoyl-hydrolase	Hen's egg	FR
Macerozyme 2S	Pectinase	*Rhiz.* sp.	YT
Maltamyl	Fungal α-amylase	*Asp. niger*	EB
Maturex	Diaacetyl reductase	*Bac.* sp.	NO
Maxamyl	Bacterial α-amylase	*Bac. licheniformis*	GB
Maxatase	Bacterial alkal. protease	*Bac. licheniformis*	GB
Maxazyme GI Immob	Glucoseisomerase, immobilized	*Actinoplanes* sp.	GB

Maxilact	β-Galactosidase	*Sacch. lactis*	GB
Maxinvert	β-Fructofuranosidase	*Sacch. lactis*	GB
Maxiren	Rennet 100%	*Kluyveromyces lactis*	
GB Megafresh	Bacterial α-amylase	*Bac. subtilis*	EB
Meicellulase	Cellulase	*Trich. reesei*	MJ
MKC-Fungal Amylase	Fungal α-amylase	*Asp. oryzae*	SO
Molsine	Mix of hydrolases	*Asp. sp.*	YT
Multifresh	Fungal α-amylase	*Asp. niger*	EB
Mycolase	Fungal α-amylase	*Asp. oryzae*	GB
Neutrase	Bacterial neutral protease	*Bac. amyloliquefaciens*	NO
Newlase	Fungal acid protease	*Rhiz. sp.*	AM
Novamyl	Maltogenic amylase	*Bac. sp.*	NO
Novoren	Rennet	*Mucor miehei*	NO
Novozyme 188	Cellobiase	*Asp. niger*	NO
Novozyme 230	Inulinase	*Asp. sp.*	NO
Novozyme 234	Cell-wall-lysing hydrolase	*Trich. harzianum*	NO
Optamyl-L	Bacterial α-amylase	*Bac. subtilis*	SO
Optidex-L	Amyloglucosidase	*Asp. niger*	SO
Optimase	Bacterial alkal.protease	*Bac. licheniformis*	SO
Optimash	Bacterial α-amylase	*Bac. subtilis*	SO
Optisprit	Amyloglucosidase	*Asp. niger*	SO
Optisweet S	Glucoseisomerase	*Strep. olivaceus*	SO
Optisweet Immob.	Glucoseisomerase, immobilized	*Strep. rubiginosus*	SO
Optitherm	Bacterial α-amylase	*Bac. licheniformis*	SO
Orientase 90	Bacterial neutral protease	*Bac. subtilis*	UA
Oropan	Proteases	Pancreatic, fungal	RM
Ovazyme	Glucoseoxidase, catalase	*Asp. niger*	FR
Palatase	Lipaesterase	*Asp. niger*	NO
Pancellase	Cellulase	*Asp. niger*	YT
Pandicase XP	Fungal neutral protease	*Asp. oryzae*	YT
Panprosin	Fungal acid protease	*Asp. niger*	YT
Panzym	Pectinase	*Asp. niger*	BÖI
Pectinex	Pectinase	*Asp. niger*	NO
Pectinol	Pectinase	*Asp. niger*	GR
Pectolase	Pectinase	*Asp. niger*	GT
Pentopan	Xylanase	*Asp. niger*	NO
Piccantase	Lipaesterase	Microbiell	GB
Plexazyme LA 1	β-Galactosidase, immobilized	*Asp. oryzae*	RM

Powerzym 5000	Hemicellulase	*Asp. niger*	SE
Proctase	Fungal acid protease	*Asp.* sp.	MJ
Prolysin	Bacterial neutral protease	*Bac. subtilis*	HU
Promen	Bacterial neutral protease	*Bac. subtilis*	DA
Pronase	Protease	*Strep. griseus*	KK
Proteinase 18	Bacterial neutral protease	*Bac. subtilis*	ABM
Protin A	Bacterial alkal. protease	*Bac. subtilis*	DA
Prozyme	Fungal protease	*Asp.* sp.	AM
Promozyme	Pullulanase	*Bac. acidopullulyticus*	NO
Pulluzyme	Pullulanase	*Klebsiella aerogenes*	ABM
Rapidase AG	Amyloglucosidase	*Asp. niger*	GB
Rapidase C	Pectinase	*Asp. niger*	GB
Rapidase CPE	Pectinesterase	*Asp. niger*	GB
Rennilase	Rennet	*Mucor* sp.	NO
Rhozyme P 11	Fungal protease	*Asp. oryzae*	GR
Rhozyme P 53	Bacterial protease	*Bac. subtilis*	GR
Rhozyme P 64	Bacterial alkal. protease	*Bac. licheniformis*	GR
Rohalase A	Bacterial α-amylase	*Bac. amyloliquefaciens*	RM
Rohalase M	Fungal α-amylase	*Asp. oryzae*	RM
Rohamalt MG	β-Glucanase	*Trich. reesei*	RM
Rohamalt P	Bacterial protease	*Bac. subtills*	RM
Rohament CA	Hemicellulase	*Asp. niger*	RM
Rohament PC	Macerase	*Asp. niger*	RM
Rohament P	Pectinglycosidase	*Asp.* sp.	RM
Rohapect D,MA, MB,C	Pectinasen	*Asp. niger*	RM
Samptase	Fungal acid protease	*Asp. niger*	HU
Sanactase	Fungal α-amylase	*Asp.* sp.	MY
Sanzyme	Fungal α-amylase	*Asp. oryzae*	SA
Sanzyme SS	Fungal acid protease	*Asp.* sp.	SA
Savinase	Bacterial alkal. protease	*Bac.* sp.	NO
Sclase	Pectinase	*Asp. usamii*	SA
Speedase	Bacterial α-amylase	*Bac. subtilis*	NG
Spezyme	Glucoseisomerase	*Strep.* sp.	FR
Spezyme GI	Glucoseisomerase	*Strep. rubiginosus*	FR
Starzyme AG	Amyloglucosidase	*Asp. niger*	SE
Sternzyme H	Hemicellulase	*Trich. reesei*	ST
Sternzyme B 2000	Fungal protease	*Asp. niger*	ST
Sternzyme BE	Hemicellulase	*Asp. niger*	ST
Sternzyme LQ	Protease, hemicellase	*Bac.* sp., *Asp.* sp.	ST
Sternzyme BK	Bacterial protease	*Bac. subtilis*	ST

Sumizyme AC	Cellulase	*Asp. niger*	SN
Sumizyme AP	Pectinase	*Asp. niger*	SN
Sumizyme C	Cellulase, hemicellulase	*Trich. reesei*	SN
Sumizyme FP	Fungal acid protease	*Asp. oryzae*	SN
Sumizyme MP	Fungal alkal. protease	*Asp. melleus*	SN
Sumizyme MC	Macerase	*Rhiz. oryzae*	SN
Sweetzyme	Glucoseisomerase, immobilized	*Bac. coagulans*	NO
Taka-Sweet	Glucoseisomerase, immobilized	*Bac.ol vaceus var.*	SOE
Taka-Therm	Bacterial α-amylase	*Bac. licheniformis*	SOE
Termamyl	Bacterial α-amylase	*Bac. licheniformis*	NO
Thermoase	Protease	*Bac. thermoproteolyticus*	DA
Thermolysin	Protease	*Bac. thermoproteolyticus*	DA
Tunicase	Cellwall lytic enzyme	*Arthrobacter* sp.	KI
Ultrazyme	Pectinase	*Asp. niger*	NO
Unicase	Amyloglucosidase	*Rhiz.* sp.	YT
Veron A, AV, AC	Fungal α-amylase	*Asp. oryzae*	RM
Veron F 25	Bacterial α-amylase	*Bac.* sp.	RM
Veron HE	Hemicellulase	*Trich. reesei*	RM
Veron P	Bacterial protease	*Bac. subtilis*	RM
Veron SX	Xylanase	Microbiell	RM
Veron GX	Xylanase	*Asp. niger*	RM
Xylanase 250	Xylanase	*Trich. viride*	UA
YL 5	Cellwall lytic enzyme	Microbiell	AM

INDEX